IN VITRO HAPLOID PRODUCTION IN HIGHER PLANTS

Current Plant Science and Biotechnology in Agriculture

VOLUME 29

Aims and Scope
The book series is intended for readers ranging from advanced students to senior research scientists and corporate directors interested in acquiring in-depth, state-of-the-art knowledge about research findings and techniques related to all aspects of agricultural biotechnology. Although the previous volumes in the series dealt with plant science and biotechnology, the aim is now to also include volumes dealing with animals science, food science and microbiology. While the subject matter will relate more particularly to agricultural applications, timely topics in basic science and biotechnology will also be explored. Some volumes will report progress in rapidly advancing disciplines through proceedings of symposia and workshops while others will detail fundamental information of an enduring nature that will be referenced repeatedly.

The titles published in this series are listed at the end of this volume.

In Vitro Haploid Production in Higher Plants

Volume 5 – Oil, Ornamental and Miscellaneous Plants

Edited by

S. MOHAN JAIN
Plant Production Department, University of Helsinki, Helsinki, Finland

S.K. SOPORY
School of Life Science, Jawaharlal Nehru University, New Delhi, India

and

R.E. VEILLEUX
Department of Horticulture, Virginia Polytechnic Institute and State University, Blacksburg, Virginia, U.S.A.

KLUWER ACADEMIC PUBLISHERS
Dordrecht/Boston/London

Library of Congress Cataloging-in-Publication Data

In vitro haploid production in higher plants / editors, S. Mohan Jain, S.K. Sopory, R.E. Veilleux.
p. cm. -- (Current plant science and biotechnology in agriculture ; v. 23)
Includes index.
Contents: v. 1. Fundamental aspects
ISBN 0-7923-3580-5 (set : alk. paper). -- ISBN 0-7923-3577-5 (hardback : alk. paper)
1. Micropropagation. 2. Haploidy. 3. Crops--Genetic engineering. 4. Plant breeding. I. Jain, S. Mohan. II. Sopory, S. K. III. Veilleux, R. E. IV. Series: Current plant science and biotechnology in agriculture ; 23.
SB123.6.I45 1996
631.5'23--dc20 95-304

ISBN 0-7923-3979-7 (Volume 5)
SET: ISBN 0.7923-3580-5

Published by Kluwer Academic Publishers,
P.O. Box 17, 3300 AA Dordrecht, The Netherlands.

Kluwer Academic Publishers incorporates
the publishing programmes of
D. Reidel, Martinus Nijhoff, Dr W. Junk and MTP Press.

Sold and distributed in the U.S.A. and Canada
by Kluwer Academic Publishers,
101 Philip Drive, Norwell, MA 02061, U.S.A.

In all other countries, sold and distributed
by Kluwer Academic Publishers Group,
P.O. Box 322, 3300 AH Dordrecht, The Netherlands.

Printed on acid-free paper

Printed in the Netherlands

Table of Contents

Haploid production in higher plants
A dedication

INDRA K. VASIL

The value of haploids in genetic analysis and plant breeding has been known for a long time. Natural haploid embryos and plants, derived from gametophytic cells, have been described in about 100 species of angiosperms. However, haploids occur only rarely in nature. To be useful, they must be produced in large numbers. Therefore, many attempts have been made over the years to increase the efficiency of *in ovulo* haploid production, but none of these has proven to be of wide practical utility. The early attempts to obtain haploid plants from the male gametophyte of gymnosperms (Tulecke, 1953) and angiosperms (Yamada *et al.*, 1963) resulted only in the production of haploid callus tissues (Vasil, 1980).

Embryo-like structures formed in cultured anthers of *Datura innoxia* were first described by Guha and Maheshwari (1964). They were considered to have originated from the somatic tissues of the anther. In a subsequent study, it was determined that the somatic embryos and the resulting plantlets were indeed derived from the developing microspores and were haploid in nature (Guha and Maheshwari, 1966). As is true of most pioneering studies, these first androgenic haploids were neither grown to maturity, nor were the experimental conditions for their production clearly defined. Their real value was in demonstrating the feasibility of the experimental production of haploids. Haploid plants were soon obtained from cultured anthers of *Nicotiana sylvestris* and *N. tabacum* by Bourgin and Nitsch (1967). These and subsequent studies by Nitsch and Nitsch (1969) clearly established that the culture of excised anthers at a precise stage of development was the most important requirement for switching the development of pollen from a gametophytic to a sporophytic phase, resulting in the formation of haploid embryos and/or plants. They also described a simple nutrient medium for the culture of anthers, and an easy procedure for obtaining dihaploid homozygous plants. The elegant, simple and reliable method of haploid production invented by Jean Pierre Nitsch and his associates provided much stimulus for future studies by many others.

During the past three decades many improved methods as well as nutrient media have been developed to increase the efficiency of production of andro-

Prof. S.C. Maheshwari

Dr. J.P. Nitsch

genic haploids, from cultured anthers as well as isolated microspores, in a wide variety of species. Success has also been achieved in obtaining gynogenic haploids from cultured ovaries or ovules. As a result, haploids are being used increasingly and profitably in breeding programmes for the development of new and improved cultivars. The various chapters in this and the companion volumes describe in detail the basic as well as many applied aspects of haploid production and utilization.

It has been my pleasure and privilege to have known the late Jean Pierre Bourgin, Sipra Guha-Mukherjee, Satish C. Maheshwari, Colette Nitsch and the late Jean Pierre Nitsch, all pioneers in haploid research. These volumes are dedicated to them for their seminal contributions to the experimental production of haploids and for creating a whole new field of basic and applied plant research.

References

Bourgin, J.P. and J.P. Nitsch, 1967. Obtention de *Nicotiana* haploïdes à partir d'étamines cultivées *in vitro*. Ann. Physiol. Veg. 9: 377–382.

Guha, S. and S.C. Maheshwari, 1964. *In vitro* production of embryos from anthers of *Datura*. Nature 204: 497.

Guha, S. and S.C. Maheshwari, 1966. Cell division and differentiation of embryos in the pollen grains of *Datura in vitro*. Nature 212: 97–98.

Nitsch, J.P. and C. Nitsch, 1969. Haploid plants from pollen grains. Science 163: 85--87.

Tulecke W., 1953. A tissue derived from the pollen of *Ginkgo biloba*. Science 117: 599–600.

Yamada, T., T. Shoji and Y. Sinoto. 1963. Formation of calli and free cells in the tissue culture of *Tradescantia reflexa*. Bot. Mag. Tokyo 76: 332–339.

Vasil, I.K., 1980. Androgenetic haploids. Int. Rev. Cytol. Suppl. 11A: 195–223.

General Preface

Since the beginning of agricultural production, there has been a continuous effort to grow more and better quality food to feed ever increasing populations. Both improved cultural practices and improved crop plants have allowed us to divert more human resources to non-agricultural activities while still increasing agricultural production. Malthusian population predictions continue to alarm agricultural researchers, especially plant breeders, to seek new technologies that will continue to allow us to produce more and better food by fewer people on less land. Both improvement of existing cultivars and development of new high-yielding cultivars are common goals for breeders of all crops. *In vitro* haploid production is among the new technologies that show great promise toward the goal of increasing crop yields by making similar germplasm available for many crops that was used to implement one of the greatest plant breeding success stories of this century, i.e., the development of hybrid maize by crosses of inbred lines. One of the main applications of anther culture has been to produce diploid homozygous pure lines in a single generation, thus saving many generations of backcrossing to reach homozygosity by traditional means or in crops where self-pollination is not possible.

Because doubled haploids are equivalent to inbred lines, their value has been appreciated by plant breeders for decades. The search for natural haploids and methods to induce them has been ongoing since the beginning of the 20th century. Blakeslee (1921) first identified naturally occurring haploids of *Datura stramonium* and subsequently, natural haploids of many other plants were reported by various researchers. However, naturally occurring haploids could not be produced in sufficient numbers by reliable techniques for their extensive use in breeding programmes. In 1964, the research group headed by Prof. S.C. Maheshwari, Department of Botany, Delhi University, India, reported haploid production in *Datura innoxia* for the first time by anther culture. Since this discovery, many of the limitations of the technique have been overcome such that it is currently employed for the production of haploids and doubled haploids of many crop species throughout the world. The early contributions of Drs. C. Nitsch and J.-P. Nitsch (France), G. Melchers (Germany), M.S. Swaminathan (India), I.K. Vasil (USA), N.

Sunderland (UK), and Hu Han (China) towards overcoming these limitations and adapting the technology to a variety of crops must be acknowledged. Their realization of the potential of anther culture and tireless pursuit of reliable techniques that would facilitate its success has led to its implementation in breeding programmes.

In addition to its practical applications, *in vitro* haploid extraction has changed our understanding of developmental processes in plants. Androgenesis can be thought of as a type of somatic embryogenesis that involves cells, i.e., microspores, that at first thought would not have been expected to be embryogenically competent. How can the natural course of microsporogenesis be diverted onto an embryogenic pathway? Why are some microspores competent for androgenic development and not others? How can the process of anther culture be a heritable trait in crosses between competent and non-competent parents? Does the process of anther culture impose some selection pressure on the population of microspores or otherwise result in some undesirable change expressed in the population of regenerated plants? Why are albinos so common among the anther-derived regenerants of some species when it is obvious that microspores must contain proplastids in order for green plants to be regenerated at all? We have only begun to answer some of these questions.

This book project was submitted with the consent of co-editors to the Kluwer Academic Publishers, Dordrecht, The Netherlands. The publisher had this project reviewed by anonymous reviewers. Finally, on the basis of the positive comments of the reviewers, the publisher gave us the contract to proceed with this book project. We have not followed any conservative format of chapters and gave all the liberty to the authors to write the way they felt appropriate. Most of the chapters are reviews of work done. However, in some cases where a lot of work has not been done in the past, the authors have been encouraged to give their own research findings in details.

In this set of volumes, we have made an attempt to assimilate detailed descriptions of various aspects of anther culture and related *in vitro* procedures. Many chapters have been written by experts in the various applications of anther culture to specific crops. In addition to crop-by-crop discussions on the progress of anther culture, we have also included chapters on other topics concerning the utilization of *in vitro* haploids in plants. Embryogenic microspores have recently been regarded as ideal targets for genetic transformation. Molecular markers such as RFLPs, RAPDs, or SSRs can be used to determine disturbed segregation ratios in haploid populations or to tag traits of interest to plant breeders. The potential of pollen protoplasts is discussed. *In vitro* selection during androgenesis, both imposed and inadvertent, is also considered.

The series is divided into five volumes. Volume 1 contains 18 chapters and primarily covers fundamental aspects of haploidy and various methods of haploid extraction, e.g., anther culture, microspore culture, ovary culture, etc. The second volume comprises 21 chapters and describes applications of haploid breeding in protoplast manipulations, mutation breeding, RFLP mapping, identification of quantitative trait loci (QTLs), cryopreservation, chromosome engineering by anther culture, molecular biology of pollen rejection, transformation of pollen/microspores, etc. The third volume has 20 chapters focussed on haploid breeding in selected important crops including vegetables (*Allium* spp., *Brassica* spp., *Capsicum*, *Cichorium*, *Cucumis*, *Solanum melongena*, *Solanum tuberosum*); fruit crops (*Malus*, *Fragaria*, *Vitis*); and other miscellanaeous crops (*Beta*, *Coffea*, *Ginkgo*, *Glycine*, *Medicago*, *Saccharum*, *Sinocalamus latiflora*). We have included 11 chapters in the fourth volume on haploid breeding in cereals (wheat, rice, barley, oat, sorghum, maize, triticale, rye, pearl millet, buckwheat) whereas this fifth and final volume has emphasized ornamentals, tobacco, tomato, cotton, linseed, sunflower, asparagus, niger, and gynogenic haploids in angiosperms.

While preparing these volumes, we were overwhelmed by the enthusiastic response and timely cooperation of invited authors and the many research scientists who gave freely of their time to review the manuscripts. The reviewers were: R.I.S. Brettel (Australia); B.S. Ahloowalia (Austria); J.M. Bonga, K.N. Kao, K. Kasha, L.K. Kott, K.P. Pauls, R. Sadashivaiah (Canada); Hu Han (China); H. Ahokas, V. Kauppinen, J. Peltonen, S. Sarvori, L. Simola, P.M.A. Tigerstedt (Finland); C. Dóre, R.S. Sangwan (France); B. Foroughi-Wehr, D. Hess, S. Deimling, H. Uhrig, G. Wenzel (Germany); S.S. Bhojwani, P.B. Kirti, S.C. Maheshwari, A.F. Mascarenhas, P.S. Nadgauda, D. Pental, S.K. Raina, P.S. Rao, N. Sarin, G. Lakhshmi Sita (India); A. Mujeeb-Kazi (Mexico); H. Dons, K. Sree Ramulu (The Netherlands); D.S. Brar, G.S. Khush (Philippines); M. Zenkteler (Poland); A.M. Vieitez (Spain); C. Bornman, K. Glimelius, A. Wallin (Sweden); W. Chang (Taiwan); J. Dunwell, V.E. Franklin-Tong, W. Powell (UK); P.S. Baenziger, E. Earle, G.J. Galletta, D.J. Gray, P.K. Gupta, J.P. Helgeson, J. Janick, S.M. Reed, H.S. Rines, G. Schaeffer, T.L. Sims, I.K. Vasil, J.M. Widholm (USA).

It is still too early to write the last chapter on *in vitro* haploids. There are those who argue that its great potential will result in improved cultivars of many of our major crops. On the other hand, there are those who think its potential has been overrated, that the severe inbreeding depression observed among primary doubled haploids and the lack of selection pressure for functional sexual flower parts during the process of androgenesis will result only in useless, fruitless plants. The first cultivars employing anther-derived doubled haploids in their pedigree have already been released for a number of crops including wheat, rice, maize, and asparagus. Whether these cultivars and future such releases will endure remains to be seen.

S. Mohan Jain
S.K. Sopory
R.E. Veilleux

Preface to Volume 5

This fifth and final volume on *In Vitro* Haploid Production in Higher Plants, comprises 13 chapters and is divided into three sections. Most chapters are crop specific and each chapter contains an introduction about the selected plant, the techniques (anther culture, microspore culture, ovary/ovule culture) that have been successfully used for haploid production, the factors that have influenced the success of these techniques, the identification and genetic characterization of haploid regenerants, the application of haploids in breeding, and a brief conclusion on the potential of haploid breeding in the specific crop. The chapters contained in Section 1 are focussed on selected oil crops, including linseed (*Linum usitatissimum*), sunflower (*Helianthus annuus*) and niger (*Guizotia abyssinica*). Section 2 deals with ornamental plants including *Petunia*, lily (*Lilium* spp.), tulip (*Tulipa* spp.), and *Camellia japonica*. The third and final Section comprises miscellaneous plants including various fruit, vegetable, or specialty crops [asparagus (*Asparagus officinalis*), tobacco (*Nicotiana* spp.), cotton (*Gossypium* spp.), gynogenic haploids in angiosperms, *Citrus*, tomato (*Lycopersicon esculentum*), and interspecific somatic hybrids of potato (*Solanum brevidens* and *S. tubersosum*)]. Some chapters have also included a discussion of the potential of protoplast manipulations and genetic transformation of the particular crop under discussion.

Our thanks to authors for their time and effort to write these chapters and to the reviewers for working to improve the quality of the manuscripts.

S. Mohan Jain
S.K. Sopory
R.E. Veilleux

Acknowledgements

Ever since I finished my Ph.D. on *In vitro* haploids in higher plants, I have wanted to edit a book on this important subject and tucked this thought in the back of my mind. On separate opportunities, I mentioned it to Profs. S.K. Sopory and Richard E. Veilleux, inviting them to co-edit such a book. They graciously accepted my invitation to become co-editors. Their critical review of manuscripts and valuable suggestions have substantially improved the quality of these volumes. I am thankful to both Sudhir and Richard for helping me on this ambitious project and it has been a great pleasure working with them.

I appreciate the invited authors for their punctuality in meeting deadlines for submission of their contributions and all the reviewers (named in the General Preface) for constructive and timely critical reviews of the manuscripts. Their comments have been extremely useful for improving the quality of these volumes.

I am thankful to my colleagues Prof. Eija Pehu, Mr. Tapio Poutala, and Mr. Matti Teittinen for their assistance.

While editing this book, I had the opportunity to visit the University of Tuscia, Italy, as a visiting professor fellow. I am thankful to Prof. Eddo Rugini for his warm hospitality. During my short stay, I managed to find the time to edit several manuscripts.

Also, with great love and affection, I want to thank my daughters Sarita and Sonia, and my wife, Marja Liisa, for their unceasing patience and understanding while I was working on this time-consuming project.

Finally, I express my deepest sense of appreciation to Adrian Plaizier of Kluwer Academic Publishers, the Netherlands, for giving us the opportunity to work on this project. Adrian has always been cooperative and helpful, encouraging me with intelligent advice.

S. Mohan Jain
Book Project Leader

1. Haploidy and related biotechnological methods in linseed (*Linum usitatissimum* L.)

R. BERGMANN and W. FRIEDT

Contents

1. Introduction

Linseed or flax (*Linum usitatissimum* L.) is one of the oldest cultivated plants in temperate regions. The two primary uses of flax are as an oilseed crop and as a fibre crop. Regarding the fatty acid composition of linseed (oil flax) a high content of linolenic acid is desirable for oleochemical purposes. For human nutrition, however, a maximum proportion of 3% linolenic acid has been prescribed because higher contents increase the sensitivity to oxidation (rancidity). Flax fibres are needed for the production of linen in the textile industry. Additionally, short fibres may be used as insulating material for buildings or the lining of cars (Kaul *et al.*, 1994).

Vavilov (1926) described six different gene centres of linseed: India, the Near East, Ethiopia, Russia, southwestern Europe and western Europe. *Linum angustifolium* L. ($2n = 2x = 30$) is the presumed ancestor of domesticated linseed, *Linum usitatissimum* L. ($2n = 2x = 30$: Hoffmann, 1961; Khrzhanovskii *et al.*, 1980). Within the genus *Linum*, a wide karyotypical heterogeneity has been documented. Chromosome numbers of wild *Linum* species have been described to vary from $2n = 12$ to 84 (Martzenitzina, 1927; Kikuchi, 1929; Rogers & Harris, 1966; Plessers, 1966; Yermanos, 1966; Harris, 1968; Bari & Godward, 1970; Robertson, 1971; Chennaveeraiah & Joshi, 1983). Linseed is an autogamous species with only about 4% cross-fertilization. Therefore, flax breeders have mainly used pedigree selection, the bulk breeding method or progeny methods to develop breeding lines and new cultivars. Mutation induction (Green & Marshall, 1984; Nichterlein, 1987) and interspecific hybridization (Yermanos & Gill, 1969; Seetharam, 1972; Nickel, 1993) have been used to expose variation in fatty acid composition. About 200 wild species have been described in the genus *Linum* (Durrant, 1979) but only a few are sexually compatible with *L. usitatissimum*. Wild *Linum* species vary not only for fatty acid composition of the oil but also possess valuable resistance genes against pathogens. The application of

S.M. Jain, S.K. Sopory & R.E. Veilleux (eds.), In Vitro *Haploid Production in Higher Plants, Vol. 5,* 1–16.

biotechnology may be helpful to accelerate breeding programmes or to improve the efficiency of selection, as demonstrated in linseed and other oil crop species (Friedt, 1988, 1992). For example, new interspecific hybrids could be produced by the use of the embryo rescue technique or by protoplast fusion, respectively.

2. Haploidy in linseed

Haploid plants can be useful in plant breeding programmes for the rapid development of recombinant homozygous lines (doubled haploids) which combine various desirable characteristics. In linseed, haploidy can be achieved through *in vivo* methods by selection of polyembryonic seedlings or through *in vitro* methods of ovule, anther or isolated microspore culture. The term polyembryony encompasses different mechanisms that can result in twin seedlings:

a) the ovule develops into a multicellular mass from which multiple embryos develop,
b) a proembryo develops after normal fertilization but the terminal cell divides into two cells, each of which gives rise to an embryo,
c) twin embryos are derived from each of the daughter cells that result from the first mitotic division of the zygote,
d) synergids, antipodes or cells of the endosperm develop into embryos,
e) two nucelli develop within the same integument,
f) more than a single embryo sac derive from one nucellus,
g) a somatic cell of the ovule differentiates without meiosis and fertilization.

According to Johansen (1951), the term polyembryony should only be applied in cases a) and b) and for twin seedlings, respectively. The origin of the twin seedlings and the heritability of polyembryony in *Linum* were first described by Kappert (1950, 1953). Later, Wricke (1954) confirmed these results and conducted additional studies on the factors that influence polyembryony. He found that the date of sowing had a strong effect on the twinning rate. In some genotypes (i.e., "RA 91" and "Rocket 4") that produced haploid-diploid twins, the rate of haploid plant production was originally about 6–8% (Plessers, 1963; Rajhaty, 1976). After hybridization and selection, the rate of twin seedlings was increased up to 32% (Green & Salisbury, 1983). The heritability of polyembryony has been studied through crosses between haploid and diploid lines (Murray, 1980, 1985) as well as between twinning and non-twinning lines of flax. Presumably, additive gene action is primarily responsible for the production of twin seedlings. Haploid-diploid twin embryos are derived from genetically different nuclei. Cell division of the second, apomictic haploid nucleus has been reported to be dependent on fertilization of the ovule. To recover DHs, Rowland & Weerasena (1986) initiated a recurrent selection program using polyembryony found in selected genotypes of *L. usitatissimum*. It was more efficient to use the diploid twin seedling as the female parent than the haploid. Moreover, no significant difference in the twinning frequency of F_2 seed of crosses between twinning and non-twinning lines (range 0.8–1.9%) was observed. The twinning frequency of individual DHs within crosses varied from 0.4 to 20.3%.

3. Anther and microspore culture

First experiments on flax anther culture have been reported by Sun (1979) and Sun and Fu (1981) using certain genotypes and media. Sun (1979) selected out of 11 induction media the B5–medium (Gamborg *et al.*, 1968) as the most suitable with an average callus regeneration rate of 6.6%. The medium contained 4 mg/l IAA, 1–2 mg/l kin, 1500 mg/l lactalalbumin hydrolysate and 200 mg/l RNA. Sun and Fu (1981) also used 2 mg/l kin but 8 mg/l IAA for their medium. In our laboratory, experiments on linseed anther culture were initiated by Nichterlein *et al.* (1989). Later, Umbach (1991) was able to regenerate plants from anther-derived callus of different genotypes. However, both the growing conditions of donor plants and the genotype had a substantial influence on the regeneration rate (Nichterlein *et al.*, 1991; Umbach & Friedt, 1991). For growing the donor plants, different temperature regimes were tested: 18/16°C, 12/10°C, 16/10°C and 14/8°C (day/night). Best *in vitro* response and shoot regeneration from anther callus were achieved when the donor plants were grown under a 16 h day at 14°C/8°C (day/night) temperature when the rate of callusing anthers was about 58%. Under the other conditions only 4–18% of the anthers started callusing and neither calluses nor shoots could be observed. With optimum conditions for donor plant cultivation, the maximal shoot regeneration was obtained from calli on modified N6 medium (Chu, 1978) containing zeatin (1 mg l^{-1}). Direct embryogenesis of microspores in anther culture, however, remains difficult. Although microspore embryos have been observed, all regenerated flax plantlets have developed through a callus phase. A schematic overview on the production of DHs of linseed by anther culture is presented in Fig. 1. The applicability of anther culture for the production of DHs as an alternative to pedigree breeding has been investigated by Bickert *et al.* (1993), Bickert (1994) and Friedt *et al.* (1995). From four cross combinations, a total of 83 DHs were produced and propagated in the nursery. Analyses of the fat content and the fatty acid composition of seeds showed remarkable variation among DHs. As an example the fat content and the linolenic acid proportion of DHs of the cross "Atalante" × "Avantgard" are presented in Table 1.

The culture of isolated microspores may further improve the efficiency of haploid extraction for the production of DHs (Fig. 2). Plant regeneration from microspores has been achieved via organogenesis (Nichterlein *et al.*, 1992; Nichterlein & Friedt, 1993). Although embryos were observed in microspore cultures, it has not yet been possible to regenerate plants directly from microspores. Developing flower buds of the donor plants were har-

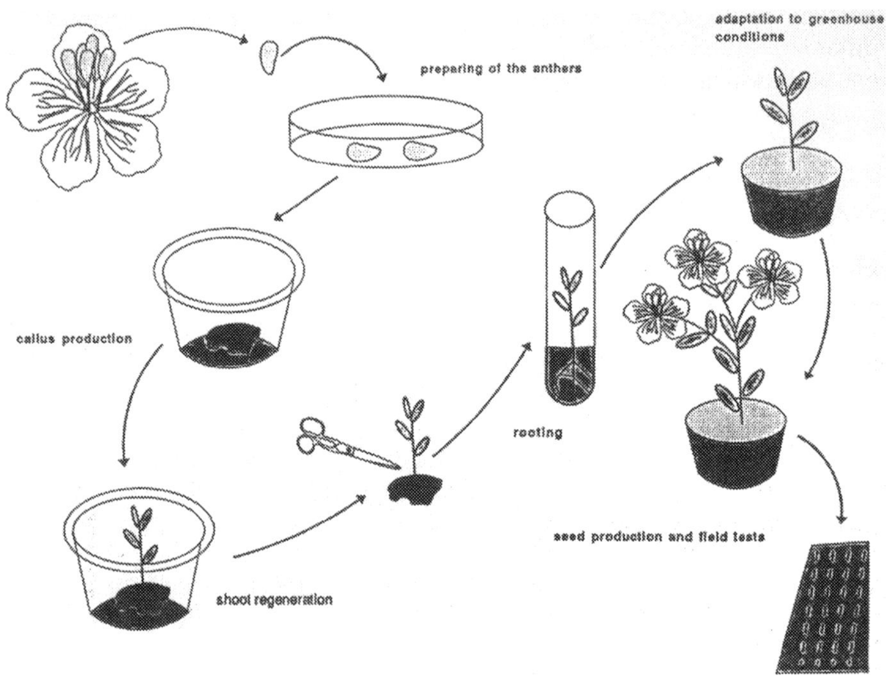

Figure 1. Production of doubled haploid lines of linseed by anther culture technique (mod. from Bickert, 1994).

vested before anthesis when microspores were in the late uninucleate stage. The microspore stage was determined by using a hematoxylin stain (Kindinger & Beckett, 1985). This stage is normally present in flower buds of 1.5–2.3 cm diameter. The viability of microspores was tested by staining in Alexander's solution (Alexander, 1969) (Fig. 4). First microspore divisions were observed in a modified NLN 82 liquid medium described by Gland *et al.* (1988) after two weeks incubation (Fig. 3). Calli larger than 1 mm diameter were transferred onto solid modified N6 medium (Chu, 1978) containing 3% maltose, 375 mg l^{-1} glutamine, 250 mg l^{-1} asparagine, 125 mg l^{-1} serine and 1 mg l^{-1} zeatin (Nichterlein *et al.*, 1991) to initiate shoots. Shoot regeneration occurred from 36–66% of the cultured calli (Fig. 5). Shoots larger than 3 cm were rooted on MS basal medium supplemented with IAA and developed into normal plants that were adapted to greenhouse conditions for seed production (Fig. 6, Table 2). Although regeneration rate was low compared to anther culture, later results were promising (Bickert *et al.*, 1993).

The use of haploidy in practical linseed breeding programmes has been limited and will require technological improvement in order to gain more general acceptance (cf. Friedt *et al.*, 1995). The direct production of homo-

Table 1. Linolenic acid content, fat content, colour of seeds and colour of blossoms of DH-lines of the cross 'Atalante' × 'Avantgard' (Bickert, 1994)

DH-line	% C 18:3	Fat (%)	Colour of blossom	Colour of seeds
"Atalante"	56.8	40.3	Blue	Brown
"Avantgard"	66.8	41.2	White	Yellow
1	54.8	37.6	Blue	Brown
3	56.5	36.9	Blue	Brown
4	58.9	38.6	Blue	Brown
5	57.6	38.2	Blue	Brown
6	50.4	37.6	Blue	Brown
8	62.8	38.0	White	Brown
9	55.8	37.9	Blue	Brown
12	61.2	38.2	Blue	Brown
13	58.8	40.5	White	Yellow-brown
14	58.7	37.5	White	Brown
15	56.7	37.3	Blue	Brown
17	59.2	38.0	White	Brown
18	59.6	39.9	Blue	Brown
20	58.3	38.5	Blue	Brown
21	57.0	39.3	Blue	Brown
26	56.7	37.5	White	Brown
30	64.4	37.7	White	Yellow
32	58.1	38.0	Blue	Brown
34	53.7	37.2	Blue	Brown-yellow
42	57.1	38.8	Blue	Brown
43	59.1	37.7	White	Yellow-brown
44	53.7	37.0	Blue	Brown
45	53.9	37.9	Blue	Brown
46	54.0	37.3	Blue	Brown
47	65.9	38.2	White	Brown-yellow
48	58.9	38.3	White	Brown-yellow
53	58.1	37.3	Blue	Brown
55	55.8	36.1	Blue	Brown
57	59.0	35.9	Blue	Brown
58	54.4	38.0	Blue	Brown
61	58.5	37.4	White	Brown
62	57.0	38.9	Blue	Brown
63	61.5	38.2	Blue	Brown
64	55.9	38.2	White	Brown
67	65.0	37.2	Blue	Brown
69	58.6	38.6	White	Brown-yellow
70	58.3	37.8	White	Yellow

zygous DHs will assist in selecting for qualitative traits like high linolenic acid content or disease resistance. Besides haploid technology, there are additional biotechnologies that can be used to increase variability in the linseed genome for subsequent genetic improvement via breeding.

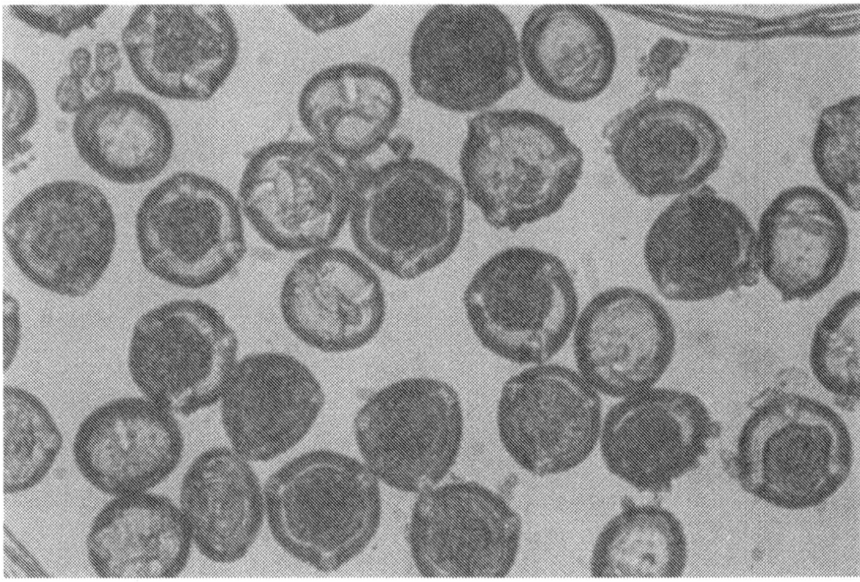

Figure 2. Isolated microspores of linseed.

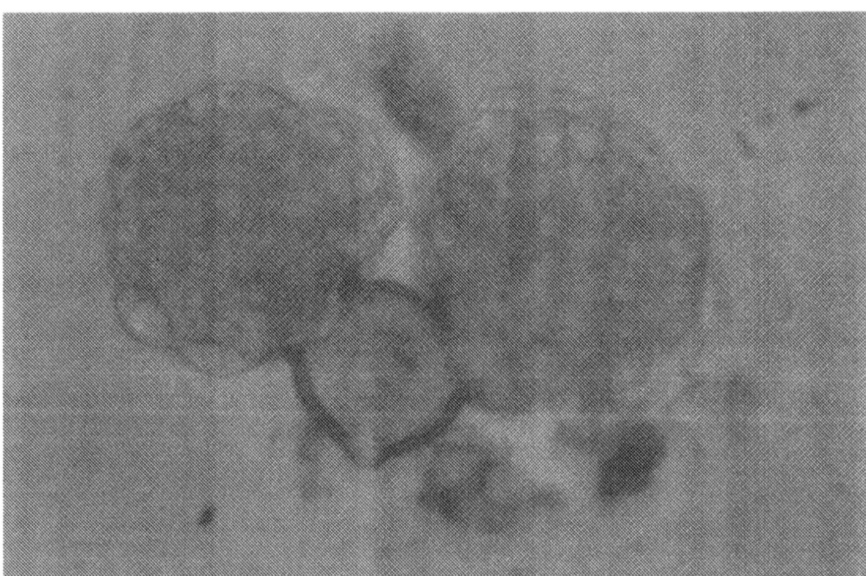

Figure 3. First cell divisions in a linseed microspore.

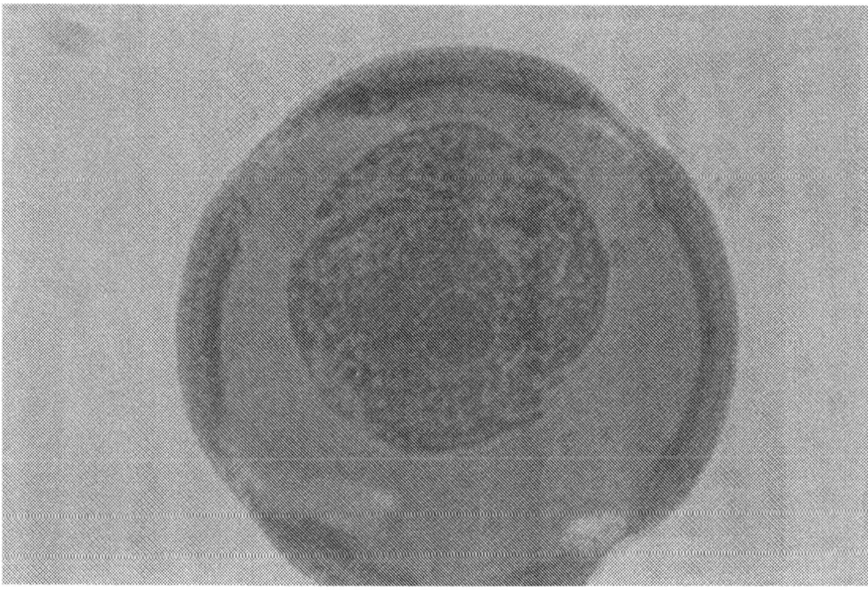

Figure 4. Microspore of linseed in the late uninucleate stage stained with Alexander solution.

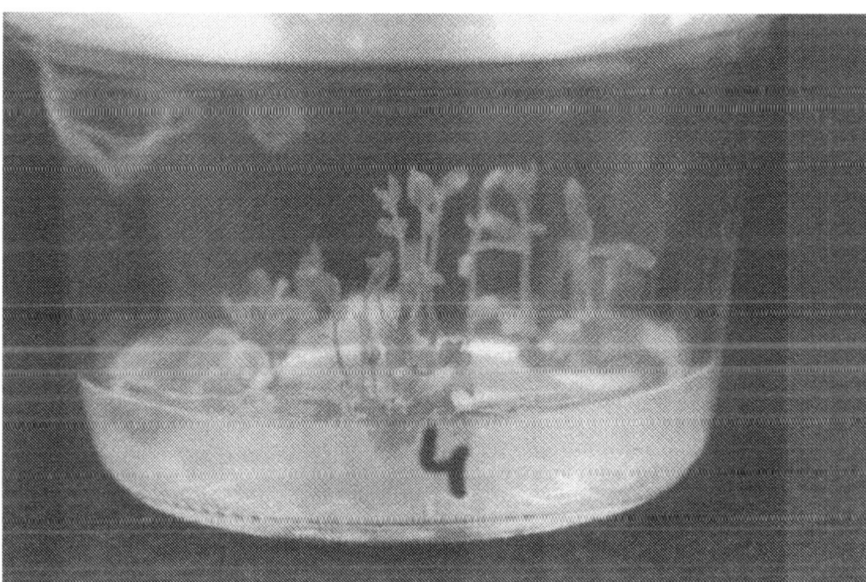

Figure 5. Differentiation of shoots from calli of linseed microspores.

Figure 6. Linseed shoot transferred onto rooting medium.

4. Embryo rescue

The search for new sources of genetic variation in wild species of the genus *Linum* has resulted in a number of interesting traits, such as increased oil quality and disease resistance. However, most wild species are not sexually compatible with *L. usitatissimum* due to pre- or post-fertilization barriers (Seetharam, 1972; Green, 1983; Gosh & Shivanna, 1984). By application of an improved embryo rescue technique, Nickel (1993) first succeeded in the regeneration of plants from globular embryos after interspecific hybridization between *L. usitatissimum* and *L. monogynum* (cf. Friedt *et al.*, 1989). But embryo rescue can only be applied in cases of post-fertilization barriers. When pre-fertilization incompatibilities exist, the genomes of the different species may be combined by somatic hybridization.

5. Protoplast fusion

A basic requirement for successful protoplast fusion is a reliable plant regeneration system from protoplast culture. First results of flax protoplast regeneration were reported by Gamborg & Shyluk (1976). They isolated protoplasts from hypocotyl segments and observed cell division and production of calli in media with various cytokinins. Organogenesis with root formation

Table 2. Composition of media for microspore culture in linseed (Nichterlein & Friedt, 1993)

Components	Mod. NLN 82 (mg/l)	Mod. N6 (mg/l)	SV 4* (mg/l)	Rooting medium (mg/l)
$Ca(NO_3)_2 \cdot 4H_2O$	500	–	–	–
$CaCl_2 \cdot 2H_2O$	–	166	440	440
KH_2PO_4	125	400	170	170
KNO_3	125	2830	1900	1900
$MgSO_4 \cdot 7H_2O$	125	185	370	370
NH_4NO_3	–	–	1650	1650
$(NH_4)_2SO_4$	–	463	–	–
$CoCl_2 \cdot 6H_2O$	0.025	–	0.025	0.025
$CuSO_4 \cdot 5H_2O$	0.025	–	0.025	0.025
H_3BO_3	10	1.6	6.2	6.2
KI	–	0.8	0.83	0.83
$MnSO_4 \cdot H_2O$	18.94	3.3	16.9	16.9
$NaMoO_4 \cdot 2H_2O$	0.25	–	0.25	0.25
$ZnSO_4 \cdot H_2O$	7.69	1.5	8.6	8.6
$FeSO_4 \cdot 7H_2O$	27.8	27.8	27.8	27.8
$FeNa_2 \cdot EDTA$	37.3	37.3	37.3	37.3
L-Asparagine	–	250	–	–
L-Glutamine	800	375	–	–
L-Glycine	2	2	2	2
L-Serine	100	125	–	–
Nicotinic acid	5	0.5	0.5	0.5
Pyridoxine-HCl	0.5	0.5	0.5	0.5
Thiamine-HCl	0.5	1.0	0.1	0.1
Biotin	0.05	–	–	–
Folic acid	0.5	–	–	–
Ca-Panthothenate	0.5	–	–	–
myo-Inositol	100	100	100	100
Sucrose	130,000	–	30,000	10,000
Maltose		31,560	–	–
BA	0.1	–	1	–
Zeatin	–	10	–	–
IAA	0.3	–	0.2	0.1
NAA	0.5	–	–	–
Kinetin	–	–	–	–
$AgNO_3$	–	–	1	1

* SV 4 = shoot propagation medium.

could, at least, be initiated. The first regenerated plants from isolated protoplasts of *L. usitatissimum* were reported by Barakat & Cocking (1983). They used hypocotyls, roots and cotyledons of young seedlings as donor tissues for protoplast isolation. With protoplasts derived from hypocotyl tissue, only rhizogenesis was observed, whereas plant regeneration was achieved from protoplasts isolated from roots and cotyledons. Based on these results, Barakat & Cocking (1985) also succeeded in regeneration from protoplasts of some wild species: *L. lewisii* and *L. strictum* produced whole plantlets via organogenesis whereas only rhizogenesis was observed for *L. alpinum*, *L.*

narbonense, *L. grandiflorum* and *L. altaicum*. Ling & Binding (1987) reported plant regeneration from five wild *Linum* species and eight cultivars of linseed; with *L. alpinum*, a somatic embryo was found in regenerating protoplast cultures. Using a similar procedure, Zhan *et al.* (1989) regenerated plants from protoplasts of *L. marginale*. Regeneration of protoplasts via somatic embryogenesis was further improved by Ling & Binding (1992) and extended to include the wild species, *L. alpinum*, *L. amurense*, *L. hologynum* and *L. perenne*. Subsequently, about 20% of the embryos could be converted into plants on MS medium containing 0.5 μM IBA and 0.25 μM each of BA, kin and zeatin.

Successful somatic hybridization between *Linum usitatissimum* (cv. Linda) and *L. monogynum* has recently been reported by Bergmann *et al.* (1993, 1994). Protoplasts of young cotyledons of both species were electrofused and incubated in agarose droplets in a modified NLN 82 medium described by Lichter (1985). For the identification of interspecific hybrids representing heterokaryotic fusion products, isoenzyme electrophoresis was used. The enzyme systems 6–phosphogluconate dehydrogenase (6–PGD) and peroxidase (PER) were sufficiently polymorphic for detection of putative somatic hybrids.

Regarding the multiple applications of plant tissue culture in *Linum* Table 3 gives a compact overview on media and inocula used by different authors.

6. Genetic transformation

In linseed and flax, genetic transformation has been achieved by two functional vector systems: a) *Agrobacterium tumefaciens*, and b) *A. rhizogenes*. Crown gall induction by wild type Ti-plasmids of *A. tumefaciens* has been demonstrated by Hepburn *et al.* (1982, 1983). Basiran *et al.* (1987) used hypocotyl segments for the production of transformed shoots by an *A. tumefaciens* strain. They succeeded in transformation and efficient expression of a functional nopaline synthase (*nos*) gene and a chimaeric nos-neomycin phosphotransferase II (*npt-II*) construct to allow selection of kanamycin resistant transformed plant cells. Transgenic shoots were easily obtained from kanamycin resistant callus developing on hypocotyl sections. By Southern hybridization T-DNA was detected in all kanamycin resistant tissues.

Herbicide resistance genes have also been successfully used in transformation experiments (McHughen *et al.*, 1986; McHughen, 1992). Stable transformation in the progenies of transgenic flax plants was reported by using an *A. tumefaciens* strain carrying a disarmed Ti-plasmid vector containing a chimeric *npt-II* gene and a glyphosate resistance gene (Jordan & McHughen, 1988a). The non-specific herbicide, glyphosate, acts by inhibiting the enzyme 5–enolpyruvylshikimate-3–phosphate synthase which is a part of the shikimate pathway involved in the synthesis of aromatic amino acids. Transformed shoots were regenerated from the inoculated tissue and their transgenic

Table 3. Summary of media used for various inocula (mod. McHughen, 1990)

Media for callus formation	Initial explant	Authors
Norstog B_{II} + 2,4–D	Hypocotyl	McHughen & Swartz, 1984
MS + 2,4–D	Hypocotyl	McHughen & Swartz, 1984
	Protoplasts	Ling & Binding, 1992
MS + BA + NAA	Stem pieces	Cullis, 1981
White + 2,4–D	Cotyledons, hypocotyl	Mathews & Narayanaswamy, 1976
MS + 2,4–D	Cotyledons, hypocotyl	Mathews & Narayanaswamy, 1976
White + kin + NAA	Cotyledon	Ibrahim, 1971
MS (Major salts) + B5	Hypocotyl, protoplasts	Gamborg & Shyluk, 1976
MS + kin + IAA	Cotyledon	Ibrahim & Phan, 1978
MS + kin + 2,4–D	Hypocotyl	Quatrano, 1968
N6 + zeatin	Microspores	Nichterlein *et al.*, 1991
Shoot or bud formation		
MS + BA + NAA	Stem pieces	Murray *et al.*, 1977
MS + BA + NAA	Protoplasts	Barakat & Cocking, 1983
MS + kin + IBA	Hypocotyl	McHughen & Swartz, 1984
MS	Hypocotyl	Mathews & Narayanaswamy, 1976
MS + IBA	Hypocotyl	Mathews & Narayanaswamy, 1976
MS + BA	Hypocotyl, apical meristem	Lane, 1979
MS + BA	Hypocotyl	Gamborg & Shyluk, 1976
LS + kin + NAA	Cotyledons	Rybczynski, 1975
MS + BA + kin + zeatin + IBA	Protoplasts	Ling & Binding, 1992
Shoot elongation		
MS	Stem pieces	Murray *et al.*, 1977
Rooting		
MS + NAA	Hypocotyl, apical meristem	Lane, 1979
MS + NAA	Cotelydons	Mathews & Narayanaswamy, 1976
B5	Stem pieces	Murray *et al.*, 1977
1/2 B5 + IAA	Protoplasts	Barakat & Cocking, 1983
	Protoplasts	Gamborg & Shyluk, 1976
B5		McHughen & Swartz, 1984
MS or B5 + NAA	Microspores	Nichterlein *et al.*, 1991
Embryos		
MS + NAA	Protoplasts	Ling & Binding, 1992

nature was confirmed by a combination of leaf-callus assay, nopaline assay and progeny tests. An inoculation method with a transformation efficiency of 10% has been developed. This rate is high enough for the successful incorporation of genes of agronomic importance and selectable markers, e.g., herbicide, disease or antibiotic resistance. After selection of transformed callus, however, careful analysis of regenerated shoots must be carried out to distinguish escapes from true transgenics (Jordan & McHughen, 1988b). The rate of cellular transformation on flax hypocotyl explants was investigated by Dong & McHughen (1991). The binary plasmid p35GUSINT containing GUS as a selectable marker in the octopine-type *Agrobacterium* strain GV2260 was used as a vector system. A prolonged co-cultivation resulted in high transformation rates. The efficiency of these two vector systems was compared by testing six flax cultivars (Zhan *et al.*, 1988). Explants inoculated with *A. tumefaciens* resulted in transformed callus only and transformed shoots could not be regenerated. However, cotyledon explants from cvs. Bombay and Precederia inoculated with *A. rhizogenes* formed transformed roots which then developed transformed shoots. Zhan *et al.* (1988) reported transformed plantlets with curled leaves, short internodes and a root system characterized by plagiotropic behaviour. The presence of T-DNA from the Ri plasmid was confirmed by Southern hybridization analysis.

McHughen & Holm (1991) tested two transgenic flax lines derived from cv. Norlin with a mutant *Arabidopsis* ALS (acetolactate synthase) gene conferring resistance to sulfonylurea herbicides in a replicated, field trial. The results showed that there were no significant differences between the transgenic lines and the parent for any agronomic trait measured in untreated soil. There were also no significant differences for performance of the transgenic lines in the untreated and the herbicide treated soils. However, the parent was devastated by the presence of the herbides in treated soil. In another field trial, five transgenic flax lines containing the ALS gene and four check cultivars were compared at three locations. Here it could also be proved that there was no detrimental effect of the T-DNA on important agronomic traits (McHughen & Rowland, 1991).

7. Prospects

For the future, merging the techniques of genetic engineering with those of haploid induction may result in the efficient production of homozygous doubled haploid transgenic linseed or flax with improved gene expression. Because of the self-pollinating nature of cultivated flax, such transgenics, if sufficiently fertile, could result in direct cultivar release after field testing for agronomic performance.

8. References

Alexander, P., 1969. Differential staining of aborted and non-aborted pollen. Stain Technol. 44: 117–122.

Barakat, M.N. & E.C. Cocking, 1983. Plant regeneration from protoplast-derived tissues of *Linum usitatissimum* L. (flax). Plant Cell Rep. 2: 314–317.

Barakat, M.N. & E.C. Cocking, 1985. An assessment of cultural capabilities of protoplasts of some wild species of *Linum*. Plant Cell Rep. 4: 164–167.

Bari, G. & M.B.E. Godward, 1970. Interspecific crosses in *Linum*. Euphytica 19: 443–446.

Basiran, N., P. Armitage, R.J. Scott & J. Draper, 1987. Genetic transformation of flax (*Linum usitatissimum*) by *Agrobacterium tumefaciens*: Regeneration of transformed shoots via a callus phase. Plant Cell Rep. 6: 396–399.

Bergmann, R., A. Penner, H. Köhler & W. Friedt, 1993. Interspezifische somatische Hybridisierung in der Gattung *Linum* mit dem Ziel extremer Fettsäurevarianten. Bericht 44. Tagung, Arbeitsgem. Saatzuchtleiter, Gumpenstein, p. 199. Bundesanstalt für alpenländ. Landwirtschaft, Gumpenstein.

Bergmann, R., A. Penner, H. Köhler & W. Friedt, 1994. Elektrofusion als eine Methode zur Erweiterung der genetischen Variabilität in der Gattung *Linum*. 2. GPZ-Tagung Quedlinburg, 2.–5. März 1994, Vortr. Pflanzenzüchtung 28: 41–43.

Bickert, C., 1994. Untersuchungen zur Züchtung von Öllein (*Linum usitatissimum* L.) mit Hilfe der Antherenkulturtechnik und der Pedigree-Selektion im Hinblick auf eine Nutzung als nachwachsender Rohstoff. Diss., Universität Giessen. Shaker Verlag, Aachen.

Bickert, C., R. Bergmann, A. Büchsenschütz-Nothdurft & W. Friedt, 1993. Einsatz der Antheren- und Mikrosporenkultur in der Züchtung von Öllein im Hinblick auf eine Verbesserung von Fettgehalt und Fettqualität. Bericht 44. Tagung, Arbeitsgem. Saatzuchtleiter, Gumpenstein, pp. 115–121. Bundesanstalt für alpenländ. Landwirtschaft, Gumpenstein.

Chennaveeraiah, M.S. & K.K. Joshi, 1983. Karyotypes in cultivated and wild species of *Linum*. Cytologia 48: 833–841.

Chu, C.-C., 1978. The N_6 medium and its application to anther culture of cereal crops. In: Proc. Symp. Plant Tissue Culture, pp. 43–50. Science Press, Beijing.

Cullis, C.A., 1981. DNA sequence organisation in the flax genome. Biochem. Biophys. Acta 652: 1–15.

Dong, J.-Z. & A. McHughen, 1991. Patterns of transformation intensity on flax hypocotyls inoculated with *Agrobacterium tumefaciens*. Plant Cell Rep. 10: 555–560.

Durrant, A., 1979. Flax and linseed – *Linum usitatissimum*. In: N.W. Simmonds (Ed.), Evolution of Crop Plants, pp. 190–193. Longman, London.

Friedt, W., 1988. Biotechnology in breeding of industrial oil crops: the present status and future prospects. Fat Sci. Technol. 90: 51–55.

Friedt, W., 1992. Breeding and agronomic development of linseed and sunflower for technical markets. In: K. Anthony (Ed.), New Crops for Temperate Regions, pp. 222–234. Chapman & Hall, London.

Friedt, W., K. Nichterlein & M. Nickel, 1989. Biotechnology in breeding of flax (*Linum usitatissimum*): In: G. Marshall (Ed.), Flax: Breeding and Utilisation, pp. 5–13. Kluwer Academic Publishers, Dordrecht.

Friedt, W., C. Bickert & H. Schaub, 1995. *In vitro* breeding of high linolenic, doubled-haploid lines of linseed (*Linum usitatissimum* L.) via androgenesis. Plant Breed. 114: 322–326.

Gamborg, O.L. & J.P. Shyluk, 1976. Tissue culture, protoplasts, and morphogenesis in flax. Bot. Gaz. 4: 301–306.

Gamborg, O.L., R.A. Miller & K. Ojima, 1968. Nutrient requirements of suspension cultures of soybean root cells. Exp. Cell. Res. 21: 359–368.

Gland, A., R. Lichter & H.-G. Schweiger, 1988. Genetic and exogenous factors affecting embryogenesis in isolated microspore cultures of *Brassica napus* L. J. Plant Physiol. 132: 613–617.

Gosh, S. & K.R. Shivanna, 1984. Interspecific incompatibility in *Linum*. Phytomorphology, Delhi 34: 128–135.

Green, A.G., 1983. Interspecific hybridization in the genus *Linum*. In: Proc. Austral. Plant Breed. Cong., Adelaide, pp. 302–304.

Green, A.G. & P.A. Salisbury, 1983. Inheritance of polyembryony in flax (*Linum usitatissimum*). Can. J. Genet. Cytol. 25: 117–121.

Green, A.G. & D.R. Marshall, 1984. Isolation of induced mutants in linseed (*Linum usitatissimum*) having reduced linolenic acid content. Euphytica 33: 321–328.

Harris, B.D., 1968. Chromosome numbers and evolution in North American species of *Linum*. Amer. J. Bot. 55: 1197–1204.

Hepburn, A.G., L.E. Clark & C.A. Cullis, 1982. *Agrobacterium tumefaciens* T-DNA inserts into the flax genome. Heredity 49: 136–137.
Hepburn, A.G., L.E. Clark, K.S. Blundy & J. White, 1983. Nopaline Ti-plasmid, pTiT37, T-A insertions into a flax genome. J. Mol. Appl. Genet. 2: 211–224.
Hoffmann, W., 1961. Lein, *Linum usitatissimum* L. In: Th. Roemer & W. Rudorf (Eds.), Handbuch der Pflanzenzüchtung Bd. 5, 2. Auflage, pp. 264–366. Verlag Paul Parey, Berlin.
Ibrahim, R.K., 1971. Media for growth of flax tissue culture. Can. J. Bot. 49: 295–298.
Ibrahim, R.K. & C.T. Phan, 1978. Phenolic synthesis in relation to chloroplast ultrastructure in flax callus and cell suspension cultures. Biochem. Physiol. Pflanz. 172: 199–212.
Johansen, D.A., 1951. Plant Embryology (Embryogeny of the Spermatophyta). Chronica Botanica Co., (A new series of plant science books, v. 24.) Waltham, Mass., USA.
Jordan, M.C. & A. McHughen, 1988a. Glyphosate tolerant flax plants from *Agrobacterium* mediated gene transfer. Plant Cell Rep. 7: 281–284.
Jordan, M.C. & A. McHughen, 1988b. Transformed callus does not necessarily regenerate transformed shoots. Plant Cell Rep. 7: 285–287.
Kappert, H., 1950. Botanische Untersuchungen zur Erblichkeit der Polyembryonie. Mod. Biol. (Festschr. f. Hans Nachtsheim), pp. 80–194.
Kappert, H., 1953. Erbliche Polyembryonie bei *Linum usitatissimum*. Biol. Zbl. 53: 276–307.
Kaul, H.-P., M. Scheer-Triebel & K.-U. Heyland, 1994. Selection criteria for short-fibre flax. Plant Breed. 113: 130–136.
Khrzhanovskii, V.G., S.F. Ponomarenko & V.A. Doguzashvilii, 1980. The problem of the origin and evolution of the genus *Linum* L. Biol. Bull. Acad. Sci. USSR 6: 580–595.
Kikuchi, M., 1929. Cytological studies of the genus *Linum* L. Jpn. J. Genet. 4: 202–212.
Kindinger, B. & J.B. Beckett, 1985. A hematoxylin staining procedure for maize pollen grain chromosomes. Stain Technol. 60: 265–269.
Lane, W.D., 1979. Influence of growth regulators on root and shoot initiation from flax meristem-tips and hypocotyls *in vitro*. Physiol. Plant 45: 260–264.
Lichter, R., 1985. From microspores to rape plants: a tentative way to low glucosinolate strains. In: World Crops Production, Utilization, Description, Vol. 11, pp. 268–277. Martinus Nijhoff/W. Junk, Dordrecht.
Ling, H.Q. & H. Binding, 1987. Plant regeneration from protoplasts in *Linum*. Plant Breed. 198: 312–317.
Ling, H.Q. & H. Binding, 1992. Improvement of plant regeneration from *Linum* protoplasts by the induction of somatic embryogenesis. J. Plant Physiol. 139: 422–426.
Linsmaier, E.M. & F. Skoog, 1965. Organic growth factor requirements of tobacco tissue cultures. Physiol. Plant 18: 100–127.
Martzenitzina, K.K., 1927. The chromosome of some species of the genus *Linum*. Bull. Appl. Bot. Genet. Plant Breed. 17: 253–264.
Mathews, V.H. & S. Narayanaswamy, 1976. Phytohormone control of regeneration in cultured tissues of flax. Z. Pflanzenphysiol. 80: 436–442.
McHughen, A., 1990. Flax (*Linum usitatissimum* L.): *In vitro* studies. In: Y.P.S. Bajaj (Ed.), Legumes and Oilseed Crops I. Biotechnology in Agriculture and Forestry, Vol. 10, pp. 502–514. Springer-Verlag, Berlin.
McHughen, A., 1992. Genetic engineering for crop improvement: the linseed/flax story. AgBiotech News Information 4: 53–58.
McHughen, A. & M. Swartz, 1984. A tissue culture derived salt-tolerant line of flax (*Linum usitatissimum* L.). J. Plant Physiol. 117: 109–117.
McHughen, A. & F. Holm, 1991. Herbicide resistant transgenic flax field test: Agronomic performance in normal and sulfonylurea-containing soils. Euphytica 55: 49–56.
McHughen, A. & G.G. Rowland, 1991. The effect of T-DNA on the agronomic performance of transgenic flax plants. Euphytica 55: 269–275.
McHughen, A., R. Browne, D. Kneesaw & M. Jordan, 1986. Ti-mediated transformation and regeneration of flax plants *in vitro*. In: 6th Int. Congr. Plant Tissue and Cell Culture, Minneapolis, Book of Abstracts, p. 130.

Murashige, T. & F. Skoog, 1962. A revised medium for rapid growth and bioassays with tobacco tissue cultures. Physiol. Plant 15: 473–497.

Murray, B., 1980. Diploid, aneuploid and triploid plants from haploid × diploid crosses in flax (*Linum usitatissimum*). Can. J. Genet. Cytol. 22: 591–596.

Murray, B., 1985. Studies of haploid-diploid twins in flax. Can. J. Genet. Cytol. 19: 177–186.

Murray, B.E., R.J. Handyside & W.A. Keller, 1977. *In vitro* regeneration of shoots on stem explants of haploid and diploid flax (*Linum usitatissimum*). Can. J. Genet. Cytol. 19: 177–186.

Nichterlein, K., 1987. Untersuchungen zur züchterischen Nutzung mutagen induzierter Variation des Fettsäuremusters beim Lein (*Linum usitatissimum* L.). Diss., Universität Giessen, Giessen.

Nichterlein, K. & W. Friedt, 1993. Plant regeneration from isolated microspores of linseed (*Linum usitatissimum* L.). Plant Cell Rep. 12: 426–430.

Nichterlein, K., H. Umbach & W. Friedt, 1989. Investigations on androgenesis in breeding of linseed (*Linum usitatissimum* L.). In: Vortr. f. Pflanzenzüchtung 15/I, Proc. 12th EUCARPIA-Congr., Göttingen, pp. 13–25.

Nichterlein, K., H. Umbach & W. Friedt, 1991. Genotypic and exogenous factors affecting shoot regeneration from anther callus of linseed (*Linum usitatissimum* L.). Euphytica 58: 157–164.

Nichterlein, K., R. Bergmann & W. Friedt, 1992. Development of doubled haploids by microspore culture of *Linum usitatissimum*. In: XIIIth EUCARPIA-Congress, Angers, Book of Poster Abstracts, pp. 201–202.

Nickel, M., 1993. Untersuchungen zur Erweiterung der genetischen Variation des Fettsäuremusters beim Lein (*Linum usitatissimum* L.) mit Hilfe von Artkreuzungen und *in vitro* Techniken. Diss., Universität Giessen. Shaker Verlag, Aachen.

Norstog, K., 1973. New synthetic medium for the culture of premature barley embryos. In Vitro 8: 307–308.

Plessers, A.G., 1963. Haploids as a tool in flax breeding. Cereal News 8: 3–6.

Plessers, A.G., 1966. The variation in fatty acid composition of the seed of *Linum* species. Can. J. Genet. Cytol. 8: 328–335.

Quatrano, R., 1968. Freeze-preservation of cultured flax cells utilizing dimethyl sulfoxide. Plant Physiol. 43: 2057–2061.

Rajhaty, T., 1976. Haploid flax revisited. Z. Pflanzenzüchtg. 76: 1–10.

Robertson, K.R., 1971. The Linaceae in the southeastern United States. J. Arnold Arboretum 52: 649–665.

Rogers, C.M. & B.D. Harris, 1966. Some new chromosome numbers in *Linum*. Madrono 18: 179–180.

Rowland, G.G. & L.A. Weerasena, 1986. Some observations on polyembryony in crosses between twinning and nontwinning lines of flax. Can. J. Plant Sci. 66: 819–824.

Rybczynsky, J.J., 1975. Callus formation and organogenesis of mature cotyledons of *Linum usitatissimum* L. var. *szokijskij in vitro* culture. Genet. Pol. 16: 1–12.

Seetharam, A., 1972. Interspecific hybridization in *Linum*. Euphytica 21: 489–495.

Sun, H., 1979. Preliminary report on anther culture of flax (*Linum usitatissimum* L.). K'o Hsueh T'ung Poa Kexue Tong bao 24: 948–950.

Sun, H. & W. Fu, 1981. Induction of pollen plants in flax (*Linum usitatissimum* L.) and preliminary observations on performance of their progenies. Acta Genet. Sin. 8: 369–374.

Umbach, H.A., 1991. Untersuchungen zur züchterischen Verbesserung des Ölleins (*Linum usitatissimum* L) als nachwachsender Rohstoff für die Oleochemie und Entwicklung einer verbesserten "Haploiden-Methode" *in vitro*. Diss., Universität Gießen, Giessen.

Umbach, H. & W. Friedt, 1991. Praxisrelevante Ergebnisse zur Anwendung der Haploidmethode im Züchtungsgang von Lein (*Linum usitatissimum* L.). Bericht 42. Tagung, Arbeitsgem. Saatzuchtleiter, Gumpenstein, pp. 155–162. Bundesanstalt für alpenländ. Landwirtschaft, Gumpenstein.

Vavilov, N.J., 1926. Studies on the origin of cultivated plants. Bull. appl. Bot. Gen. Pl. Br. 16: 1–248.

White, P.R., 1963. The Cultivation of Animal and Plant Cells. 2nd edn. Ronald Press, New York.

Wricke, G., 1954. Die Manifestierung der erblich bedingten Polyembryonie beim Lein in ihrer Abhängigkeit von Außenfaktoren. Sonderdr. aus Biol. Zentralblatt 73: 49–88.

Yermanos, D.M., 1966. Variability in seed oil composition of 43 *Linum* species. J. Am. Oil. Chem. Soc. 43: 546–549.

Yermanos, D.M. & K.S. Gill, 1969. Cytology of autotetraploids of *L. usitatissimum* L. and *Linum angustifolium* Huds. and their amphidiploid hybrids. Crop Sci. 9: 249–250.

Zhan, X.-C., D.A. Jones & A. Kerr, 1988. Regeneration of flax plants transformed by *Agrobacterium rhizogenes*. Plant Mol. Biol. 11: 551–559.

Zhan, X.-C., D.A. Jones & A. Kerr, 1989. *In vitro* plantlet formation in *Linum marginale* from cotyledons, hypocotyls, leaves, roots and protoplasts. Aust. J. Plant Physiol. 16: 315–320.

2. Haploid production and application of molecular methods in sunflower (*Helianthus annuus* L.)

W. FRIEDT, T. NURHIDAYAH, T. RÖCHER, H. KÖHLER, R. BERGMANN and R. HORN

Contents

1. Introduction

Cultivated sunflower (*Helianthus annuus* L.) is one of the most important oil crops of the world. Its seed oil can be used for human consumption as well as a raw material for oleochemistry. It can also be used as a substitute for mineral oil in various applications, such as fuels, lubricants or oils for hydraulic systems.

The genus *Helianthus* comprises 67 species native to the Americas. Two species, *H. annuus* and *H. tuberosus* are cultivated as food plants and several species are grown as ornamentals. The basic chromosome number of the genus *Helianthus* is $n = 17$. Although diploid, tetraploid, and hexaploid species occur, the commercial sunflower is a diploid (Heiser, 1978).

Sunflower is a predominantly cross-pollinated species, with pollination primarily accomplished by insects and only to a limited degree by wind. A system of genetic self-incompatibility exists in certain lines, but generally a wide range of self-fertility occurs among individual plants in breeding populations.

Most breeding procedures utilized in cross-fertilized species like maize (*Zea mays* L.) are suitable for sunflower, with certain modifications required due to floral morphology. Development of hybrids is a primary objective of most breeding programmes, although improved open-pollinated cultivars and synthetic cultivars may also have value, especially in countries where hybrid seed production is not feasible for technical or economic reasons (Fick, 1989). For commercial hybrid production, homozygous lines are of great importance. In contrast to conventional inbreeding, haploid techniques allow

S.M. Jain, S.K. Sopory & R.E. Veilleux (eds.), In Vitro *Haploid Production in Higher Plants, Vol. 5,* 17–35.

the production of doubled haploids, i.e., homozygous plants, within a comparatively short period of time.

The genetic basis of modern hybrids is narrow (Arnaud, 1986). Hybrid sunflower cultivars have been derived almost exclusively from a single cytoplasmic male sterile (CMS) source, discovered by Leclercq (1969) in the progeny of a cross between *Helianthus petiolaris* and *H. annuus* and transferred to sunflower germplasm. Due to the exclusive use of female lines with this cytoplasm for hybrid seed production, all hybrids cultivated world-wide are closely related – at least with regard to their cytoplasm (Hahn & Friedt, 1994; Horn *et al.*, 1994). Therefore, increased breeding efforts are required for broadening the genetic variability of cultivated sunflower. Successful applications of new methods of biotechnology can contribute to overcome problems of a narrow genetic base, and at the same time accelerate the breeding process. Sunflower has already been successfully manipulated with various *in vitro* techniques, e.g., embryo culture ("embryo rescue"), meristem culture and protoplast culture. In particular, interspecific hybrids, produced mainly by embryo rescue, have provided a unique source of "new" traits, like cytoplasmic male sterility (CMS) or resistance to devastating diseases, like *Sclerotinia* rot. The improvement and application of anther and microspore culture would allow the rapid production of homozygous sunflower lines. Isozyme markers and molecular techniques, such as restriction fragment length polymorphism (RFLPs), DNA fingerprinting and polymerase chain reaction (PCR), can be implemented to characterize and identify the regenerated plant material (Friedt, 1992; Fig. 1).

2. Application of haploid techniques in sunflower

One of the applications of a technique that produces haploids or spontaneously doubled haploids is to accelerate the breeding process. In addition, the technique may facilitate the selection of quantitative traits, characters controlled by individual recessive genes, or genes incorporated from alien species.

2.1. *Anther culture*

Although the production of microspore-derived embryos from cultured anthers has been well-established for many crop plants, in sunflower there is still a need to improve haploid techniques before its application in practical breeding is possible. So far, success in anther culture of cultivated sunflower has been poor, whereas better results have been obtained for interspecific hybrids. Several publications have described extensive callusing induced from anthers of various interspecific sunflower hybrids cultured *in vitro* (Bohorova *et al.*, 1985; Mezzarobba & Jonard, 1986; Vasiljevic *et al.*, 1991). Successful regeneration of shoots was also reported by Mezzarobba & Jonard (1988)

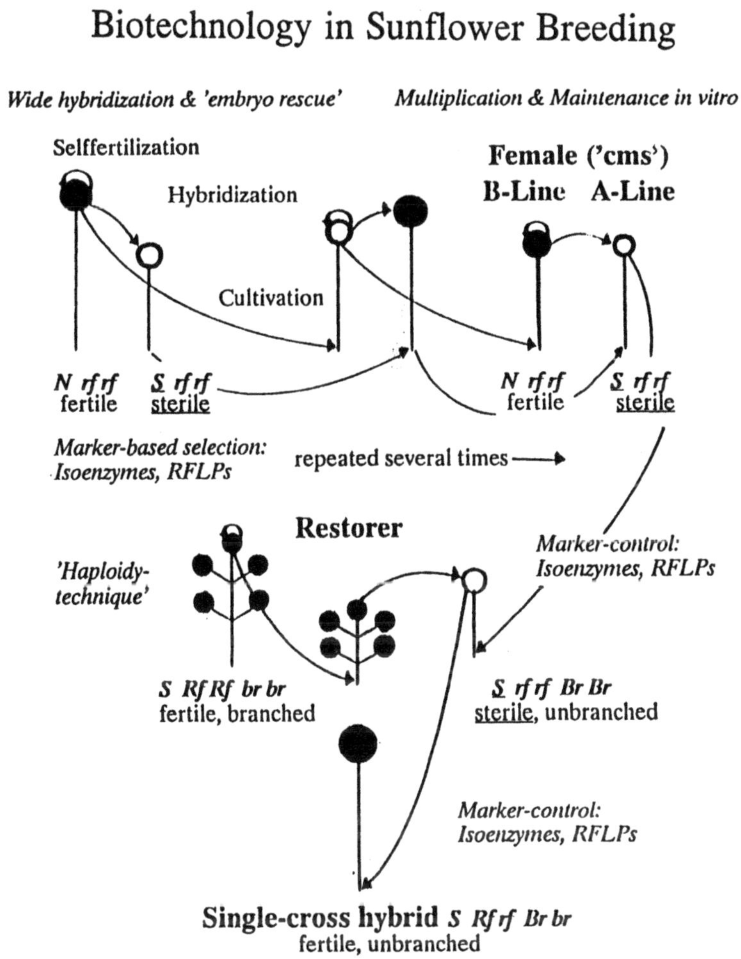

Figure 1. Scheme of inbred line development and hybrid breeding in sunflower with indications for potential applications of cellular and molecular methods.

and Gürel *et al.* (1991a). In the latter report, an M2 medium with half-strength MS macronutrients and complete MS micronutrients, supplemented with amino acids and high sucrose (120 g l^{-1}) proved to yield the most regenerated shoots for most of the genotypes tested.

Direct shoot formation from anthers of *H. divaricatus* and a *H. annuus* × *H. decapetalus* hybrid cultured on medium with 5 mg l^{-1} zeatin was described by Bohorova *et al.* (1985). Other combinations of growth regulators including kin (0.2 mg l^{-1}) and 2,4–D (1.0 mg l^{-1}), tended to promote callus formation but failed to induce shoot regeneration. In addition, secondary callus and shoots developed on stem explants cultured on a medium containing 2 mg l^{-1} BA, 0.2 mg l^{-1} IAA and 20 mg l^{-1} adenine. Shoots developed

Table 1. Culture media used in anther culture experiments according to different references

Components	Media			
	M1[1]	M2[2]	M3[1]	B1[3]
Macro nutrients	$\frac{1}{2}$ MS	$\frac{1}{2}$ MS	$\frac{1}{2}$ MS	MS
Micro nutrients	$\frac{1}{2}$ MS*	MS	$\frac{1}{2}$ MS	MS
Fe-solution	$\frac{1}{2}$ MS	MS	$\frac{1}{2}$ MS	MS
Vitamins (mg l^{-1})	M&W*	M&W*	M&W*	W**
B12	0.2	0.2	0.2	–
Myoinositol	–	–	–	500
Amino acids (mg l^{-1})				
Glycine	1.0	2.0	2.0	–
L-Glutamine	–	40.0	40.0	–
L-Serine	–	25.0	25.0	–
L-Tryptophan	–	1.0	1.0	–
L-Cysteine	–	2.5	2.5	–
Casein hydrolysate	–	–	–	500
Growth regulators				
NAA	0.5	0.5	0.5	–
BA	0.5	0.5	0.5	–
Zeatin	–	–	–	5.0
Sucrose (g l^{-1})	120.0	120.0	120.0	30.0
Gelrite (g l^{-1})	3.3	3.3	–	3.3
Agar (g l^{-1})	–	–	6.0	–
pH	5.9	5.8	5.9	5.8

[1] Mezzarobba & Jonard (1988), mod.
[2] Mezzarobba, pers. comm.
[3] Bohorova *et al.* (1985).
*,** According to Morel & White (1951) and White (1963), cit. in Gürel *et al.* (1991a).

roots on White's medium, even after several transfers and subcultures. However, on repeated subculture, an increase in chromosome number ($2n = 45$–51, 68 and 102) above the diploid number ($2n = 34$) was observed (Bohorova *et al.*, 1985). An overview of culture media used in anther culture experiments by Mezzarobba & Jonard (1988), Bohorova *et al.* (1985) and Gürel *et al.* (1991a) is given in Table 1.

Regeneration of six androgenic plants, including two haploids, from anthers of the French cv. "Inra" was reported by Mix (1985). Another haploid plant was recovered from anthers of cv. Luciole. Alissa *et al.* (1985) regenerated haploid, polyploid and aneuploid plants from four hexaploid and one diploid wild *Helianthus* species and their interspecific hybrids with different sunflower lines. The best results were observed for a *H. annuus* × *H. resinosus* hybrid (53% of anthers produced plants), while no plants could be regenerated from a *H. annuus* × *H. tuberosus* hybrid. The medium supplemented with 0.5 mg l^{-1} NAA and 0.5 mg l^{-1} BA proved most effective. Nenova *et al.* (1992) obtained plants from anthers of interspecific hybrids (*H. annuus* × *H. smithii*, *H. annuus* × *H. eggertii*) on B5 medium. A culture

Table 2. Composition of induction, regeneration and rooting media (mg l^{-1}) with basal medium MS (Nurhidayah *et al.*, 1996)

Components (mg l^{-1})	Induction medium	Regeneration medium	Rooting medium
Glycine	2.000	2.000	2.000
Myoinositol	100.000	100.000	100.000
Nicotinic acid	1.000	0.500	0.500
Thiamine-HCl	1.000	0.500	0.100
Pyridoxine-HCl	1.000	0.500	0.500
Biotin	0.010	–	–
BA	0.500	–	–
NAA	0.500	–	–
Sucrose	30,000	30,000	10,000
Gelrite	3,000	3,500	4,000
pH	5.8	5.8	5.8

method for embryo induction and plant regeneration from anthers of four sunflower genotypes has recently been described by Thengane *et al.* (1994). The MS basal medium induced more embryos over all cultivars tested. To optimize the response, MS basal medium was supplemented with 0.2–2.0 mg l^{-1} 2,4–D and 0.5 and 1.0 mg l^{-1} BA. The effects of cold pretreatment, growth regulator regime and sucrose concentration were tested for embryogenic efficiency. However, genotype was the most significant factor for androgenic capacity and only a few plantlets were obtained from embryos of one genotype.

Nurhidayah *et al.* (1996) cultured anthers of interspecific F_1–hybrids between *H. annuus* [HA89(cms)] and three wild species (*H.* × *laetiflorus*, *H. resinosus* and *H. tuberosus*) on modified MS medium (Table 2). Anthers containing microspores at the early uninucleate stage were cultured under different dark conditions and then under light (1500–2000 lux) at 30°C and 35°C. Embryos of regenerable genotypes were observed 10 to 30 days after plating the anthers. The highest direct embryo induction rates (up to 90% responsive anthers) were observed at an incubation temperature of 30°C (Table 3). A genotype by temperature interaction was evident. *H. tuberosus* hybrid genotype 10/25 exhibited direct embryogenesis only when cultured at 35°C whereas genotype 40/3 yielded more direct embryos at 30°C compared to 35°C. A total of more than 2,000 shoots was regenerated and transferred to the field (Fig. 2). Isozyme electrophoresis was employed to demonstrate the androgenic origin of at least some of the plants (see "Biochemical and molecular methods for identification of genotypes").

However, the regeneration of diploid plants from the somatic anther tissue remains a major problem. For example, Zhong *et al.* (1995) have recently reported the recovery of embryo-derived plants from sunflower anthers: all plants proved to be diploid ($2n = 34$).

Table 3. Response of different sunflower genotypes to anther culture under different dark conditions at 30°C and 35°C (Nurhidayah *et al.*, 1996)

Genotype	Dark period (days)	Number of anthers plated		% Anthers with direct embryogenesis		% Anthers with callus formation		% Anthers with root formation		Average number of direct embryos per anther	
		30°C	35°C	30°C	35°C	30°C	35°C	30°C	35°C	30°C	35°C
40/3	0	220	120	93	22	6	0	0	0	8	1
	6	220	120	69	13	19	0	0	0	2	1
	12	220	120	61	10	34	0	0	0	2	1
	18	220	120	46	7	41	0	0	0	2	1
	Mean	220	120	67	12	25	0	0	0		
11/51	0	360	200	4	4	50	5	1	0	1	1
	6	360	200	6	6	77	9	3	0	1	1
	12	360	200	6	9	82	19	7	1	1	1
	18	360	200	1	5	90	28	9	1	1	1
	Mean	360	200	4	6	74	15	5	1		
10/25	0	160	120	0	2	57	27	0	0	0	1
	6	160	120	0	2	63	29	0	0	0	1
	12	160	120	0	3	73	31	0	0	0	1
	18	160	120	0	2	81	37	0	0	0	1
	Mean	160	120	0	2	68	30	0	0		
30/61	0	120	100	6	7	67	11	0	0	2	1
	6	120	100	3	8	72	33	0	0	1	1
	12	120	100	2	5	82	35	1	0	1	1
	18	120	100	0	2	93	38	1	0	0	1
	Mean	120	100	3	5	79	29	2	0		

* Interspecific F_1 hybrids: 40/3 (*H. annuus* × *H. resinosus* 1545); 11/51 (*H. annuus* × *H. tuberosus* 1705); 10/25 (*H. annuus* × *H. tuberosus* 5), 30/61 (*H. annuus* × *H. laetiflorus*).

2.2. *Microspore culture*

An alternative method to anther culture for recovery of doubled haploid lines is microspore culture. Division of cultured microspores was first reported by Gürel *et al.* (1991b). Three hybrids and an inbred line were used as donor plants in their experiments. The flower buds were surface-sterilized by immersion in 70% ethanol followed by 2.5% NaOCl for 2 min each and then the buds were rinsed twice in sterile distilled water. Flower buds were selected that mainly contained microspores at the uninucleate stage. Microspores were isolated and purified by crushing the buds followed by several centrifugation steps. Modified NLN medium (Gland *et al.*, 1988) was used for culture induction, whereas MS medium (Murashige & Skoog, 1962) was used to try to induce regeneration. Microspore density was adjusted to 40,000 per ml liquid culture medium. Isolated microspores were incubated at 30°C under continuous light. Embryos were obtained from one hybrid and an inbred line; however, they died on transfer to MS medium.

Table 4. Solid culture media used for microspore culture of sunflowers (mg l^{-1}) (Todorova *et al.*, 1993)

Component	NLN	AA1	AA2	L4M1	L4M2
$Ca(NO_3)_2 \cdot 4H_2O$	500	–	–	–	–
$CaCl_2 \cdot 2H_2O$	–	440	440	440	440
KH_2PO_4	125	170	170	68	68
$MgSO_4 \cdot 7H_2O$	125	370	370	738	738
KNO_3	125	1,900	1,900	–	–
NH_4NO_3	–	1,650	1,650	–	–
KCl	–	–	–	1,177	1,177
$CoCl_2 \cdot 6H_2O$	0.025	0.025	0.025	0.024	0.024
$CuSO_4 \cdot 5H_2O$	0.025	0.025	0.025	0.025	0.025
H_3BO_3	10	6.2	6.2	6.2	6.2
$MnSO_4 \cdot 7H_2O$	18.94	16.95	16.95	0.17	0.17
$NaMoO_4 \cdot 2H_2O$	0.5	0.25	0.25	0.024	0.024
$ZnSO_4 \cdot 7H_2O$	12.31	8.6	8.6	0.28	0.28
KI	–	0.83	0.83	–	–
$FeSO_4 \cdot 7H_2O$	27.8	27.8	27.8	27.85	27.85
$Na_2EDTA \cdot 2H_2O$	37.3	37.3	37.3	37.25	37.25
Myoinositol	100	100	100	100	100
Thiamine-HCl	0.5	1	1	1	1
Pyridoxine-HCl	0.5	1	1	1	1
Nicotinic acid	5	1	1	1	1
Biotin	0.05	0.1	0.1	0.01	0.01
Folic acid	0.5	–	–	–	–
Ca-pantothenate	–	–	–	1	1
Glutamine	800	–	800	1,095	1,095
Glycine	2	1	1	–	–
L-Serine	100	–	100	–	–
Casein hydrolysate	–	–	–	1,000	1,000
BA	0.05	0.5	0.5	1	1
IAA	0.03	0.5	0.5	–	–
NAA	0.5	–	–	3	3
2,4 D	–	–	–	0.1	0.1
Sucrose	120,000	60,000	120,000	20,000	40,000
Mannitol	–	–	–	–	80,000
Gelrite	7,000	7,000	7,000	7,000	7,000
Charcoal	8,000	8,000	8,000	8,000	8,000
pH	6.2	5.7	6.2	5.7	6.2

Todorova *et al.* (1993) developed a system for isolation of pure, viable microspores by the use of centrifugation steps and ascorbic acid (vitamin C). They used eight lines derived from interspecific hybridization and three sunflower hybrids as donor plants. The flower buds contained microspores at the mid to late uninucleate stages. Microspores were plated at an optimal density of 6×10^4 per ml. Three basic media and their modifications were tested: NLN (Lichter, 1985), L4M1/L4M2 (Lenee & Chupeau, 1985) and a modified MS, AA1/AA2 (Table 4). The modifications to MS consisted of addition of glutamine (800 mg l^{-1}), serine (100 mg l^{-1}), BA (0.5 mg l^{-1}),

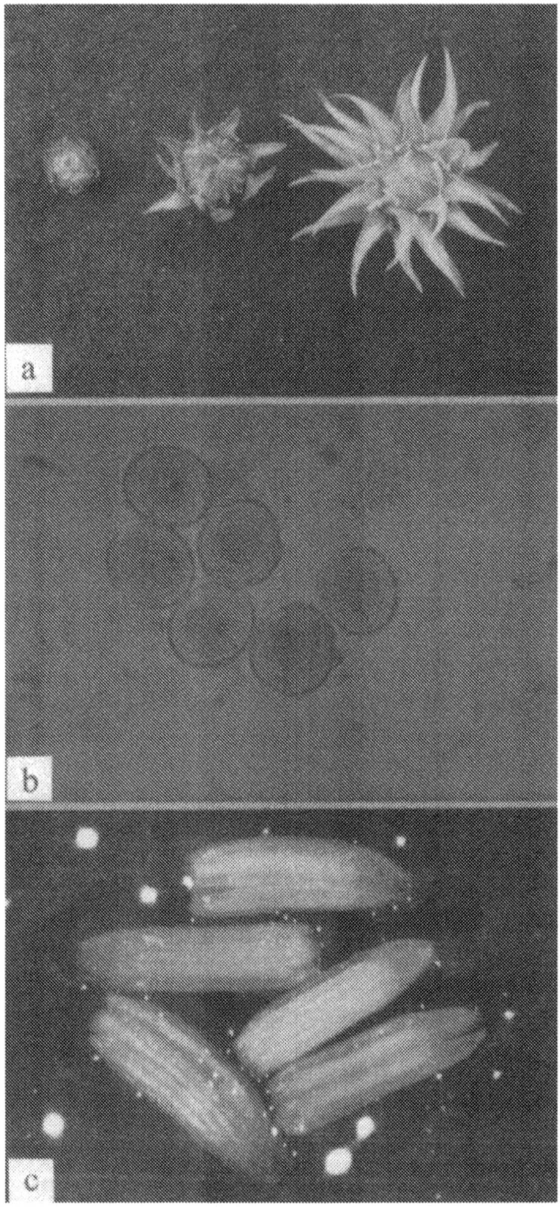

Figure 2. (a) Optimal size of flower buds for anther culture of F_1 donor plants, right: *H. annuus* × *H. resinosus*, middle: *H. annuus* × *H. tuberosus*, left: *H. annuus* × *H. laetiflorus*. (b) Microspores of sunflower at the early uninucleate stage. (c) Anthers excised from the donor plants. (d) Anthers with emerging callus and the beginning of plant differentiation on induction medium. (e) *In vitro* regenerated sunflower plantlet from anther culture. (f) Sunflower plantlets on rooting medium.

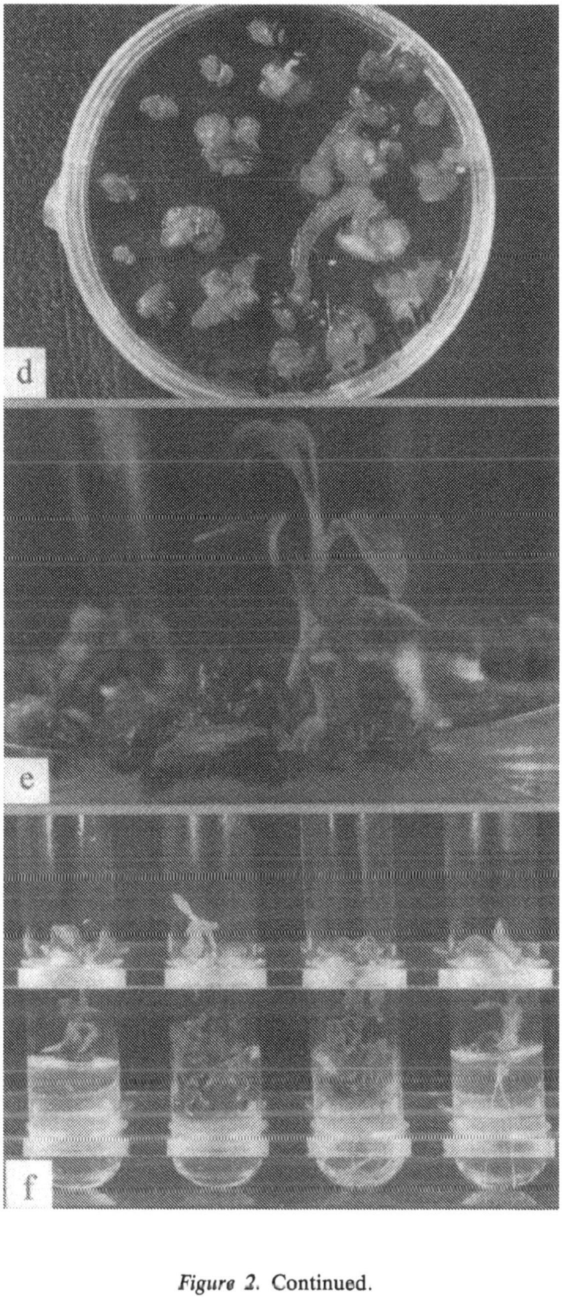

Figure 2. Continued.

IAA (0.5 mg l^{-1}), modified vitamin contents, increased sucrose concentration to 12%, and elevated pH to 6.2 (Table 4, medium AA2). The highest induction of microspore response was achieved on medium AA2. However, only a few cell divisions occurred and no further development was observed. Corresponding results were obtained by Coumans & Zhong (1995) by using liquid N6–medium supplemented with NAA (1 mg l^{-1}), BA (0.2 mg l^{-1}), and maltose (0.44 M), which resulted in an increased microspore division rate and microcallus formation.

Dahlhoff (1994) attempted to remove various inhibitory factors in sunflower microspore culture. He demonstrated that phenols produced by the plant tissue during the isolation process could be reduced by adding ascorbic acid to the culture medium. He also demonstrated that activated charcoal was necessary to promote cell division of the microspores. However, only a few cell divisions were observed and the development of the cultures stopped after five days. Substantial technical improvements to both anther and microspore culture are still needed, particularly with regard to genotypic and environmental (physiological) effects on the androgenic response.

3. Biochemical and molecular methods for identification of genotypes

Biochemical and molecular methods have been useful for genotypic analysis in many crop species. These methods have been applied in sunflower as a tool for early identification of important agronomic traits, such as quality characteristics, disease resistance, and stress tolerance. Isozymes have been utilized in both basic and applied research in many species (Goodman & Stuber, 1980; Tanksley, 1983; Cooke, 1984).

One recurring problem with several tissue culture systems has been the difficulty of determining the androgenic origin of regenerated plants. Biochemical and molecular markers represent powerful tools to determine precisely the origin of plants derived from microspore or anther culture. With improvements in technique, populations of doubled haploid lines have been produced in several major crop species. Doubled haploid populations have proven particularly useful in the production of molecular maps and in tagging important agronomic traits (Cloutier & Landry, 1994).

3.1. *Isozymes*

Isozyme electrophoresis has been a comparatively simple, fast and inexpensive method for an identification of *Helianthus* species, sunflower lines, hybrids and doubled haploid lines. Anisimova *et al.* (1991) used isozyme markers (in addition to helianthinine) for the identification of sunflower lines. Rieseberg & Seiler (1990) applied the isozyme technique in an evolutionary analysis of cultivated sunflower in relation to wild *H. annuus* populations. They concluded that domesticated sunflower must have evolved from a limited gene pool. Isozyme markers have also been successfully used to identify *Helianthus* species, their interspecific crosses and progenies (Dahlhoff *et al.*, 1992; Köhler *et al.*, 1992; Köhler & Friedt, 1992).

Another application of isozymes has been to determine androgenic vs. somatic origin of plants regenerated from anther culture. Cytological investigations cannot distinguish spontaneously doubled haploids from diploids derived from somatic tissue. Therefore, methods are required that allow somatic and androgenic plants to be distinguished, preferably in early developmental stages.

The first reports on the verification of haploids with the help of isozyme electrophoresis were published on asparagus (Colby & Peirce, 1988) and potato (Liu & Douches, 1994). In the latter case, haploids were produced by interspecific hybridization of eight tetraploid potato cultivars with a haploid-inducing pollinator. Isozyme analysis and visual examinations were used independently to compare the efficiency of discriminating hybrids from haploids. Approximately 80% of the hybrids could be identified by isozyme analysis, while 77% could be distinguished through visual examination. However, a combination of both visual and electrophoretic methods made hybrid identification even more efficient.

Nurhidayah *et al.* (1993) used isozyme markers (glucosephosphate isomerase (GPI), phosphoglucomutase (PGM) and menadione reductase (MNR) as well as DNA and PCR fingerprint techniques, for the identification of regenerated plants derived from anther culture of sunflower interspecific hybrids. Unfortunately, the isozyme patterns of the donor plants were expressed in the regenerated plants as well, indicating somatic origin. These results were confirmed by PCR and DNA fingerprinting, where a clear differentiation between the donor plant and the regenerated plants was not possible. Consequently, origin of the investigated plants from reduced microspores could not be demonstrated. However, recent investigations (Table 5) have shown that isozyme markers are a useful tool for the verification of sunflower haploids. Three of four isozyme systems (GPI, PGM, and EST) did not reveal any polymorphism. However, with the MNR system, it was possible to infer the androgenic origin of nearly 2,000 regenerated plants. Figure 3 shows the MNR zymogram of a hybrid anther donor plant, its parents, and several corresponding regenerated plants. Novel patterns that differed from the anther donor observed for some of the regenerated plants indicates their possible androgenic origin. A total of 10 of 13 anther derived regenerants were observed to differ from the anther donor by their MNR zymogram. It was possible to identify 10% of the plants to be of possible androgenic origin by the isozyme technique (Röcher, pers. comm.). However, these results must be confirmed by ploidy analysis and/or molecular methods.

3.2. *Molecular methods*

Molecular methods will likely make an essential contribution to the future of sunflower breeding. RFLPs have already been employed for identification

Table 5. Results of the investigated regenerants from the genotypes 11/51* and 40/3* characterized by different isozyme systems: menadione reductase (MNR), esterase (EST), glucosephosphate isomerase (GPI) and phosphoglucomutase (PGM)

Genotype	Number of regenerants examined	Isozyme system	Plants showing the same pattern as the F_1 hybrid	Plants showing a divergent pattern from the F_1 hybrid	% plants with a different pattern from the F_1 hybrid
40/3	1211	MNR	1196	15	1.2
		EST	1211	0	
		PGM	1211	0	
		GPI	1211	0	
11/51	70	MNR	70	1	1.4
		EST	70	0	
		PGM	70	0	
		GPI	70	0	

* Interspecific F_1 hybrids: 40/3 (*H. annuus* × *H. resinosus* 1545); 11/51 (*H. annuus* × *H. tuberosus* 1705).

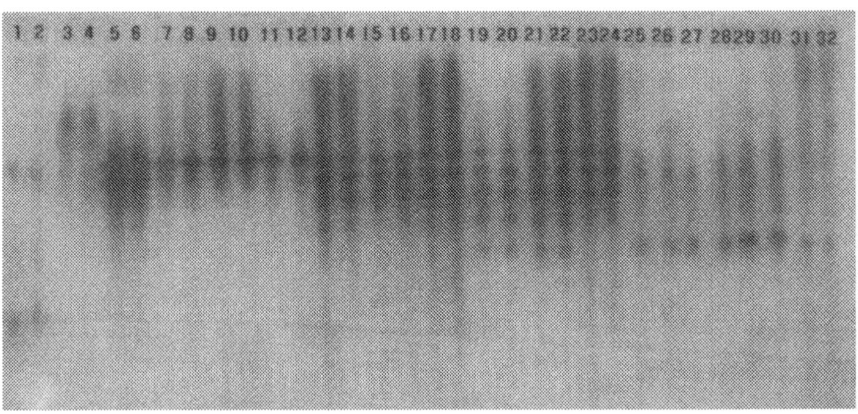

Figure 3. MNR zymogram of anther-derived plants of the interspecific hybrid *H. annuus* cv. HA 89 × *H. resinosus*: Lanes 1 and 2, *H. annuus* cv. HA 89; lanes 3 and 4, *H. resinosus*; lanes 5 and 6, F_1 hybrid anther donor; lanes 7–32, regenerated plants derived from anther culture (13 regenerated plants with each sample occurring in duplicate lanes, side by side). Anther-derived plants in lanes 13–18 show the same pattern as the anther donor whereas those in lanes 7–12 and 19–32 exhibit different patterns (Nurhidayah *et al.*, 1996).

of genotypes and species in numerous plant taxa. Techniques based on PCR, such as RAPDs (randomly amplified polymorphic DNA; Williams *et al.*, 1990) and SSRs (simple sequence repeats or microsatellites; Jeffreys *et al.*, 1985), offer considerable potential for the identification and isolation of markers linked to valuable characters such as disease resistance, cytoplasmic male sterility (CMS), oil quality and stress tolerance. Most of the published results on marker-assisted selection have been obtained by application of RFLPs or RAPDs, e.g., in barley (Graner & Bauer, 1993), rice (Ahn *et al.*, 1992; Mohan *et al.*, 1994), and tomato (Klein-Lankhorst *et al.*, 1991; Williamson *et al.*, 1994). The identification of valuable characters using RFLPs would facilitate selection at early growth stages and at the initiation of a sunflower breeding programme. First results of a genetic linkage analysis in sunflower were published by Leon *et al.* (1995). The authors used RFLP markers for a genetic analysis of oil percentage in sunflower seed. The RFLP markers (201 codominant loci) detected six regions representing 57% of the genetic variation of seed oil percentage; two of these regions were associated to seed oil percentage, two with kernel percentage, and two to both.

Gentzbittel *et al.* (1994) and Berry *et al.* (1994) have used RFLPs to determine genetic relationships among inbred lines of cultivated sunflower. They were able to show a clear differentiation between restorer and maintainer lines. Estimations of genetic diversity showed that the available genetic variability in sunflower is among the smallest described for crop plants.

The construction of genetic linkage maps of sunflower can be successfully engineered by RAPDs, as demonstrated by Rieseberg *et al.* (1993) for *Helianthus anomalus*, or by RFLPs. Recently, Gentzbittel *et al.* (1995) published a composite linkage map of sunflower based on 180 RFLP markers that identified 18 linkage groups, covering a total of 650 cM. A more detailed linkage map of *H. annuus* has been constructed by Berry *et al.* (1995) based on the segregation of 234 RFLP loci. The genetic markers, covering 1,380 cM of the sunflower genome, were arranged in 17 linkage groups, corresponding to the haploid chromosome number.

Another promising application of molecular methods is the identification of *Helianthus* genotypes, species, interspecific hybrids and their progenies derived either by conventional breeding or haploid techniques. Probably the best-suited technique for this aim concerning the more closely related taxonomic levels is oligo fingerprinting, especially using $(GATA)_4$ as a probe. This allows an accelerated construction of near isogenic lines (NILs), examination of the homogeneity of lines, a detailed determination of the genetic relatedness or distance of interspecific hybrid progenies after several generations of self-fertilization, haploid breeding or backcrossing, and selection for desired traits (Dehmer *et al.*, 1992; Dehmer & Friedt, 1993, 1994).

Molecular methods can be used to identify doubled haploid lines derived from anther culture. Schmidt *et al.* (1993) first reported results on the verification of doubled haploid breeding lines in sugar beet (*Beta vulgaris*) by using oligonucleotide fingerprinting. They performed a DNA fingerprint analysis of four selected DH lines. The DNA of individuals of each line was digested with either *Rsa*I or *Hae*III. After transferring the resulting fragments onto a nylon membrane, the DNA was hybridized to the oligonucleotides $(GACA)_4$ and $(GATA)_4$, respectively. The different probes revealed different levels of polymorphism. The highest degree of variability was observed after hybridization with the oligonucleotide $(GATA)_4$. The abundance and the high level of polymorphism of its target sequences made the $(GATA)_4$ probe optimally suited for the characterization and identification of doubled haploid lines of cultivated beet. In contrast to isozymes, DNA fingerprinting should also make it feasible to estimate the degree of homozygosity in a particular sunflower breeding line by comparison of individual plants.

RAPDs and AP-PCRs (Arbitrarily Primed PCR; Welsh *et al.*, 1991) can also be employed successfully for the discrimination of *Helianthus* species (Bervillé, 1993; Dehmer & Friedt, 1992), as demonstrated for other species (Demeke *et al.*, 1992; Van Heusden & Bachmann, 1992).

The discrimination of near isogenic sunflower lines (NILs) with biochemical and molecular methods has been described by Mösges & Friedt (1994). Two sets of NILs, which differed by the presence or absence of a downy mildew (*Plasmopara halstedii*) resistance gene were successfully distinguished by comparative application of isozymes, oligo fingerprints and PCR fingerprints, and it was concluded that isozyme electrophoresis was a suitable, cheap and simple alternative to discriminate genotypes for this purpose.

4. Prospects for the application of "genetic engineering"

Basic research for the incorporation of foreign genes into the *H. annuus* genome has already been conducted. Kempf & Hall (1981), Matzke *et al.* (1984), and Helmer *et al.* (1984) obtained genetically transformed calluses using disarmed *Agrobacterium tumefaciens* plasmid vectors, whereas Everett *et al.* (1987) obtained transgenic plants from hypocotyl callus. More recently, Schrammeijer *et al.* (1990, 1993) attempted transformation of sunflower cv. Zebulon via co-cultivation of dissected shoot apical meristems from seeds with a disarmed *Agrobacterium tumefaciens* strain, harbouring a binary vector with the genes encoding GUS and NPTII activity. Chimaeric expression of the two genes was observed in transformed plants, and integration of foreign DNA into the sunflower genome was confirmed by PCR. However, transformation of shoot meristem cells occurred at low frequencies. Moyne *et al.* (1989) succeeded in direct transformation of sunflower protoplasts as demonstrated by the presence of the NPTII marker gene in kanamycin-resistant calli. However, no plants could be regenerated from these calli. Stable transformants from split embryonic axes of sunflower treated with *A. tumefaciens* were reported by Bidney *et al.* (1992). Each T_0 plant was mapped for the presence of NPT II activity in leaf tissue and in the corollas of the disc florets. Stable transgenics were recovered for seven different promoter/gene constructs from nine different binary constructs, each relying on the NPT II selectable marker.

A more reliable protocol for transformation in sunflower has been worked out by Knittel *et al.* (1994). Based on microprojectile bombardment of half shoot apices in combination with *A. tumefaciens* co-cultivation, transgenic shoots were obtained from five inbred lines. Up to 7% of the explants expressed the transgenes. A minority of plants was shown to be chimaeric for expression of β-glucuronidase activity, while most appeared to be uniformly transformed. The integration of foreign genes was also demonstrated by Southern analysis. Finally, genetic transformation of an interspecific hybrid (*H. annuus* × *H. tuberosus*) has been achieved by using *A. tumefaciens* as a vector (Pugliesi *et al.*, 1993).

The recovery of intact transformed plants by manipulating single cells from cultivated sunflower remains the major bottleneck for the application of genetic engineering via protoplasts. Further improvements will be necessary for practical use of this method in sunflower breeding, where the treatment of haploid targets, either single cells or simple tissues would be highly attractive for the immediate creation of homozygous diploid transformants.

5. Conclusions

Sunflower has long been considered a recalcitrant species to various methods of biotechnology. However, substantial progress has been made in recent years:

- The recovery of haploid and doubled haploid sunflower plants has been demonstrated. However, pronounced genotypic differences have been observed for androgenic response, with most genotypes characterized as non-responsive. Therefore, the technique cannot be considered to be generally applicable or reproducible yet in sunflower.
- Progress has been made in the field of genome characterization and analysis, both by using biochemical and molecular markers as a basis for indirect "marker-assisted selection" for important agronomic traits, i.e., disease resistance, stress tolerance or quality characteristics.
- Genome analysis also forms the basis for identification and isolation of specific genes as candidates for "genetic engineering" of cultivated sunflowers, e.g., for specific traits as mentioned above.
- Haploid plants or explants would be optimal receptors for gene transfer, since the results of transformation regarding a specific trait (e.g., pathogen resistance) would be immediately expressed phenotypically, even in the case of recessive genes encoding an agronomic character. This fact would be of even greater importance in the case of quantitative traits, where doubled haploids allow reliable tests (e.g., regarding yield characteristics) due to their complete homozygosity and absence of dominance effects.

6. References

Ahn, S.N., C.N. Bollich & S.D. Tanksley, 1992. RFLP tagging of a gene for aroma in rice. Theor. Appl. Genet. 84: 825–828.

Alissa, A., H. Serieys & R. Jonard, 1985. Sur les possibilités de régénération d'espèces sauvages

et d'hybrides interspécifiques du genre *Helianthus* par androgenèse *in vitro*. C.R. Acad. Sci. Paris 300, Série III, no. I: 25–30.
Anisimova, I.N., A.V. Loskutov & I.G. Borovkova, 1991. Identification of sunflower lines by electrophoresis of helianthinine and isozymes. Sov. Agric. Sci. 6: 11–13.
Arnaud, F., 1986. Plante-Selection. Cahier Technique Tournesol. CETIOM, Paris.
Berry, S.T., R.J. Allen, S.R. Barnes & P.D.S. Caligari, 1994. Molecular marker analysis of *Helianthus annuus* L. 1. Restriction fragment length polymorphism between inbred lines of cultivated sunflower. Theor. Appl. Genet. 89: 435–441.
Berry, S.T., A.J. Leon, C.C. Hanfrey, P. Challis, A. Burkholtz, S.R. Barnes, G.K. Rufener, M. Lee & P.D.S. Caligari, 1995. Molecular marker analysis of *Helianthus annuus* L. 2. Construction of an RFLP linkage map for cultivated sunflower. Theor. Appl. Genet. 91: 195–199.
Bervillé, A., 1993. The use of RAPD markers to discriminate sections and species belonging to *Helianthus*. Biotechnol. Biotechnol. Eq. 7(4): 21–31.
Bidney, D.L., C.J. Scelonge & J.B. Malone-Schoneberg, 1992. In: Proc. 13th International Sunflower Conference, Pisa, Vol. II, pp. 1408–1412.
Bohorova, N.E., A. Atanassov & J. Georgieva-Todorova, 1985. *In vitro* organogenesis, androgenesis and embryo-culture in the genus *Helianthus* L. Z. Pflanzenzüchtg. 95: 34–44.
Cloutier, S. & B.S. Landry, 1994. Molecular markers applied to plant tissue culture. In Vitro Cell. Dev. Biol.-Plant 30P: 32–39.
Colby, L.W. & L.C. Peirce, 1988. Using isozyme markers to identify doubled haploids from anther culture of asparagus. HortScience 23: 761–763.
Cooke, R.J., 1984. The characterisation and identification of crop cultivars by electrophoresis. Electrophoresis 5: 59–72.
Coumans, M. & D. Zhong, 1995. Doubled haploid sunflower (*Helianthus annuus*) plant production by androgenesis: fact or artifact ? Part 2. *In vitro* isolated microspore culture. Plant Cell Tiss. Organ Cult. 41: 203–209.
Dahlhoff, M., 1994. Optimierung verschiedener Biotechniken bei der Sonnenblume unter besonderer Berücksichtigung des "Embryo rescue" zur Erstellung von Arthybriden in der Gattung *Helianthus*. Ph.D. Dissertation, Universität Giessen, Giessen.
Dahlhoff, M., H. Köhler & W. Friedt, 1992. New interspecific hybrids of sunflower. In: Proc. 13th International Sunflower Conference, Pisa, Vol. II, pp. 1438–1442.
Dehmer, K.J. & W. Friedt, 1992. PCR fingerprinting in the genus *Helianthus*. Abstracts XIIIth Eucarpia Congress, Angers, France, pp. 623–624.
Dehmer, K.J. & W. Friedt, 1993. Oligo-fingerprinting and genetic relationships in the genus *Helianthus*. Biotechnol. Biotechnol. Eq. 7(4): 57–59.
Dehmer, K.J. & W. Friedt, 1994. Einsatz von DNA-Fingerprints zur Charakterisierung von Verwandtschaftsverhältnissen in der Gattung *Helianthus*. Vortr. Pflanzenzüchtg. 28: 238–240.
Dehmer, K.J., G. Mösges & W. Friedt, 1992. DNA-Fingerprinting bei der Sonnenblume. Vortr. Pflanzenzüchtg. 22: 365–366.
Demeke, T., R.P. Adams & R. Chibbar, 1992. Potential taxonomic use of random amplified polymorphic DNA (RAPD): a case study in *Brassica*. Theor. Appl. Genet. 84: 990–994.
Everett, N.P., K.E.P. Robinson & D. Mascarenhas, 1987. Genetic engineering of sunflower (*Helianthus annuus* L.). Bio/Technology 5: 1201–1204.
Fick, G.N., 1989. Sunflower. In: G. Robbelen, R.K. Downey & A. Ashri (Eds.), Oil Crops of the World, their Breeding and Utilization, pp. 301–318. McGraw-Hill Publishing Company, New York.
Friedt, W., 1992. Present state and future prospects of biotechnology in sunflower breeding. In: G. Seiler (Ed.), Field Crops Research 30, pp. 425–442. Elsevier Science Publishing Co., Inc., Amsterdam.
Gentzbittel, L., F. Vear, Y.X. Zhang, A. Berville & P. Nicolas, 1995. Development of a consensus linkage RFLP map of cultivated sunflower (*Helianthus annuus* L.). Theor. Appl. Genet. 90: 1079–1086.
Gentzbittel, L., Y.X. Zhang, F. Vear, B. Griveau & P. Nicolas, 1994. RFLP studies of genetic

relationships among inbred lines of the cultivated sunflower, *Helianthus annuus* L.: evidence for distinct restorer and maintainer germplasm pools. Theor. Appl. Genet. 89: 419–425.

Gland, A., R. Richter & H.G. Schweiger, 1988. Genetic and exogenous factors affecting embryogenesis in isolated microspore cultures of *Brassica napus* L. Plant Physiol. 132: 613–617.

Goodman, M.M. & C.W. Stuber, 1980. Genetic identification of lines and crosses using isozyme electrophoresis. Annu. Corn Sorghum Res. Conf. Proc. 35: 10–31.

Graner, A. & E. Bauer, 1993. RFLP mapping of the *ym4* virus resistance gene in barley. Theor. Appl. Genet. 86: 689–693.

Gürel, A., K. Nichterlein & W. Friedt, 1991a. Shoot regeneration from anther culture of sunflower (*Helianthus annuus*) and some interspecific hybrids as affected by genotype and culture procedure. Plant Breed. 106: 68–76.

Gürel, A., S. Kontowski, K. Nichterlein & W. Friedt, 1991b. Embryogenesis in microspore culture of sunflower (*Helianthus annuus*). Helia 14: 123–128.

Hahn, V. & W. Friedt, 1994. Molecular analysis of the CMS-inducing MAX 1 cytoplasm in sunflower. Theor. Appl. Genet. 89, 379–385.

Heiser, C.B., 1978. Taxonomy of *Helianthus* and origin of domesticated sunflower. In: J.F. Carter (Ed.), Sunflower Science and Technology, pp. 31–52. American Society of Agronomy, Madison, Wisconsin.

Helmer, G., M. Casadaban, M. Bevan, L. Kayes & M.D. Chilton, 1984. A new chimeric gene as a marker for plant transformation. Bio/Technology 2: 520–537.

Horn, R., R.H. Köhler, V. Hahn, R. Kräuter, T. Hain, A. Lössl, J. Gerlach, J. Hustedt, K. Zetsche & W. Friedt, 1994. Development and analysis of alloplasmic male sterility in sunflower. Adv. Plant Breed. 18: 89–110.

Jeffreys, A.J., V. Wilson & S.L. Thein, 1985. Hypervariable "minisatellite" regions in human DNA. Nature (London) 314: 67–73.

Kempf, J.D. & T.C. Hall, 1981. Bean gene moved to sunflower cell. Agric. Res., U.S. Dept. Agric. 30: 4–5.

Klein-Lankhorst, R.M., A. Vermunt, R. Weide, T. Liharska & P. Zabel, 1991. Isolation of molecular markers for tomato (*L. esculentum*) using random amplified polymorphic DNA (RAPD). Theor. Appl. Genet. 83: 108–114.

Knittel, N., V. Gruber, G. Hahne & P. Lènèe, 1994. Transformation of sunflower (*Helianthus annuus* L.): a reliable protocol. Plant Cell Rep. 14: 81–86.

Köhler, H. & W. Friedt, 1992. Charakterisierung von *Helianthus*-Arten und Identifizierung interspezifischer Hybriden mit Hilfe der Isoenzym-Elektrophorese. Vortr. Pflanzenzüchtg. 22: 363–364.

Köhler, H., J. Weyen & W. Friedt, 1992: Genotypic characterization of *Helianthus* species, interspecific hybrids and their progeny by isoenzyme electrophoresis. In: Abstracts XIIIth Eucarpia Congress, Angers, pp. 663–664.

Leclercq, P., 1969. Une stérilité mâle cytoplasmique chez le tournesol. Ann. Amélior. Plantes 19: 99–106.

Lenee, P. & Y. Chupeau, 1985. Isolation and culture of sunflower protoplasts (*Helianthus annuus* L.): Factors influencing the viability of cell colonies derived from protoplasts. Plant Sci. 43: 69–75.

Leon, A.J., M. Lee, G.K. Rufener, S.T. Berry & R.P. Mowers, 1995. Use of RFLP markers for genetic linkage analysis of oil percentage in sunflower seed. Crop. Sci. 35: 558–564.

Lichter, R., 1985. From microspores to rape plants: a tentative way to low glucosinolate strains. In: World Crops Production, Utilization, Description, Vol. 11., pp. 268–277. Martinus Nijhoff/W. Junk, Dordrecht.

Liu, C.A. & D.S. Douches, 1994: Production of haploids of potato (*Solanum tuberosum* subsp. *tuberosum*) and their identification with electrophoretic analysis. Euphytica 70: 113–126.

Matzke, M.A., M. Susani, A.N. Binns, E.D. Lewis, I. Rubenstein & A.J.M. Matzke, 1984. Transcription of a zein gene introduced into sunflower using a Ti plasmid vector. EMBO J. 3: 1525–1531.

Mezzarobba, A. & R. Jonard, 1986. Éffets du stade de prélèvement et des prétraitements sur le développement *in vitro* d'anthères prélevées sur le tournesol cultivé (*Helianthus annuus* L.). C.R. Acad. Sci. Paris 303: 181–186.
Mezzarobba, A. & R. Jonard, 1988. L'androgenèse *in vitro* chez le tournesol cultivé (*Helianthus annuus* L.). In: Proc. 12th International Sunflower Conference, Novi Sad, pp. 562–567.
Mix, G., 1985. Antheren- und Ovarienkultur von Sonnenblumen (*Helianthus annuus* L.). Landbauforsch. Völkenrode 35(3): 153–156.
Mohan, M., S. Nair, J.S. Bentur, U.P. Rao & J. Bennett, 1994. RFLP and RAPD mapping of the rice *Gm2* gene that confers resistance to biotype 1 of gall midge (*Orseolia oryzae*). Theor. Appl. Genet. 87: 782–788.
Mösges, G. & W. Friedt, 1994. Genetic "fingerprinting" of sunflower lines and F_1 hybrids using isozymes, simple and repetitive sequences as hybridization probes, and random primers for PCR. Plant Breed. 113: 114–124.
Moyne, A.L., D. Tagu, V. Thor, C. Bergounioux, G. Freyssinet & P. Gadal, 1989. Transformed calli obtained by direct gene transfer into sunflower protoplasts. Plant Cell Rep. 8: 97–100.
Murashige, T. & F. Skoog, 1962. A revised medium for rapid growth and bioassays with tobacco tissue cultures. Physiol. Plant. 15: 473–497.
Nenova, N., P. Ivanov & M. Christov, 1992. Anther culture regeneration of F_1 hybrids of *Helianthus annuus* × *Helianthus smithii* and *Helianthus annuus* × *Helianthus eggertii*. In: Proc. 13th International Sunflower Conference, Pisa, Vol. I, pp. 1509–1514.
Nurhidayah, T., R. Horn, T. Röcher & W. Friedt, 1996. High regeneration rates in anther culture of interspecific sunflower hybrids. Plant Cell Reports (in press).
Nurhidayah, T., H. Köhler, M. Dahlhoff & W. Friedt, 1993. Anther culture of interspecific sunflower hybrids and examination of regenerants by biochemical and molecular methods. Biotechnol. Biotechnol. Eq. 7(4): 113–116.
Pugliesi, C., M.G. Biasini, M. Fambrini & S. Baroncelli, 1993. Genetic transformation by *Agrobacterium tumefaciens* in the interspecific hybrid *Helianthus annuus* × *Helianthus tuberosus*. Plant Sci. 93: 105–115.
Rieseberg, L.H. & G.J. Seiler, 1990. Molecular evidence and the origin and development of the domesticated sunflower (*Helianthus annuus*, Asteraceae). Econ. Bot. 44: 79–91.
Rieseberg, L.H., H. Choi, R. Chan & C. Spore, 1993. Genomic map of a diploid hybrid species. Heredity 70: 285–293.
Schmidt, T., K. Boblenz, M Metzlaff, D. Kaemmer, D. Weisung & G. Kahl, 1993. DNA fingerprinting in sugar beet (*Beta vulgaris*) – identification of doubled haploid breeding lines. Theor. Appl. Genet. 85: 653–657.
Schrammeijer, B., A. Hoekema & P.C. Sumons, 1993. Transformation in *Helianthus annuus* L. In: Y.P.S. Bajaj (Ed.), Biotechnology in Agriculture and Forestry, Vol. 22, pp. 213–219. Springer-Verlag, Berlin.
Schrammeijer, B., P.C. Sumons, P.J.M. van den Elzen & A. Hoekema, 1990. Meristem transformation of sunflower via *Agrobacterium*. Plant Cell Rep. 9: 5–60.
Tanksley, S.D., 1983. Gene mapping. In: S.D. Tanksley & T.J. Orton (Eds.), Isozymes in Plant Genetics and Breeding, pp. 109–138. Elsevier Science Publishers, Amsterdam.
Thengane, S.R., M.S. Joshi, S.S. Khuspe & A.F. Mascarenhas, 1994. Anther culture in *Helianthus annuus* L., influence of genotype and culture conditions on embryo induction and plant regeneration. Plant Cell Rep. 13: 222–226.
Todorova, M., M. Dahlhoff & W. Friedt, 1993. Microspore culture in sunflower (*Helianthus annuus* L.). Biotechnol. Biotechnol. Eq. 7: 83–90.
Van Heusden, A.W. & K. Bachmann, 1992. Genotype relationships in *Microseris elegans* (Asteraceae, Lactuceae) revealed by DNA amplification from arbitrary primers (RAPDs). Plant Syst. Evol. 179: 221–2133.
Vasiljevic, L., J. Atlagic & D. Skoric, 1991. Applicability of new biotechnology methods in sunflower breeding (Consultation Meeting of the FAO-Sunflower Network, Pisa 1991, unpublished manuscript).
Welsh, J., C. Petersen & M. McClelland, 1991. Polymorphisms generated by arbitrarily primed

PCR in the mouse: application to strain identification and genetic mapping. Nucl. Acids Res. 19: 303–306.

Williams, J.G.K., A.R. Kubelik, K.J. Livak, J.A. Rafalski & S.V. Tingey, 1990. DNA polymorphisms amplified by arbitrary primers are useful as genetic markers. Nucl. Acids Res. 18: 6531–6535.

Williamson, V.M., J.-Y. Ho, F.F. Wu, N. Miller & I. Kaloshian, 1994. A PCR-based marker tightly linked to the nematode resistance gene, *Mi*, in tomato. Theor. Appl. Genet. 87: 757–763.

Zhong, D., N. Michaux-Ferriére & M. Coumans, 1995. Assay for doubled haploid sunflower (*Helianthus annuus*) plant production by androgenesis: fact or artifact ? Part 1. *In vitro* anther culture. Plant Cell Tiss. Organ Cult. 41: 91–97.

3. Haploidy in niger (*Guizotia abyssinica* Cass)

P.B. KAVI KISHOR, T.P. REDDY, A. SARVESH and
G. VENKATESHAM

Contents

1. Origin and history of niger

Niger (*Guizotia abyssinica* (L. f) Cass) occurs within the Asteraceae and is known by various local names, such as Ram-til, Kala-til, Khurashi, Uchch-ellu, Gurellu, Verrinuvulu, etc., in different parts of India. It is a native of tropical Africa, and is widely grown in India, Ethiopia, East Africa, the West Indies and Zimbabwe. It is an important hardy, rainfed, herbaceous crop plant, and has received little attention with regard to agronomic practices, such as fertilization or irrigation. The crop is generally sown on: a) light lateritic or sandy soils, b) poor soils with coarse texture, and c) hilly slopes with very shallow soils. It may be grown as an intercrop with cotton, millet and pulses. This crop is widely grown in the tribal areas of India and Ethiopia as an edible and industrial oil crop. In India, it is grown over an area of 0.6 million hectares with 0.17 million tons total production. The major niger growing states are Andhra Pradesh, Bihar, Karnataka, Madhya Pradesh, Maharashtra, Orissa and West Bengal. The crop accounts for almost 2% of total seed oil with an average seed yield of 285 kg/ha. Niger occupies the fifth position in total production of oilseeds, and India accounts for 79% of total niger cultivation and 75% of total niger production in the world (Joshi, 1990).

S.M. Jain, S.K. Sopory & R.E. Veilleux (eds.), In Vitro *Haploid Production in Higher Plants, Vol. 5,* 37–51.

2. Uses of niger

Niger seeds are mainly used for extraction of oil which is about 30–50% of the total seed weight (Seegler, 1983). It is also used for making pickles and as a culinary herb (Nikam & Shitole, 1993). The oil is bluish-white in colour, with a faint odour and nutty, slightly sweet taste. In India, niger oil is used mostly in cooking, either in pure form, or mixed with other vegetable oils. It has also been used as a body ointment, in paints, soaps and for illumination. Niger oil absorbs flower fragrances making it useful as a base-oil in perfumes (Sharma, 1990). The oilcake is used as a cattle feed, especially for dairy cows, and is occasionally used as a soil amendment. Niger oil may also be used for pharmaceutical purposes similar to *Sesamum* oil.

3. Genetics and breeding of niger

Niger oil is of a high quality but average seed yield per hectare in India is lower than that of rape seed. Since it has traditionally been grown as a minor oil crop on marginal land and also as a mixed crop, it has only recently caught the attention of scientists. Certain problems preclude the realization of the full potential of niger in India, e.g., poor seed yield due to self-incompatibility, inherent low yielding capacity of the local cultivars, and susceptibility to biotic and abiotic stress. When conditions for vegetative growth are favourable, niger is prone to lodging; hence, dwarf plants with stiff stems and superior harvest index are needed for these situations (Riley & Belayneh, 1989). The cultivars evolved so far in India are selections from local collections which have low yield potential compared to Ethiopian niger (Ranganatha, 1988). Therefore, there is a need to widen the genetic diversity in this crop so that breeders can develop more productive cultivars.

Low and unpredictable yield is the main problem in niger production. Information on restructuring plant architecture towards maximal photosynthetic efficiency to improve sink capacity is meagre. There is a need to breed this crop plant for increased seed yield, fertilizer responsiveness, non-shattering, high harvest index, and for desirable maturity. As the local cultivars have been derived from a narrow genetic base, an intensive hybridization program should be undertaken to overcome this situation. Niger is an obligately cross-pollinated plant; selfing generally produces no seed due to self-incompatibility. It may be desirable to develop self-compatible lines either by mutagenesis or by selection from different germplasm collections (Sharma, 1990). Niger is sensitive to both temperature and photoperiod; hence, beneficial variants that are photo- and thermoperiod-insensitive may be derived through various techniques including tissue culture.

Genetic male sterility governed by recessive alleles has been identified in this crop. All bisexual tubular disc florets were modified into pistillate florets; therefore, male sterility is gynomonoecious. There is no phenotypic marker to identify male sterility in the vegetative stage; however, it could be identified in later stages by the large size of flower buds (Trivedi, 1990).

The primary aim of niger breeding programmes in India is to improve the seed yield and oil content, especially in the wake of large scale imports of edible oils. Since the oil content ranges from 30 to 50%, an increase in oil content could probably be achieved by selection of androclonal variants followed by breeding and selection for thin hull types. Alemau & Belayneh (1990) identified certain accessions that had low hull to embryo proportions and correspondingly high oil content. These relationships indicate that thin hulls result in higher proportion of embryo which in turn results in higher oil/protein content as well as low crude fibre. The fatty acid composition of niger oil is similar to that of sunflower oil except that sunflower oil contains 2% lignoceric acid (Seegler, 1983). Hence, breeding and subsequent selection for improved fatty acid balance has a high potential in niger. So far, limited attempts have been made on the production of haploids using anther culture for the genetic improvement of this crop. In this paper, some of our recent results on anther culture and isolation of androclonal variants are reviewed.

4. Importance of haploidy in niger

Modern production technologies capable of producing immediate tangible results should be adopted for achieving increased productivity and oil content. Large-scale production of elite clones through micropropagation or somatic embryogenesis, production of disease-free propagules, and development of cultivars utilizing tissue culture, molecular, and plant breeding techniques hold some promise of solving these problems (Prakash, 1990). Tissue culture techniques, especially, anther or pollen culture have unique potential in supplementing traditional breeding programmes using homozygous diploids. Once established, homozygous lines have the potential for crop improvement in a number of ways: a) lines can be more readily selected because additive gene effects can be more easily observed and undesirable recessive traits can be immediately eliminated, b) among F_1 hybrids, haploidization enhances the study of meiotic recombination by fixing any favourable transgression effects, and c) doubled haploids are useful in the exploitation of heterosis in heterogametic species as the homozygosity of doubled haploid lines may be greater than that of inbred lines (Chase, 1969). Such purity of doubled haploids makes it possible to produce hybrids with maximum heterosis in self-incompatible species, such as niger and turnip (Keller & Stringham, 1978). In 1984, scientists of the International Development Research Centre (IDRC), Canada, identified anther or pollen culture as powerful tools for the production of haploids to hasten breeding efforts in various oilseed crops, such as niger, sesame, and safflower. Thus, haploidy in niger is of utmost importance, not only to produce isogenic lines, but also for generating androclonal variation.

Androgenic haploids have been induced in more than 200 species, but the emphasis has been primarily on cereals and vegetable crops. In oilseed crops, only a few reports are available on the production of haploid plants through microspore culture. Keller *et al.* (1975) induced embryogenesis in the anthers of *Brassica campestris* and the anther-derived plants were found to be of pollen origin. Wenzel *et al.* (1977) and Keller & Armstrong (1978) obtained plantlets of various ploidy levels through microspore culture of rapeseed. Multicellular pollen grains and callus have been described in anthers of *Arachis correntiana*, *A. villosa* (Mroginski & Fernandez, 1980), *A. hypogaea* and *A. glabrata* (Bajaj *et al.*, 1981). Androgenic haploid production has also been achieved in safflower (*Carthamus tinctorius* L.; Prasad *et al.*, 1991) and sunflower (*Helianthus annuus* L.; Thengane *et al.*, 1994).

The use of anther culture for haploid production in niger and most other Asteraceae has been generally ineffective. Since the characteristically fused anthers are difficult to isolate without causing damage, only ovule culture has been previously used for haploid production in niger (Keller *et al.*, 1987). Regeneration of whole plants via organogenesis and somatic embryogenesis from seedling explants and cultured anthers has been reported in niger (Sarvesh *et al.*, 1993a,b, 1994a). Also, we recently reported induction of haploid embryos, *in vitro* flowering (Sarvesh *et al.*, 1996) as well as andro-clonal variation (Sarvesh *et al.*, 1994b).

5. Embryogenesis in cultured anthers of niger

5.1. *Effect of cold pretreatment on anther callus induction*

One of the important factors which can increase anther culture response is the use of various pretreatments before inoculation. The optimal duration of cold treatment usually varies with different species. Young capitula containing whitish-green anthers or uninucleate pollen grains were pretreated at 4 ± 2°C for 1 to 10 days. Before inoculation, the microspore stage was determined by acetocarmine staining. Callus initiation from anthers was not observed without cold treatment. With one day cold pretreatment, the frequency of anthers forming callus was 70 to 90% after 7–9 days. By increasing cold pretreatment to 5–10 days, the frequency of callusing was reduced to 8–10%, and initiation of callus was delayed by 13–16 days. The most effective cold treatment varied from 1 to 3 days. Callusing response was best (89.6%) when the capitula of niger cv. Ootacamund were cold-treated for one day at 4 ± 2°C. For safflower, the highest callusing from anthers was observed after a 3–day cold treatment of capitula (Prasad *et al.*, 1991), whereas, for *Sesamum indicum* L., a 6–day cold treatment at 10°C resulted in the highest rate of callus induction (80.2%) (Lee *et al.*, 1988). Our results infer that the cold pretreatment was useful not only for high frequency callus induction but also to expedite callus development in niger.

The effect of cold treatment may be indirect and increased androgenesis or callusing from cold-treated anthers has been attributed to increased pollen viability, delayed anther wall senescence, and decreased pollen abortion (Dieu & Dunwell, 1988). Therefore, cold pretreatment may be vital to alter the programme of microsporogenesis towards embryogenesis.

5.2. *Genotypic variation*

Genotype generally affects anther callus response in many plants and niger is no exception (Kavi Kishor *et al.*, 1989). Embryogenic callus formation was observed for all tested genotypes on Linsmaier & Skoog (LS) medium (Linsmaier & Skoog, 1965) fortified with 2 mg l^{-1} 2,4–dichlorophenoxyacetic acid (2,4–D) and 0.3 mg l^{-1} kinetin (kin). However, embryogenic callus initiation varied with the cultivar: 90% for cv. Ootacamund, 80% for No. 71, with sequential decreases for IET-11, RCR-18, RCR-140, GA-1, RCR-317, until only 30% was observed for IGP-76. On B_5 medium (Gamborg *et al.*, 1968), containing α-naphthaleneacetic acid (NAA) and kin, and Murashige & Skoog (MS) medium (Murashige & Skoog, 1962), supplemented with 6–benzyladenine (BA) and indole-3–acetic acid (IAA), only non-embryogenic callus was observed for any of the nine genotypes (Table 1). The time required for callus induction also varied among genotypes (five days for No. 71 to 21 days for Ootacamund) depending on the growth regulators used (Table 1). Callus obtained on medium containing 2,4–D had the potential to develop heart-shaped and bipolar embryos when transferred to a suitable medium. Different concentrations of 2,4–D, NAA, kin, BA, used individually in LS medium, failed to evoke embryogenic callus; however, non-embryogenic callus was observed from anthers with a frequency ranging from 4 to 20% (data not shown). Only a combination of 2,4–D (1–2 mg l^{-1}) and kin (0.3 mg l^{-1}) favoured a high frequency of embryogenic callus (30–90%). In the presence of 2 mg l^{-1} 2,4–D and 0.3 mg l^{-1} kin, uninucleate pollen grains produced multicellular bodies within 6–8 days. These bodies later developed into a dense callus mass. This callus first turned brownish and subsequently, embryos at various developmental stages emerged. The process of embryogenesis therefore, appeared to be asynchronous. Genotype is certainly a controlling factor for anther callus initiation in many species (Kavi Kishor *et al.*, 1989; Prasad *et al.*, 1991; Thengane *et al.*, 1994) and for plant regeneration ability (Kavi Kishor *et al.*, 1989). Zamir *et al.* (1980) attempted to understand the basis of genotypic differences for anther response in tomato and indicated that a single gene mutation was responsible. However, the precise reasons for genotypic differences in pollen plant production are largely obscure.

Table 1. Genotypic variation for embryogenic callus induction

Genotype	Media + growth regulators*	No. of anthers cultured	Frequency of callus induction (%)	Time taken for callus induction (days)	Nature of response**
Ooctacamund	B_5 NAA + kin	174	33	17	NE
	LS 2,4–D + kin	206	90	10	E
	MS BA + IAA	184	32	21	NE
No.71	B_5 NAA + kin	146	65	5	NE
	LS 2,4–D + kin	162	80	10	E
	MS BA + IAA	200	30	18	NE
GA-1	B_5 NAA + kin	126	44	15	NE
	LS 2,4–D + kin	172	49	14	E
	MS BA + IAA	144	32	20	NE with shoots
IET-11	B_5 NAA + kin	182	63	16	NE
	LS 2,4–D + kin	248	67	10	E
	MS BA + IAA	207	29	20	NE
IET-19	B_5 NAA + kin	142	20	18	NE (Poor)
	LS 2,4–D + kin	164	45	18	E
	MS BA + IAA	156	0	0	No callus
RCR-18	B_5 NAA + kin	172	32	18	NE
	LS 2,4–D + kin	160	67	15	E
	MS BA + IAA	162	15	20	NE (Poor)
RCR-140	B_5 NAA + kin	142	21	14	NE (Poor)
	LS 2,4–D + kin	180	55	15	E
	MS BA + IAA	160	14	20	NE
RCR-317	B_5 NAA + kin	184	29	18	NE
	LS 2,4–D + kin	150	33	18	E
	MS BA + IAA	158	10	20	NE (Poor)
IGP-7	B_5 NAA + kin	158	19	14	NE (Poor)
	LS 2,4–D + kin	140	30	20	E
	MS BA + IAA	165	11	20	NE (Poor)

* All media contain $2\,mg\,l^{-1}$ NAA or $2\,mg\,l^{-1}$ 2,4–D or $2\,mg\,l^{-1}$ BA and $0.3\,mg\,l^{-1}$ kin or IAA.
** E = Embryogenic callus, NE = Non-embryogenic callus.

5.3. *Influence of different media and plant growth regulators on embryogenic callus induction*

Five different media, viz. MS, B_5, Chaleff's R-2 (Chaleff & Stolarz, 1981), N_6 (Chu, 1978) and LS, differing in their organic and inorganic constituents, and plant growth regulators, were evaluated to study their influence on embryogenic and non-embryogenic callus induction and subsequent plant regeneration.

Of the five media tested, LS was found to be superior, followed by N_6, MS, B_5 and Chaleff's R-2 for callus induction from niger anthers (cv. Ootacamund). The time required for the induction of callus varied from 10 days on LS medium to 20 days on Chaleff's R-2 media (Table 2). Embryogenic callus from anthers occurred mainly on 2,4–D containing medium and, to a lesser

Table 2. Influence of different media and plant growth regulators on anther callus induction in cv. Ootacamund

Media + growth regulators (mg l^{-1})	No. of anthers inoculated	No. of anthers responded	Callus induction (%)	Time taken for callus (days)	Nature of response*
MS					
2.0 2,4–D + 0.3 kin	292	152	52	18	Embryogenic callus
2.0 NAA + 0.3 kin	200	36	18	18	Direct rooting and shoots
B_5					
2.0 2,4–D + 0.3 kin	284	121	43	17	Green friable callus
2.0 NAA + 0.3 kin	294	95	32	17	Callusing with shoots & roots
Chaleff's R-2					
2.0 2,4–D + 0.3 kin	240	39	29	20	Embryogenic callus
2.0 NAA + 0.3 kin	228	66	16	20	NE green callus
N_6					
2.0 2,4–D + 0.3 kin	286	170	59	12	Embryogenic callus
2.0 NAA + 0.3 kin	270	45	17	12	NE green callus
LS					
2.0 2,4–D + 0.3 kin	275	248	90	10	Embryogenic callus
2.0 NAA + 0.3 kin	240	110	46	10	NE green friable callus

* NE = Non-embryogenic; Immature capitula were pretreated for 1 day at 4° ± 2°C.

extent, on medium fortified with 2 mg l^{-1} NAA and 0.3 mg l^{-1} kin. Callus was creamy white on 2,4–D medium, but slightly brown to tan-coloured with anthocyanin pigments on NAA medium. Anthocyanin producing callus became necrotic over time. Thus, considerable variation was observed in the degree of pigment production, dependent on the growth regulator used. Incorporation of 0.1% activated charcoal or 0.1% ascorbic acid, or weekly subculture to fresh medium helped to circumvent the problem of necrosis. Replacing 2,4–D by NAA, BA or kin resulted either in slightly brown and friable (with NAA) or green and compact calli (with BA and kin) (Table 2).

5.4. *Maturation of anther-derived embryos*

Transfer of embryogenic callus to basal medium failed to convert globular and heart-shaped embryos (but not cotyledonary embryos) into whole plantlets. Therefore, Chaleff's R-2 medium supplemented with kin, NAA

Figure 1. (A) Cluster of embryos derived from anther callus of cv. Ootacamund. (B) Germination of haploid embryos on Chaleff's R-2 basal medium. (C) Differentiation of shoots from non-embryogenic callus of cv. Ootacamund on MS + 1 mg/l BA + 0.1 mg/l NAA. (D) Plantlet of Ootacamund showing differentiation of a capitulum. (E) Root tip of regenerated plant showing haploid number ($n = 15$) of chromosomes. (F) Capitula variants of Ootacamund recovered from androclones.

plus kin, and abscisic acid (ABA) were used for further proliferation and maturation of embryos of cv. Ootacamund. Often the embryos were fused together or appeared in clusters making it difficult to count the number of differentiated embryos per callus mass (Fig. 1A). Embryos germinated on transfer to basal medium devoid of growth regulators (Fig. 1B). The potential of callus to proliferate embryos on Chaleff's R-2 medium was dependent on the concentration of auxin, cytokinin and ABA (Table 3). Globular embryos developed into heart and bipolar embryos in the presence of ABA and NAA plus kin. Heart-shaped embryos were predominant on ABA but mostly

Table 3. Effect of plant growth regulators on pollen embryo proliferation and maturation from anther callus of cv. Ootacamund

Growth regulators (mg l^{-1})	Frequency of embryo proliferation (%)	Mean number of embryos/callus
0.1 kin	0	0
0.3 kin	8	4.4
0.5 kin	5	3.7
1 NAA + 0.3 kin	72	12.0
2 NAA + 0.3 kin	80	15.3
3 NAA + 0.3 kin	24	5.6
2 ABA	32	7.4
4 ABA	56	12.1
5 ABA	84	19.5
7 ABA	16	4.0

Embryo proliferation medium = Chaleff's R-2 + 2% sucrose + specified growth regulators.

bipolar embryos occurred in the NAA plus kin combination. Higher concentrations of ABA (4 and 5 mg l^{-1}) were better than 2 mg l^{-1}. Conversely, ABA and gibberellic acid increased only callusing from the haploid embryos of sunflower (Thengane *et al.*, 1994). However, Ammirato (1983) emphasized the role of ABA in maturation of embryos both *in vivo* and *in vitro*. In other species, ABA was used to aid maturation of somatic embryos, for example in *Pennisetum americanum* (Vasil & Vasil, 1981) and *Picea abies* (Arnold & Hakman, 1988). Thus, induction of embryogenic callus and maturation of embryos needed two different media in niger.

6. Genotypic variation for non-embryogenic callus induction

Calluses produced on B_5 medium containing 2 mg l^{-1} NAA and 0.3 mg l^{-1} kin from anthers of different genotypes were soft, friable and non-embryogenic. Marked genotypic differences for such callus induction was observed (Table 4).

7. Organogenesis from anther callus

Sectors of non-embryogenic calluses obtained on LS medium with 2 mg l^{-1} 2,4–D plus 0.3 mg l^{-1} kin and also on B_5 medium supplemented with 2 mg l^{-1} NAA and 0.3 mg l^{-1} kin were isolated and maintained on B_5 medium (2 mg l^{-1} NAA + 0.3 mg l^{-1} kin). This callus was subsequently transferred onto MS medium supplemented with 1 mg l^{-1} BA plus 0.1 mg l^{-1} NAA or 2 mg l^{-1} BA plus 0.2 mg l^{-1} NAA for shoot organogenesis. BA (1 mg l^{-1}) was best for shoot regeneration as well as for the number of shoots formed per callus mass in all the genotypes. The frequency of response was highest

Table 4. Genotypic variation for non-embryogenic callus induction

Genotype	No. of anthers cultured	Frequency of non-embryogenic callus (%)
Ootacamund	206	33
No.71	162	65
GA-1	172	44
IET-11	248	63
IET-19	168	20
RCR-18	160	32
RCR-140	180	21
RCR-137	150	29
IGP-76	140	19

B_5 medium containing 2 mg l^{-1} NAA and 0.3 mg l^{-1} kin was used.

Table 5. Genotypic variation for shoot regeneration from non-embryogenic callus*

Genotype	Growth regulators (mg l^{-1})	No. of calluses inoculated	No. of calluses with shoots	Frequency of response	No. of shoots/callus
Ootacamund	1 BA + 0.1 NAA	41	15	37	8–10
	2 BA + 0.2 NAA	38	11	29	5–8
No.71	1 BA + 0.1 NAA	38	13	34	7–9
	2 BA + 0.2 NAA	42	10	24	5–7
GA-11	1 BA + 0.1 NAA	40	7	18	3–6
	2 BA + 0.2 NAA	40	5	13	3–5
IET-11	1 BA + 0.1 NAA	37	8	21	4–6
	2 BA + 0.2 NAA	39	6	15	3–5
IET-19	1 BA + 0.1 NAA	44	7	15	3–4
	2 BA + 0.2 NAA	35	4	11	3–4
RCR-18	1 BA + 0.1 NAA	43	10	23	4–7
	2 BA + 0.2 NAA	40	7	18	3–5
RCR-140	1 BA + 0.1 NAA	39	8	20	3–5
	2 BA + 0.2 NAA	36	5	13	2–4
RCR-317	1 BA + 0.1 NAA	34	5	15	3–4
	2 BA + 0.2 NAA	42	4	9	2–4
IGP-76	1 BA + 0.1 NAA	38	6	16	3–5
	2 BA + 0.2 NAA	42	5	12	2–4

* Data were scored after 30 days on regeneration medium.

$$\text{Frequency of response (\%)} = \frac{\text{No. of calluses with shoots}}{\text{No. of calluses inoculated}} \times 100$$

in cv. Ootacamund (37%) followed by No. 71, IET-19 and RCR-317 (Table 5). Many adventitious shoot bud primordia and shoots were differentiated from anther callus on MS medium containing 1 mg l^{-1} BA and 0.1 mg l^{-1} NAA for all the genotypes (Fig. 1C). Sporadic differentiation of roots was also recorded on this medium in some cultures. When callus of cv. Ootacamund with shoot buds was subcultured to MS basal medium devoid of growth regulators, it produced many tiny shoot buds, shoots and roots after 30 days.

These shoots grew to 10–15 cm in 15 days, and 23% of them developed small capitula after 30–35 days. Our results indicate that a basal medium without growth regulators was best for further growth and rooting of the shoots. The shoot forming potential of the callus on MS medium containing 1 mg l^{-1} BA and 0.1 mg l^{-1} NAA could be maintained through 5–6 subcultures, and thereafter the morphogenetic capacity was lost. Shoot bud differentiation was also noticed on MS medium containing 2 mg l^{-1} BA and 0.2 mg l^{-1} NAA, but the frequency was reduced invariably in all cultivars.

8. *In vitro* flowering in niger

Protocols have been developed to induce flowering from plantlets derived from niger anther callus. Callus cultures of cv. Ootacamund cultured on MS medium supplemented with 1 mg l^{-1} BA plus 0.1 mg l^{-1} NAA produced capitula directly with 10% frequency. Aggregated florets without a peduncle (sessile) were noticed with flowers that were abnormal and lacking sexual organs. Shoots were excised from the haploid plantlets obtained from germinating embryos and planted on MS basal medium. These shoots rooted with 100% frequency and small capitula differentiated in 23% of cultures within 4–5 weeks (Fig. 1D). Five to six capitula arising from a single stalk could also be differentiated sporadically on this medium. Capitula formation was inconsistent and seldom noticed in other genotypes. These flowers, however, contained sex organs unlike the inflorescences that developed directly from callus.

Induction of flowers has been observed in other oil crops, such as *Arachis hypogaea* (Narsimhulu & Reddy, 1984), *Carthamus tinctorius* (Tejovathi & Anwar, 1984) and *Brassica nigra* (Mehta *et al.*, 1993). Flower formation in these cases was either from differentiated shoots or directly from the cotyledons. Niger may thus be valuable for understanding physiological and biochemical factors controlling flower formation and also the molecular changes associated with it.

9. Androclonal variation in niger

Plants regenerated from anther callus of cv. Ootacamund were transferred to soil : vermiculite mix (3 : 1) and grown to maturity with 70–90% survival. One hundred and fifty plants regenerated from anther-derived embryogenic callus of cv. Ootacamund were analyzed for chromosome number. Of these, 127 were haploid ($2n = x = 15$) (Fig. 1E), eight were diploid, and the remaining 15 were polyploid (mostly tetraploid and hexaploid). Seed-raised as well as tissue culture-derived plants served as controls to score the morphological variation among the androclones. A total of 30 plants was scored for several traits measured in the second generation. From each plant, three leaves for

Table 6. Androclonal variation for some plant traits of Ootacamund in the second generation

Genotype	Plant height (cm)	Leaf length (cm)	Leaf width (cm)	No. of branches per plant	Width of recepticle (mm)	Depth of recepticle (mm)	No. of seeds/capitulum
Seed raised plants	165.4 (±8.0)	10.6 (±0.6)	2.2 (±0.2)	13.9 (±2.0)	8.9 (±1.2)	7.8 (±1.0)	43.1 (±2.3)
AC1	50.3* (±3.1)	12.1* (±0.3)	1.2* (±0.2)	10.8 (±0.9)	8.2 (±0.3)	7.4 (±0.5)	40.4 (±1.5)
AC2	112.3* (±4.9)	12.0* (±0.5)	1.6 (±0.2)	9.8* (±0.7)	7.3 (±0.5)	6.8 (±0.7)	29.5* (±2.5)
AC3	172.2 (±2.3)	10.7 (±0.4)	1.2* (±0.2)	8.6* (±1.0)	8.4 (±0.5)	8.0 (±0.5)	34.0* (±3.1)
AC4	124.4 (±2.6)	18.6* (±0.5)	2.2 (±0.4)	16.0 (±1.2)	14.5* (±1.0)	13.5* (±1.0)	44.5 (±3.8)
AC5	139.5* (±4.0)	11.3 (±0.5)	2.3 (±0.3)	20.8* (±0.3)	9.8 (±0.9)	8.5 (±0.6)	42.0 (±1.6)
AC6	39.0 (±1.9)	9.8 (±0.2)	2.3 (±0.3)	10.3* (±1.0)	9.3 (±0.9)	7.8 (±0.9)	28.5* (±2.2)
AC7	122.3* (±2.6)	13.6* (±0.4)	1.3* (±0.3)	14.0 (±0.4)	17.6* (±0.6)	16.0* (±0.4)	52.0* (±1.4)
AC8	111.8* (±3.0)	10.0 (±0.3)	1.5* (±0.3)	10.3* (±1.3)	8.8 (±0.5)	7.3 (±0.5)	39.3 (±1.9)

[a] Data represent an average of 30 replicates for all the traits shown above.
[b] Asterisk indicates significant at p = 0.05. Other values are not significantly different.

length and width, 2–3 receptacles for width and depth, and two capitula for seed setting were scored. Self-compatibility of the plants was measured by bagging an inflorescence and counting the seed set per capitulum. All eight diploids were fertile and displayed marked morphological variation. By comparison, the range of variation observed in seed-raised plants of Ootacamund, including plant height (155–170 cm), width of receptacle (7.5–10.1 mm), and depth of recepticle (6–9.0 mm), was not significant. Based on the true-breeding nature of the anther-derived diploids, it was presumed that they were doubled haploids. Except clone AC3, all other clones showed a significant decrease in plant height, especially clones AC1 and AC6, in the first as well as second generations. The number of branches (18.0) and seeds/capitulum (50.8) were higher in clone AC7, with a maximum width (18.2 mm) and depth (17.6 mm) of the receptacles in the first generation. This is the only clone that produced more seeds per capitulum upon selfing compared with other clones, and hence this clone may be more self-compatible (Table 6).

Some of the capitula variations observed in the second generation are shown in Fig. 1F. The progeny of these eight plants were evaluated for seven plant characters and all the plants bred true for these traits in the second generation (Table 6). Dwarfs and large flower head types were some of the useful variants isolated in the present study. Since seed-derived and non-variant tissue culture-derived plants were compared with variants, the observed phenotypic differences should not have been due to culture or growth conditions of plants. Therefore, it is likely that these are true genetic variants. Peloquin (1981) suggested that chromosome substitutions may be responsible for the changes observed in plants regenerated from tissue cultures. Self-incompatibility is a constraint for increasing yield in niger. A putative self-compatible clone AC7 was detected in this study and this trait was heritable. However, these results do not clearly demonstrate the origin of these variants. The recovered diploid plants might have been derived either from the anther wall or maternal tissue. Haploid plants were treated with colchicine and some homozygous diploids were obtained but androclonal variation has not been detected among these doubled haploids.

10. Conclusions

Anther culture of niger produced both embryogenic and non-embryogenic calli on appropriate medium containing plant growth regulators. Plants derived via embryogenesis or organogenesis could be transferred successfully to pots and grown to maturity. Self-incompatibility has been described as a major constraint for increasing seed yields in niger. To the best of our knowledge, using conventional mutagenesis or somaclonal variation, no new variants for the S-locus governing self-incompatibility have previously been obtained in niger. In the present study, a single self-compatible clone was generated from anther callus of cv. Ootacamund, suggesting that anther culture can be exploited for generating beneficial androclones in niger.

11. References

Alemau, G. & H. Belayneh, 1990. Investigations on some biochemical characteristics of niger seeds (*Guizotia abyssinica* Cass). In: Abbas Omran (Ed.), Oil Crops: Proceedings of the Three Meetings Held at Pantnagar and Hyderabad, pp. 229–232. International Development Research Centre, Ottawa.

Ammirato, P.V., 1983. The regulation of somatic embryo development in plant cell cultures: suspension culture techniques and hormone requirements. Bio/Technology 1: 68–74.

Arnold, S.V. & I. Hakman, 1988. Regulation of somatic embryo development in *Picea abies* by abscisic acid. J. Plant Physiol. 132: 164–169.

Bajaj, Y.P.S., A.K. Ram, K.S. Labana & H. Singh, 1981. Regeneration of genetically variable plants from anther derived callus of *Arachis hypogaea* and *Arachis villosa*. Plant Sci. Lett. 23: 35–39.

Chaleff, R.S. & A. Stolarz, 1981. Factors influencing the frequency of callus formation among cultured rice (*Oryza sativa*) anthers. Physiol. Plant. 51: 201–206.

Chase, S.S., 1969. Monoploids and monoploid derivatives of maize (*Zea mays* L.). Bot. Rev. 5: 117–167.

Chu, C.C., 1978. The N_6 medium and its application to anther culture of cereal crops. In: Proc. Symp. Plant Tissue Culture, pp. 45–50. Science Press, Peking.

Dieu, P. & J.M. Dunwell, 1988. Anther culture with different genotypes of opium poppy (*Papaver somniferum* L.): Effect of cold treatment. Plant Cell Tiss. Organ Cult. 12: 263–271.
Gamborg, O.L., R.A. Miller & K. Ojima, 1968. Nutrient requirements of suspension cultures of soybean root cells. Exp. Cell Res. 50: 151–158.
Johsi, A.V. 1990. Present production, research and future strategy for niger in Maharashtra. In: Abbas Omran (Ed.), Oil Crops: Proceedings of the Three Meetings Held at Pantnagar and Hyderabad, pp. 171–175. International Development Reserach Centre, Ottawa.
Kavi Kishor, P.B., M. Aruna & G.M. Reddy, 1989. Plant regeneration from haploid callus of Indica rice. Proc. Ind. Nat. Sci. Acad. 55B: 193–201.
Keller, W.A. & K.C. Armstrong, 1978. High frequency production of microspore derived plants from *Brassica napus* anther cultures. Z. Pflanzenzüchtg. 80: 100–108.
Keller, W.A., P.G. Arnison & J. Cardy, 1987. Haploids from gametophytic cells – recent developments and future prospects. In: C.E. Green, D.A. Somers, W.P. Hackett & D.D. Biesboer (Eds.), Plant Tissue and Cell Culture, pp. 223–241. Alan R. Liss Inc., New York.
Keller, W.A., T. Rajhathy & J. Lacapra, 1975. *In vitro* production of plants from pollen in *Brassica campestris*. Can. J. Genet. Cytol. 17: 655–666.
Keller, W.A. & G.R. Stringham, 1978. Production and utilization of microspore-derived haploid plants. In: T.A. Thorpe (Ed.), Frontiers of Plant Tissue Culture, pp. 113–122. Calgary University Press, Calgary.
Lee, S.K., H.S. Kim, Y.T. Lee & C.H. Park, 1988. Effects of growth regulator, cold treatment and genotype in anther culture of sesame (*Sesamum indicum* L.). Research Report, Rural Develop. Administration, Biotechnology, Korea Republic 30: 74–79.
Linsmaier, E.M. & F. Skoog, 1965. Organic growth factor requirements of tobacco tissue cultures. Physiol. Plant. 18: 100–127.
Mehta, U.J., S. Hazra & A.F. Mascarenhas, 1993. Somatic embryogenesis and *in vitro* flowering in *Brassica nigra*. In Vitro Cell Dev. Biol. 29P: 14–17.
Mroginski, L.A. & A. Fernandez, 1980. Obtainment of plantlets by *in vitro* culture of anthers of wild species of *Arachis* (Leguminosae). Obagineus 35: 89–92.
Murashige, T. & F. Skoog, 1962. A revised medium for rapid growth and bioassays with tobacco tissue cultures. Physiol. Plant. 15: 473–497.
Narsimhulu, S.B. & G.M. Reddy, 1984. *In vitro* flowering and pod formation from cotyledons of groundnut. Theor. Appl. Genet. 69: 87–91.
Nikam, T.D. & M.G. Shitole, 1993. Regeneration of niger (*Guizotia abyssinica* Cass.) cv. Sahyadri from seedling explants. Plant Cell Tiss. Organ Cult. 32: 345–349.
Peloquin, S.J., 1981. Chromosomal and cytological manipulations. In: K.J. Frey (Ed.), Plant Breeding, pp. 117–150. Iowa State University Press, Ames, Iowa.
Prakash, J., 1990. Commercialisation of agricultural biotechnology in India: Opportunities, progress and problems. In: Proc. Sixth Intl. Conference, Amsterdam, pp. 789–794.
Prasad, B.R., M.A. Khadeer, P. Seeta & S.Y. Anwar, 1991. *In vitro* induction of androgenic haploids in safflower (*Carthamus tinctorius* L.). Plant Cell Rep. 10: 48–51.
Ranganatha, A.R.G., 1988. High yielding varieties of niger. Workshop on Oilseed Production Technology at National Communication and Training Centre for Oilseeds, Directorate of Oilseeds Research, Hyderabad.
Riley, K.W. & P. Belayneh, 1989. Niger seed (*Guizotia abyssinica* Cass). In: G. Robbelen, R.K. Downy & A. Ashri (Eds.), Oil Crops of the World: Their Breeding and Utilization, pp. 394–403. MacMillan Publishing Co., New York.
Sarvesh, A., T.P. Reddy & P.B. Kavi Kishor, 1993a. Plant regeneration from cotyledons of niger. Plant Cell Tiss. Organ Cult. 32: 131–135.
Sarvesh, A., T.P. Reddy & P.B. Kavi Kishor, 1993b. Embryogenesis and organogenesis in cultured anthers of an oil yielding crop niger (*Guizotia abyssinica* Cass). Plant Cell Tiss. Organ Cult. 35: 75–80.
Sarvesh, A., T.P. Reddy & P.B. Kavi Kishor, 1994a. Somatic embryogenesis and organogenesis in *Guizotia abyssinica*. In Vitro Cell Dev. Biol. 30P: 104–107.

Sarvesh, A., T.P. Reddy & P.B. Kavi Kishor, 1994b. Androclonal variation in niger (*Guizotia abyssinica* Cass). Euphytica 79: 59–64.

Sarvesh, A., T.P. Reddy & P.B. Kavi Kishor, 1996. *In vitro* anther culture and flowering in *Guizotia abyssinica* Cass. Ind. J. Exp. Biol. 34: 565–568.

Seegler, C.J.P., 1983. Oil Plants in Ethiopia, Their Taxonomy and Agricultural Significance. Centre for Agricultural Publishing and Documentation, PUDOC, Wageningen.

Sharma, S.M., 1990. Niger seed in India : Present status of cultivation, research achievements and strategies. In: Abbas Omran (Ed.), Oil Crops: Proceedings of the Three Meetings held at Pantnagar and Hyderabad, pp. 159–165. International Development Research Centre, Ottawa.

Tejovathi, G. & S.Y. Anwar, 1984. *In vitro* induction of capitula from cotyledons of *Carthamus tinctorius* (safflower). Plant Sci. Lett. 36: 165–168.

Thengane, S.R., M.S. Joshi, S.S. Khuspe & A.F. Mascarenhas, 1994. Anther culture in *Helianthus annuus* L., influence of genotype and culture conditions on embryo induction and plant regeneration. Plant Cell Rep. 13: 222–226.

Trivedi, H.B.P., 1990. Niger in tribal Bihar. In: Abbas Omran (Ed.), Oil Crops: Proceedings of the Three Meetings Held at Pantnagar and Hyderabad, pp. 176–179. International Development Research Centre, Ottawa.

Vasil, V. & I.K. Vasil, 1981. Somatic embryogenesis and plant regeneration from suspension cultures of pearl millet (*Pennisetum purpureum* Schum.). Z. Pflanzenphysiol. 111: 233–239.

Wenzel, G., I. Hoffman, I. Potrykus & E. Thomas, 1977. Increased induction and chromosome doubling of androgenetic haploid rye. Theor. Appl. Genet. 51: 81–86.

Zamir, D., R.A. Jones & N. Kedar, 1980. Anther culture of male sterile tomato (*Lycopersicon esculentum* Mill.) mutants. Plant Sci. Lett. 17: 353–361.

4. Haploidy in *Petunia*

S. MOHAN JAIN and N. BHALLA-SARIN

Contents

1. Introduction

The genus *Petunia* was first established in 1803 and since then, 30 *Petunia* species have been described, originated from South America (Sink, 1984). Because of inconsistency in taxonomic data of different *Petunia* species, the exact origin of the cultivated *Petunia hybrida* (Fig. 1) is still unknown. Van der Krol & Chua (1993) speculated that *P. hybrida* might have resulted from a cross between a member of the *Petunia axillaris* or *P. parodii* group possessing white flowers and a member of *P. violacea*/*P. inflata*/*P. integrifolia*/*P. parviflora* group having coloured flowers. Despite the lack of reliable data from chemical and protein analysis of putative ancestors, the indications are that *P. hybrida* was derived from a cross between *P. axillaris* and *P. inflata* (Sink, 1984). Restriction fragment length polymorphism (RFLP) studies have indicated that the putative ancestral lines contributed differently to the cultivated hybrid lines of *Petunia* (Van Tunen *et al.*, 1990).

Petunia has numerous cultivars and shows a range of variation in flower colour which is the result of differences in flower pigments. A new brick red *Petunia* flower colour has been generated as a result of transformation of a *Petunia* mutant RL01, which shows no flower pigmentation, with the maize *A*1 gene encoding dihydroflavonol reductase, which confers pelargonidin

S.M. Jain, S.K. Sopory & R.E. Veilleux (eds.), In Vitro *Haploid Production in Higher Plants, Vol. 5,* 53–71.

Figure 1. Petunia hybrida with white flowers growing in a pot.

coloration (Meyer *et al.*, 1987). However, this flower colour was unstable under field conditions and plants with flowers showing white or variegated pigmentation were obtained (Meyer *et al.*, 1992).

Petunia is much used as an annual ornamental species. It is a seed-propagated solanaceous plant; hundreds of seeds are readily obtained by crossing after a four or five week maturation period, and 3–4 generations per annum can be obtained under greenhouse conditions with supplementary lights. Although *Petunia* has a gametophytic self-incompatibility system and has been used extensively in research on self-incompatibility, self-pollination at the bud stage can result in adequate seed set for most selections (Ascher, 1984; Sims, 1994). In addition, *Petunia* can be propagated by cuttings or grafts allowing clones to be maintained *in vitro* or *in vivo* (Bianchi & Dommergues, 1979), has a small chromosome number ($2n = 2x = 14$; $n = 7$; Fig. 2), and is ideal for cytological and mutation studies. *Petunia hybrida* cultivars are readily transformed with *Agrobacterium*-mediated transformation system and give rise to fertile plants (Horsch *et al.*, 1988). Furthermore, male sterility (ms) occurs in *Petunia*, characterized by the lack of pollen grains in otherwise normal anthers (Evenor *et al.*, 1984), and it has three types of genetic control systems: a) a multiple gene system, b) a cytoplasmic male sterility (cms) or nuclear-cytoplasmic system, and c) a single nuclear gene system (see Evenor & Izhar, 1984).

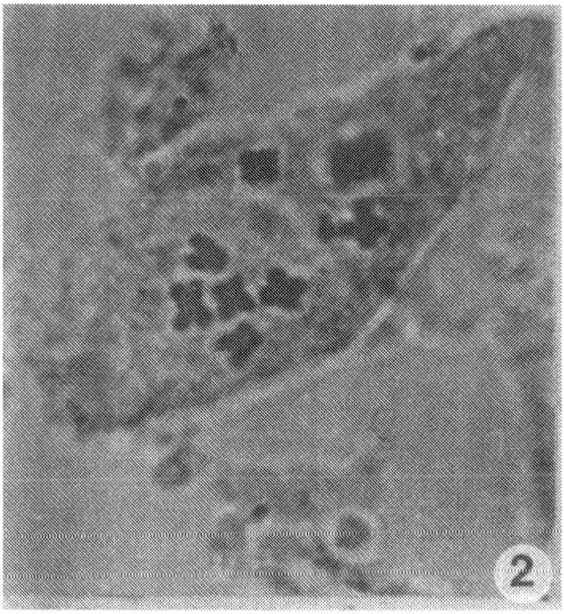

Figure 2. Petunia hybrida haploid chromosome number ($n = 7$).

2. Haploid production

Although the usefulness of haploids has long been recognized, no appropriate method has been available to produce haploids in large numbers in a short period. The origin of haploids may be either spontaneous or induced by different experimental approaches (Lacadena, 1974). Natural haploids were first discovered in *Datura stramonium* (Blakeslee *et al.*, 1922), *Nicotiana tabacum* (Clausen & Mann, 1924), and *Triticum compactum* (Gains & Aase, 1926). Since these discoveries, natural haploids have been reported in many plant species (Kimber & Riley, 1963; Magoon & Khanna, 1963; Katayama & Nei, 1964). However, their rarity and sporadic occurrence has effectively prevented the use of natural haploids in plant breeding programmes. Guha & Maheshwari (1964) in Delhi University, India, reported for the first time haploids in *Datura innoxia* by anther culture. Since this discovery, a barrage of publications appeared on anther culture of various plant species (Sunderland, 1971, 1974; Kasha, 1974; Vasil & Nitsch, 1975; Acharya & Ramji, 1977; Nitzsche & Wenzel, 1977; Maheshwari *et al.*, 1980, 1982; Dunwell, 1985; Hu & Yang, 1986; Bajaj, 1990; Raghavan, 1990; Wenzel *et al.*, 1995). Whether it was a conscientious effort or coincidence, the solanaceae were favoured for *in vitro* cultures, not only of anthers but of all explants. Now that anther culture is an established technique, haploids can be produced in a wide range of plants. However, besides the anther culture technique, several other techniques have also been developed for the production of haploids (Rao & Suprasanna, 1996). In the genus *Petunia*, haploids have been produced by ionizing radiation (gamma rays), by crossing lines with specific genetic markers and then screening the progeny for their expression, ovary and ovule culture, and anther or microspore culture (Straub, 1973; Sangwan & Norreel, 1975; Singh & Cornu, 1976; DeVerna & Collins, 1984; Jain, 1978; Mitchell *et al.*, 1980). Although anther culture has been the main approach for haploid production, we will review various methods of haploid production in *Petunia*.

2.1. *Ovary and ovule culture*

Maheshwari & Rangaswamy (1965) suggested that ovule culture may prove useful in the artificial induction of parthenogenesis. Ovary and ovule cultures have been used for the production of haploids over a wide range of genera (Keller & Korzun, 1996; Lakshmi Sita, 1996; Yang & Zhou, 1982). The first report on haploid plants produced by ovary culture was with *Hordeum vulgare* (San Noeum, 1976, 1979). DeVerna & Collins (1984) considered that the generation of maternally derived haploids by ovary or ovule culture may be advantageous for a number of reasons. It may be the only efficient means of producing haploids in *Beta vulgaris* (Bornman, 1985), *Gerbera jamesonii* (Cappadocia *et al.*, 1988; Cappadocia & Veith, 1990), *Allium* spp. and other plant species (Keller & Korzun, 1996), and may also be useful for comparison with paternally derived haploids. By culturing placenta-attached ovules prior to anthesis, DeVerna & Collins (1984) produced significantly more maternal haploids of *P. axillaris* than from similar cultures after anthesis. They suggested placenta-attached ovule culture as an alternative to anther culture for haploid production.

2.2. *Use of ionizing radiation for haploid production*

Because of interest in haploidy for both basic and applied genetics, and their low occurrence in natural populations, several different approaches have been used to produce haploids in large number (Lacadena, 1974). Physical agents such as ionizing and non-ionizing radiation have been used for pollen irradiation to induce *in situ* haploid plants. Radiation destroys the generative function of pollen without affecting its capability in egg-cell stimulation, thus allowing parthenogenic embryos to develop (Sestili & Ficcadenti, 1996). The successful use of pollen irradiation has been rare. However, low numbers of haploids have been obtained in apple, cacao, melon, *Petunia*, barley, and onion (Sestili & Ficcadenti, 1996). Although initially successful in obtaining haploids through irradiated pollen, Straub (1973) later obtained the first spontaneous *Petunia* haploid in 1965. Maizonnier (1976) also recovered a few haploid *Petunia* plants by the use of irradiated pollen. Raquin (1985) pollinated three different F_1 genotypes of *P. hybrida* with irradiated pollen (gamma ray dose 6–100 kR) and ovaries were harvested 9–14 days after pollination and cultured *in vitro*. At doses up to 30 kR, hybrid plants (mainly diploids) were produced, whereas only gynogenic haploids were produced at doses of 60 kR and above. Using *P. hybrida* and *P. parodii*, Raquin *et al.* (1989) reported induction of androgenesis by *in vitro* culture of ovaries irradiated with gamma ray doses ranging from 50–1000 Gy before pollination. Their results indicated: (1) gamma ray doses ranging from 200–1000 Gy led to the development of two types of plants: haploids $2n = x = 7$ and overdiploids $2n > 2x = 14$, (2) the androgenic origin of haploids was confirmed with genetic markers, (3) androgenic haploids contained the chloroplasts of the irradiated female parent without visible change to the cp DNA (chloroplast) pattern after irradiation, and (4) four possible androgenic events had occurred between the two *Petunia* species: *hybrida* haploids with *hybrida* or *parodii* cytoplasm, and *parodii* haploids with *parodii* or *hybrida* cytoplasm. The production of overdiploids may have been due to spontaneous chromosome doubling of the male gamete. Haploid plants with an additional chromosome and hyperploid plants were also obtained, suggesting participation of irradiated gametes in the fertilization process (Raquin, 1990). Gamma-induced androgenesis and spontaneous androgenesis have been studied by using genetic markers and lethal mutants (Singh & Cornu, 1976; Horlow *et al.*, 1996).

2.3. *Anther culture*

The first attempt to produce *Petunia* haploid embryos through anther culture was made by Bernard (1971). This was followed by the success in obtaining androgenic plants of *Petunia* by Raquin & Pilet (1972). They, however, obtained predominantly triploids via callus formation from anther cultures with microspores at the mitotic stage, excised from the flower buds of hybrids between *P. axillaris* × *P. hybrida*. They proposed the origin of triploids was as a result of either endomitosis or fusion of three nuclei of microspores. Sopory & Maheshwari (1973) obtained plantlets directly from anther culture of *P. hybrida* on Nitsch medium supplemented with kinetin (kin), naphthaleneacetic acid (NAA) and 15% coconut water. However, they did not indicate the ploidy of regenerated plantlets. Engvild (1973) obtained nearly all triploid plants by anther culture of *P. axillaris*, originated from diploid anthers ($2n = 2x = 14$) with immature pollen grains. Gupta (1982) reported predominantly triploid (80.3%) microspore-derived plants of *P. parodii* as a result of fusion of interphase nuclei in two different pathways. Engvild *et al.* (1972) confirmed that plantlet ploidy was a function of the developmental stage of the flower bud in *Datura innoxia*. Anthers with young mononucleate microspores gave haploids, while anthers with binucleate microspores gave diploids and anthers with older binucleate microspores gave triploids, probably because DNA synthesis in the generative nucleus, prior to the second

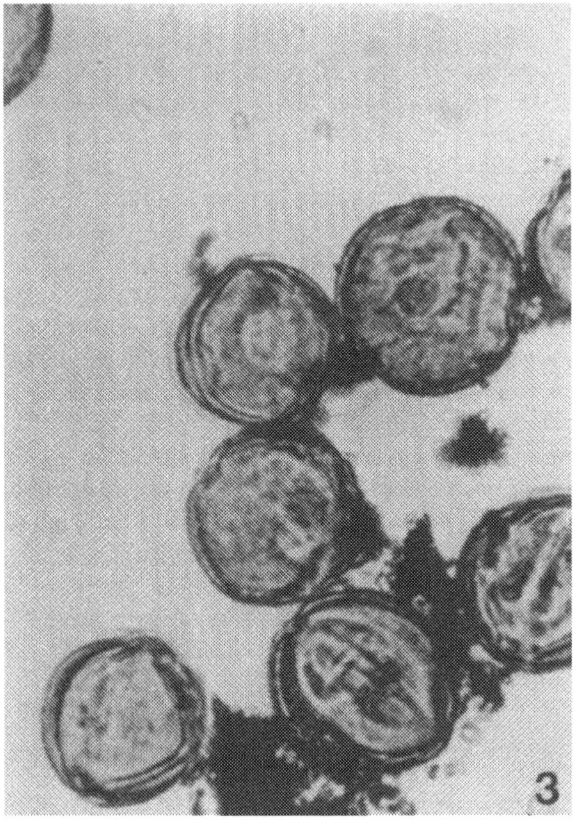

Figure 3. Uninucleate stage of *Petunia hybrida* microspores.

pollen mitosis, had already occurred. Gangupati *et al.* (1976) suggested that microspores initially produced callus in which endomitosis and/or nuclear fusion occurred before embryo development. Growth regulators required for microspore development may have induced endomitosis or nuclear fusion in the developing cells or selected for development of non-haploid microspores (Mitchell *et al.*, 1980). Swamy & Chacko (1973), Wagner & Hess (1974) and Jain (1978) regenerated haploid, diploid and triploid *P. hybrida* plantlets from anther cultures having pollen at the uninucleate stage (Fig. 3). Mitchell *et al.* (1980) observed genotype-dependent anther culture response for microspore-derived plantlets. A few haploid plants were obtained, but most of the plants derived from anthers were diploid and triploid or tetraploid. Gupta (1983) produced microspore-derived haploid, diploid and triploid plants in *P. violacea* by culturing anthers containing early binucleate microspores. Cytological studies revealed that non-haploid embryos developed by nuclear fusion at early stages of pollen embryogenesis.

2.3.1. *Factors affecting anther culture*
Several researchers have succeeded in obtaining *Petunia* haploids and plants with different ploidy levels by anther culture (Gupta, 1983; Santos *et al.*, 1983; Liscum & Hangarter, 1991). However, the percentage of responding *Petunia* anthers for haploid production has been rather low, up to 2–3%. We will discuss different factors affecting the efficiency of *Petunia* anther culture for haploid production.

2.3.1.1. *Basal medium.* The most common media used for *Petunia* anther culture have been MS (Murashige & Skoog, 1962), Nitsch (1969), Sangwan & Norreel (1975) and Nitsch & Nitsch (1969). Mitchell *et al.* (1980) used modified MS medium salts with vitamins of Gamborg *et al.* (1968) for *Petunia* callus proliferation and regeneration. However, qualitative and quantitative variations were made in the carbon source, growth regulators and adjuvants. Babbar & Gupta (1980) showed that external supply of nutrients was not required for *Petunia* anthers to initiate a switch from the gametophytic to the sporophytic mode of development.

2.3.1.2. *Carbon source.* The presence of sugars, especially sucrose has been essential in the culture medium for optimal anther culture response. Jain (1978) showed that 2% sucrose in the culture medium induced the greatest embryo formation in *P. hybrida* anther cultures and that higher sucrose concentrations were deleterious to cultured anthers. Mitchell *et al.* (1980) also used 2% sucrose in anther culture of *Petunia*. Raquin (1983) tested the effect of various sugars – sucrose, lactose, maltose, glucose, fructose, and galactose – as carbon source for *in vitro* differentiation in anther cultures of a hybrid clone of *P. axillaris* × *P. hybrida*. The results indicated that embryos were obtained with all the sugars tested, and a metabolizable sugar in the culture medium was necessary to allow the growth of androgenic embryos. Galactose was not utilized and appeared to be toxic. Maltose yielded better results when compared with glucose and sucrose.

2.3.1.3. *Growth regulators and adjuvants.* Various combinations and concentrations of auxins and cytokinins have been supplemented in different basal media to obtain callus, embryos and plantlets from cultured *Petunia* anthers (Raquin & Pilet, 1972; Engvild, 1973; Sopory & Maheshwari, 1973; Gangupati *et al.*, 1976; Jain, 1978; Mitchell *et al.*, 1980; Jain *et al.*, 1980; Martineau *et al.*, 1981). Except for the first two reports, differentiation from callus has been reported by all other researchers. Either auxin [NAA, 2,4-dichlorophenoxyacetic acid (2,4-D), indolebutyric acid (IBA)] alone or in combination with a low concentration of cytokinin [kin, benzyladenine (BA) or zeatin (zea)] could induce callus formation from the anthers. The effect of including 1% activated charcoal with and without growth regulators for *Petunia* anther culture was tested by Martineau *et al.* (1981). Activated charcoal completely inhibited callus and root formation from anthers cultured on medium containing 0.1 mg/l NAA and 0.1 mg/l BA but did not stimulate microspore development on medium without growth regulators. On the contrary, Jain (1978) reported an increase in the response of *P. hybrida* anthers producing embryos by adding 1% activated charcoal to the medium. Bajaj *et al.* (1976) also found in *Nicotiana tabacum* that the percentage of anthers producing embryos could be enhanced by adding 2% activated charcoal in the medium. They also observed a small increase in the formation of diploid plants by activated charcoal. Activated charcoal adsorbs from the culture media both toxic substances produced by the tissues and exogenously-supplied growth regulators. The positive effect of activated charcoal on anther culture may be dependent on plant type and genotype used.

Callus differentiation could be achieved by lowering the concentration of growth regulators in the medium. In general, regeneration of anther-derived callus was more difficult compared with stem or leaf-derived callus (Mitchell *et al.*, 1980; Sopory & Maheshwari, 1973; Swamy & Chacko, 1973). However, a combination of auxin and cytokinin caused an interaction resulting in improved growth and differentiation of *Petunia* anther-derived callus (Jain, 1978).

Direct embryo formation from *Petunia* anthers (Fig. 4) has been reported by Engvild (1973), Sopory & Maheshwari (1973), Gangupati *et al.* (1976), Malhotra & Maheshwari (1977), Jain *et al.* (1980) and Gupta (1982). Better efficiency of anthers producing embryos was obtained in *P. axillaris* when in the culture medium organic N was supplemented in the form of bacto-casamino acids, L-glutamine, L-asparagine, L-arginine, L-tryptophan, and L-cysteine (Engvild, 1973). Gupta (1982) obtained embryos from *Petunia* microspores by culturing anthers on Nitsch & Nitsch (1969) medium amended with 100 mg/l L-serine and 800 mg/l L-glutamine along with 1.0 mg/l kin. Addition of 10–15% coconut water in the culture medium was beneficial for *Petunia* anthers to produce embryos (Engvild, 1973; Sopory & Maheshwari, 1973; Jain, 1978). Gupta & Carlson (1972) reported that calluses can be preferentially maintained in the haploid state in *Nicotiana tabacum* by adding para-flurophenylalanine (PFA) to the culture medium. Jain (1978) observed that 15 μg/ml PFA was most effective in increasing the percentage of anthers producing embryos compared to the control. However, this increase was not equivalent to stimulation obtained by manipulating growth regulators in the medium.

2.3.1.4. *Physical factors*. There is little information available on the effect of pH on *Petunia* anther culture. Most studies have used a pH range between 5.5–5.8 for *Petunia* anther culture. Jain (1978) studied the effect of pH (4.5, 5.0, 5.5, 5.8, 6.0, 6.5) on *P. hybrida* anther culture response on Nitsch medium containing 10^{-7} M kin, and found the highest anther culture response for embryo formation at pH 5.8.

Temperature. Malhotra & Maheshwari (1977) worked on the enhance-

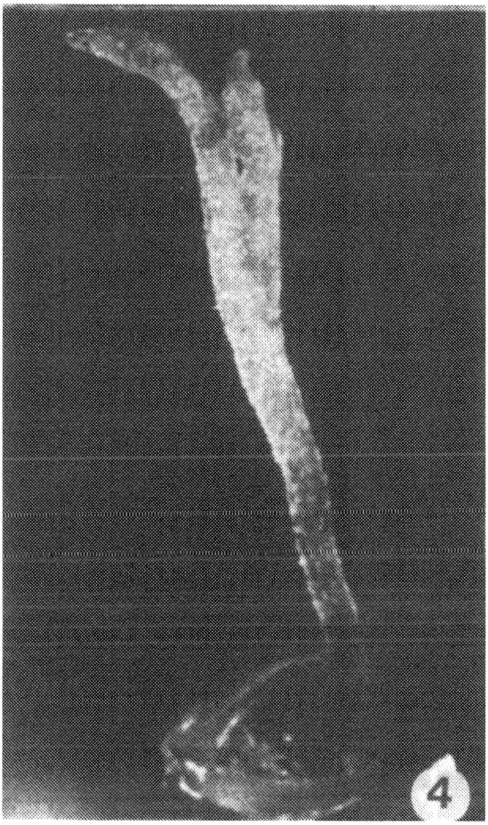

Figure 4. Direct embryo formation from *Petunia hybrida* cultured anther.

ment of pollen embryo development in *P. hybrida* and reported the stimulation of division in pollen grains by cold treatment. Depending upon the medium used, cold treatment at 6°C for 48 h lead to direct or indirect embryo formation. Addition of kin to the Nitsch basal medium induced direct embryo formation from *Petunia* anthers; when 15% coconut water was supplemented in this medium, anthers formed callus which later differentiated into embryos. The cold treatment did not affect the number of anthers forming embryos. Nonetheless, it resulted in an increased percentage of anthers forming direct embryos. The percentage of embryo formation rose to 6.9 from 4.4. Regardless of the medium, the number of developing embryos per anther was 3–4 compared to one or rarely two in the controls. Enhanced production of embryos per anther indicated that cold treatment diverted more pollen grains towards direct production of embryos, whereas in earlier investigations, the percentage of anthers forming direct embryos was less than one (Raquin *et al.*, 1982; Binding, 1972a,b; Sopory & Maheshwari, 1973; Engvild, 1973; Wagner & Hess, 1974; Gangupati *et al.*, 1976). However, this increase in embryo formation was much lower than that obtained by manipulation of growth regulators in the culture medium (Jain, 1978; Jain *et al.*, 1980).

Gupta (1982) also reported a significant influence of low temperature treatment on the embryogenic response of microspores of *Petunia*. Though the anthers responded to form embryos and plantlets without cold treatment, the frequency of embryogenic anthers was low. Embryogenesis was stimulated significantly when pre-excised anthers were subjected to cold treatment at 5°C for 3 and 6 days or at 7–9°C for 6 and 12 days at the stage just before or during first pollen mitosis. In this study, the most effective temperature pretreatment was 7–9°C for 6 days which produced an average 2.6 plantlets (range 1–6 plantlets) from an embryogenic anther. Low temperature treatment increased the percentage of embryogenic anthers to 5.8 as compared to 2.4 for the controls. Babbar & Gupta (1980) reported that cold treatment (0 ± 1°C for five days) followed by 1 day high temperature (12 ± 1°C or 25 ± 1°C) treatment of flower buds induced mitotic activity in *P. hybrida* pollen. The authors presumed that prechilling of floral buds induced the pollen to switch to sporophytic development. This was manifested when buds were incubated at relatively higher temperatures without external supply of nutrients.

Light conditions. There is no literature available on the effect of the quality of light on anthers forming embryos. However, the cultures have been incubated under constant light (Engvild, 1973), 1500 lux light intensity for 12 h (Gangupati *et al.*, 1976), 16 h photoperiod (Malhotra & Maheshwari, 1977), continuous 150–250 lux (Jain, 1978), 12 h per day 500 lux light intensity (Gupta, 1983). Gupta (1982) also reported initial storage of cultures in the dark at 28°C for the induction and development of embryos.

2.3.1.5. *Role of genotype*. Mitchell *et al.* (1980) made first detailed study on the role of genotype in the induction of callus/embryo and plantlet formation in *Petunia* by culturing anthers of several cultivars of *P. hybrida* (Rose du Ciel), *P. axillaris* and of F_1 hybrids, and the backcross generations of these two species. Among the inbred lines, *P. axillaris* produced a high frequency of callus compared to Rose du Ciel cultivar. In the progeny of crosses between *P. axillaris* and Rose du Ciel, hybrids having *P. axillaris* as the female parent produced anther callus at a higher rate than when Rose du Ciel was used as a female parent. Maternal genotype did not influence the backcross generation as much as it did the F_1 hybrids. The frequency of plantlet formation from anthers of plants from backcrosses ranged from 0.5–5.5% depending upon the particular cross and the medium used. The rates of plantlet formation obtained by Mitchell *et al.* (1980) were comparable to those obtained by Malhotra & Maheshwari (1977) using an unidentified *Petunia* line. Mitchell *et al.* (1980) predicted that creating a hybrid population of anther donor plants from genotypes exhibiting high protoplast or tissue culture regeneration capacity might become a general method for obtaining haploids from recalcitrant species for anther culture because these lines are more likely amenable to tissue culture conditions compared with recalcitrant genotypes.

Raquin (1982) carried out studies on the genetic control of embryo formation and their quality from cultured anthers of *Petunia* using hybrids at different levels of heterozygosity under similar culture conditions. The yield of embryos was directly linked to the heterozygosity of the anther donor plants. The quality of embryos was expressed as the capacity to give viable plants, which was carried by certain lines and transmitted to their progeny and did not appear as a monogenic factor. These results might have been due to the nurse effect provided by anther wall in hybrid genotypes. Pelletier & Ilami (1972) established the importance of conditioning brought about by somatic tissue of tobacco and *Petunia* in the androgenic process. Therefore, the embryo production seemed to be essentially under sporophytic control, and heterozygosity was important for producing embryos (Raquin, 1982). A significant albeit slight difference was noticed in the response of anthers of the reciprocal hybrids. This was attributed to an inherent advantage of *P. axillaris* cytoplasm (Raquin, 1982; Mitchell *et al.*, 1980).

Gupta (1983) compared data on androgenic efficiency of *P. violacea* with *P. axillaris* (Engvild, 1973), *P. hybrida* (Wagner & Hess, 1974), and *P. axillaris* × *P. hybrida* (Raquin & Pilet, 1972). Both the frequency of productive anthers and the number of plantlets per productive anther were much higher for *P. violacea*. However, under the same cultural conditions, *P. parodii* showed significantly higher productivity compared to *P. violacea* (Gupta, 1982). Genotype seems to play an important role in the production of androgenic plants. Jain (1978) and Jain *et al.* (1980) obtained a higher frequency of androgenic plants than previously reported for *Petunia* using an unidentified *P. hybrida* and manipulating the growth regulators in the culture medium.

2.3.1.6. *Developmental stage of the microspores at the time of culture.* The developmental stage of microspores in anthers seems to be a crucial factor for the induction and development of androgenic *Petunia* embryos. Raquin & Pilet (1972) reported predominantly triploids via anther culture of *P. axillaris* and *P. hybrida* hybrids, although haploids and diploids were also obtained. Later on Raquin *et al.* (1982) showed that the homozygous polyploid of *Petunia* were produced by endoreduplication in uninucleate microspores, prior to the first embryonic division. Engvild (1973) observed that anthers containing only binucleate microspores formed predominantly triploid plantlets in *P. axillaris*, which was probably due to the completion of DNA synthesis prior to second pollen mitosis. All other reports indicated few haploids among androgenic *Petunia* plants (Wagner & Hess, 1974; Gangupati *et al.*, 1976; Mitchell *et al.*, 1980; Gupta, 1982, 1983). However, Jain

(1978) and Jain *et al.* (1980) obtained predominantly haploids from anthers cultured at the uninucleate stage compared to tetrad, binucleate or early meiotic stage in *P. hybrida*. In other genera also, success has also been achieved by culturing anthers at the uninucleate stage (Vasil & Nitsch, 1975; Maheshwari *et al.*, 1982; Bajaj, 1990; Raghavan, 1990). The question is why uninucleate pollen grains are most responsive. There is no unequivocal answer to this question yet. Gupta (1982) suggested the origin of triploids was due to the fusion of interphase nuclei in two different pathways. In the majority of embryogenic microspores, a vegetative nucleus of 1C DNA content fused with an endoreduplicated 2C DNA generative nucleus at the trinucleate stage and produced true triploid embryos and plantlets (A pathway). In the failure of fusion, both the vegetative and the generative nuclei divided separately and, in multinucleate microspores two or more daughter nuclei fused to form a mixoploid embryo. Such mixoploid embryos produced mixoploid plants and ploidy chimerism within individual plants. Haploid plant formation occurred when vegetative and/or generative nuclei divided repeatedly and failed to fuse during embryo formation (Gupta, 1983). According to Sopory (1972), uninucleate pollen grains may have sufficient endogenous growth regulators for the initiation of growth and development of pollen for haploid production; such conditions may be absent during other microspore developmental stages.

2.4. *Microspore culture*

Binding (1972a,b) first cultured isolated pollen from several cultivars of *P. hybrida* and obtained multinucleate structures containing as many as 30 nuclei. Pollen inoculated either at the uninucleate or mature stages did not divide in any medium tested, but responded if inoculated just after mitosis on different culture media. Auxins and cytokinins were essential for pollen division *in vitro*. Sucrose, though essential, could be replaced with mannitol. Sangwan & Norreel (1975) first regenerated plantlets from microspore culture of *P. hybrida*. They determined the precise developmental stage of immature pollen grains which gave the best androgenic response in order to increase the yield of haploid plants. The microspores cultured at early, mid, or late uninucleate stages did not divide in any medium or combination tested. Pollen inoculated just before or just after the first mitosis, however, responded to the various media tested. The microspore cultures were initiated on medium containing half-strength MS macroelements and microelements (Murashige & Skoog, 1962) with vitamins of Nitsch & Nitsch (1969) (Sangwan & Norreel, 1975), NAA, BA, boric acid, sucrose and glucose, and incubated at 25°C for 7 days in the dark after which they were illuminated under 1000 lux at 28 ± 1°C. Only a small percentage of pollen grains developed into plantlets (approximately one plant per eight anthers) (Sangwan & Norreel, 1975).

Table 1. Morphological differences between androgenic haploid and colchicine diploidized *Petunia hybrida*

Characters of plants	Haploids (cm)	Diploids (cm)
Girth of stem	0.2–0.3	0.6–0.7
Length of internode	1.0–1.3	2.4–2.8
Length of leaves	0.5–1.1	1.0–2.1
Height of plants	4.5–5.1	10.2–12.3

Figure 5. Colchicine treated diploidized *Petunia hybrida* flowering plants.

3. Haploid characteristics and determination

Haploids, in most cases have only half the genome of the sporophyte from which they were derived, a condition that results in altered phenotype at both the cellular and whole plant levels. Therefore, the general characteristics of haploids compared to their parents are: diminished vigour, stunted growth, reduced height, narrower leaves, smaller flowers with varying degrees of sterility and poor seed set (Magoon & Shambulingappa, 1962). Alternatively, haploids may be as good in appearance as their progenitors or sometimes even more vigorous (Magoon & Khanna, 1963). Jain (1978) compared morphology of haploids and diploidized *P. hybrida* anther-derived plants and found that the girth of stem, length of internode, length of leaf and height of plants were almost half for the haploids (Table 1; Fig. 5). Straub (1973) observed marked variation among haploid *Petunia* for flower colour and this characteristic remained unchanged in the diploidized plantlets.

Ploidy determination of anther-derived *Petunia* and other plants has been done with cytology (Rao & Suprasanna, 1996). Mitchell *et al.* (1980) found that counting chloroplasts in guard cell pairs from leaflets of anther-derived plants was suitable for indicating ploidy of small shoots in culture. Santos & Handro (1983) demonstrated that leaf size was not a good characteristic for discriminating among plants of different ploidy – flower and stomata characteristics were more adequate for this purpose. Furthermore, they showed that stepwise discriminant analysis of chloroplast number, leaf length/width ratio, petal length was appropriate for the discrimination of ploidy classes. Kamo & Griesbach (1989) identified haploids with microfluorimetry, chloroplasts per guard cell pair and chromosome root tip counts. Haploids and chimeric haploid/diploid plants could not be distinguished by chloroplast counts, but were identified by either chromosome counts or microfluorimetry. Altman *et al.* (1994) measured pollen size to determine ploidy of anther-derived *Arabidopsis thaliana* plants.

4. Diploidization of haploid lines

Ploidy level in androgenic haploid *Petunia* plants varies depending on the auxin and cytokinin in the medium (Liscum & Hangarter, 1991). Treatment of haploid leaf tissue with 30–100 μM BA for 9 to 12 days prior to culture on regeneration medium had a stabilizing effect on the genome such that 80% of the regenerants were haploids. In contrast, pretreatment with 50 μM IAA for two days induced endomitosis and resulted in 90% diploid regenerants (Liscum & Hangarter, 1991). The stability of haploidy in regenerated plantlets may also be dependent on the type of explant used. Jain (1978) cultured leaf, stem, callus, and axillary buds from androgenic haploid *P. hybrida* plants, and obtained highest number of regenerated haploid plants from axillary buds on culture medium containing 10^{-7} M each of kin and indoleacetic acid (IAA). However, when callus was used as an explant on the same medium, the percentage of haploid plants was slightly higher. Kamo & Griesbach (1993) propagated haploid *Petunia* of cv. Mitchell by tissue culture in the greenhouse and the resulting majority of plants (80%) was chimaeric rather than purely diploid or haploid. Protoplasts isolated from mesophyll cells of diploid plants were initially purly diploid, but later increased levels of ploidy. Auxins and cytokinin did not affect ploidy changes at the initial cell divisions. Knowledge of ploidy variation in haploid plants and cultures can be exploited to micropropagate selected haploid clones or to double the chromosome number of haploids to derive homozygous lines.

5. Conclusions

The recovery of haploid plants in *Petunia* has been low compared to diploid and triploid plants, irrespective of the method of haploid production. This variation in ploidy has been attributed to endomitosis or nuclear fusion, growth regulators, genotype, and the developmental stage of microspores. Cold pretreatment of anthers did not dramatically enhance the frequency of haploid plantlets. Jain (1978) and Jain *et al.* (1980) were the only ones to recover a preponderance of haploids by: a) culturing *P. hybrida* anthers with microspores at the uninucleate stage, and b) modifying the growth regulators (auxin and cytokinin) in the culture medium. These results underscore the importance of both genotype and culture conditions for optimizing haploid induction.

Binding (1974a) utilized haploid cultures of *P. hybrida* for the selection of streptomycin resistant lines. Two selections were auxin autotrophs. The haploid tissues later became diploid, tetraploid and even octoploid. Surprisingly, streptomycin resistance was not found in diploid *Petunia*.

The success of Binding (1974b) in obtaining divisions from leaf protoplasts of aseptically-grown haploid *P. hybrida* paved the way for transformation with *Agrobacterium tumefaciens* (Jia *et al.*, 1983), isolation of nitrate reductase (NR) deficient mutants of *Petunia* as well as their genetic and biochemical characterization (Steffen & Schieder, 1984), later led to intraspecific hybridization by fusing haploid tetrad protoplasts with cell suspension protoplasts resulting in triploid gametosomatic hybrids (Lee & Power, 1988). These achievements have made *Petunia* a model ornamental plant. However, work is needed to improve the recovery of haploids in anther culture and to identify genes for flower colour or other important ornamental traits. The use of haploids for genetic transformation and gene mapping may facilitate molecular solutions to problems in the ornamental industry.

6. Acknowledgements

We are grateful to Dr. Yogesh Sharma, History Centre, JNU, New Delhi, for the translation of French research papers into English.

7. References

Acharya, B.C. & M.V. Ramji, 1977. Experimental androgenesis in plants – a review. Proc. Ind. Acad. Sci. 86(B): 337–360.

Altman, T., B. Damm, W.B. Frommer, T. Martin, D. Schewizer, L. Willmitzer & R. Schmidt, 1994. Easy determination of ploidy level in *Arabidopsis thaliana* plants by means of pollen size measurement. Plant Cell Rep. 13: 652–656.

Ascher, P.D., 1984. Self-incompatibility. In: K.C. Sink (Ed.), *Petunia*: Monographs on Theoretical and Applied Genetics, Vol. 9, pp. 92–109. Springer-Verlag, Berlin.

Babbar, S.B. & S.C. Gupta, 1980. Chilling induced androgenesis in anthers of *Petunia hybrida* without any culture medium. Z. Pflanzenphysiol. 100: 279–283.

Bajaj, Y.P.S., J. Reinert & E. Heberle, 1976. Factors enhancing *in vitro* production of haploid plants in anthers and isolated microspores. In: R.J. Gautheret (Ed.), A La Mémoire de Georges Morel, pp. 212–228. Masson and Cie, Paris.

Bajaj, Y.P.S. (Ed.), 1990. Biotechnology in Agriculture and Forestry, Vol. 12: Haploids in Crop Improvement. Part I. Springer-Verlag, Berlin.

Bernard, S., 1971. Développement d'embryons haploïdes à partir d'anthères cultivées *in vitro*. Étude cytologique comparée chez le tabac et le *Petunia*. Rep. Cytol. Biol. Veg. 34: 165–188.

Bianchi, F. & P. Dommergues, 1979. *Petunia* genetics. *Petunia* as a model for plant research genetics and mutagenesis. Ann. Amélior. Plantes 29: 607–610.

Binding, H., 1972a. Nuclear and cell-division in isolated pollen of *Petunia hybrida* in agar suspension cultures. Nature New Biol. 237: 283–285.

Binding, H., 1972b. Selection in kallus Kulturen mit haploiden Zellen. Z. Pflanzenzüchtg. 67: 33–38.

Binding, H., 1974a. Mutation in haploid cell cultures. In: K.J. Kasha (Ed.), Haploids in Higher Plants: Advances and Potential, pp. 323–337. University of Guelph, Guelph.

Binding, H., 1974b. Cell cluster formation by leaf protoplasts from axenic cultures of haploid *Petunia hybrida* L. Plant Sci. Lett. 2: 185–188.

Blakeslee, A.F., J. Belling, M.E. Farnham & A.D. Bergner, 1922. A haploid mutant in the Jimson weed, *Datura stramonium*. Science 55: 646–647.

Bornman, C.H., 1985. Haploidization of sugar beet (*Beta vulgaris*) via gynogenesis. In Vitro 1: 3.

Cappadocia, M., L. Chretien & G. Laublin, 1988. Production of haploids in *Gerbera jamesonii* via ovule culture: influence of fall versus spring sampling on callus formation and shoot regeneration. Can. J. Bot. 66: 1107–1110.

Cappadocia, M. & J. Veith, 1990. *Gerbera jamesonii* H. Bolus ex Hook: *In vitro* production of haploids. In: Y.P.S. Bajaj (Ed.), Biotechnology in Agriculture and Forestry, Vol. 12, Haploids in Crop Improvement, pp. 417–427. Springer-Verlag, Berlin.

Clausen, R.E. & M.C. Mann, 1924. Inheritance in *Nicotiana tabacum*. V. The occurrence of haploid plants in interspecific progenies. Proc. Natl. Acad. Sci. 10: 121–124.

DeVerna, J.W. & G.B. Collins, 1984. Maternal haploids of *Petunia axillaris* (Lam.) B.S.P. via culture of placenta attached ovaries. Theor. Appl. Genet. 69: 187–192.

Dunwell, J.M., 1985. Embryogenesis from pollen *in vitro*. In: M. Zaitlin, P. Day, A. Hollaender & C.M. Wilson (Eds.), Biotechnology in Plant Science: Relevance to Agriculture in the Eighties, pp. 49–76. Academic Press, Inc., Orlando, FL.

Engvild, K.C., 1973. Triploid petunias from anther cultures. Hereditas 74: 144–147.

Engvild, K.C., I. Linde-Laursen & A. Lundqvist, 1972. Anther cultures of *Datura innoxia*: flower bud stage and embryoid level of ploidy. Hereditas 72: 331–332.

Evenor, D. & S. Izhar, 1984. Coexistence of cytoplasmic and nuclear genes for male sterility in *Petunia*. Mol. Gen. Genet. 194: 523–527.

Evenor, D., A. Guri & S. Izhar, 1984. The relationship between a nuclear restorer gene and the nuclear gene for male sterility in *Petunia*. Theor. Appl. Genet. 67: 475–478.

Gains, E.F. & H.C. Aase, 1926. A haploid wheat plant. Am. J. Bot. 13: 373–328.

Gamborg, O.L., R.A. Miller & K. Ojima, 1968. Nutrient requirements of suspension cultures of soybean root cells. Exp. Cell Res. 50: 157–158.

Gangupati, L.S., R. Doreswamy, E.K. Chacko & R.N. Bhat, 1976. Studies on anther culture. 1. Induction of androgenic plants and callus in *Petunia hybrida* Vilm. Curr. Sci. 45: 714–715.

Guha, S. & S.C. Maheshwari, 1964. *In vitro* production of embryos from anthers of *Datura*. Nature 204: 497.

Gupta, N. & P.S. Carlson, 1972. Preferential growth of haploid plant cells *in vitro*. Nature New Biol. 239: 86.

Gupta, P.P., 1982. Genesis of microspore derived triploid *Petunias*. Theor. Appl. Genet. 61: 327–331.

Gupta, P.P., 1983. Microspore-derived haploid, diploid, and triploid plants in *Petunia violacea* Lindl. Plant Cell Rep. 2: 255–256.

Horlow, C., S. Hamza, Y. Chupeau & G. Pelletier, 1996. Conditional lethal markers: Spontaneous haploid selection in plants. In: S.M. Jain, S.K. Sopory & R.E. Veilleux (Eds.), *In Vitro* Haploid Production in Higher Plants, Vol. 1, Kluwer Academic Publishers, Dordrecht, pp. 297–315.

Horsch, R.B., J. Fry, N. Hoffman, J. Niedermeyer, S.G. Rogers & R.T. Fraley, 1988. Leaf disk transformation. In: S.B. Gelvin & R.A. Schilperoort (Eds.), Plant Molecular Biology Manual, A5, pp. 1–9. Kluwer Academic Publishers, Dordrecht.

Hu, H. & H.Y. Yang, 1986. Haploids in Higher Plants *in Vitro*. China Acad. Publ., Beijing, Springer-Verlag, Berlin.

Jain, S.M., 1978. *In vitro* production of haploids in higher plants. Ph.D. Thesis submitted to Jawaharlal Nehru University, New Delhi.

Jain, S.M., S. Bagga, N. Bhalla-Sarin, S. Guha-Mukherjee & S.K. Sopory, 1980. *In vitro* culture of anthers in *Petunia hybrida* and *Brassica oleracea*. In: P.S. Rao, M.R. Heble & M.S. Chadha (Eds.), Plant Tissue Culture Genetic Manipulation and Somatic Hybridization of Plant Cells, Proc. Natl. Symp., Bhabha Atomic Research Center, Bombay, pp. 26–32.

Jia, J.F., R.D. Shillito & I. Potrykus, 1983. Crown gall transformation of regenerating protoplasts of haploid and diploid *Petunia hybrida* var. Mitchell by *Agrobacterium tumefaciens*. Z. Pflanzenphysiol. 112: 1–6.

Kamo, K.K. & R.J. Griesbach, 1989. Determination of ploidy level in "Mitchell" petunias. Plant Sci. 65: 119–124.

Kamo, K. & R.J. Griesbach, 1993. Ploidy changes in "Mitchell" petunia. Acta Hort. 336: 307–314.

Kasha, K.J., 1974. Haploids in Higher Plants: Advances and Potential. The University of Guelph, Guelph, Ontario.

Katayama, Y. & M. Nei, 1964. Studies on the haploidy in higher plants. Rep. Lab. Plant Breed., Fac. Agr., Miyazaki Univ. No. 22: 1–78.

Keller, E.R.J. & L. Korzun, 1996. Ovary and ovule culture for haploid production. In: S.M. Jain, S.K. Sopory & R.E. Veilleux (Eds.), *In Vitro* Haploid Production in Higher Plants, Vol. 1, Fundamental Aspects and Methods, pp. 217–235. Kluwer Academic Publishers, Dordrecht.

Kimber, G. & R. Riley, 1963. Haploid angiosperms. Bot. Rev. 29: 480–531.

Lacadena, J.R., 1974. Spontaneous and induced parthenogenesis and androgenesis. In: K.J. Kasha (Ed.), Haploids in Higher Plants: Advances and Potential. The University of Guelph, Guelph, Ontario, pp. 13–32.

Lakshmi Sita, L., 1996. Gynogenic haploids *in vitro*. In: S.M. Jain. S.K. Sopory & R.E. Veilleux (Eds.), *In Vitro* Haploid Production in Higher Plants, Vol. 4. Kluwer Academic Publishers, Dordrecht (in press).

Lee, C.H. & J.B. Power, 1988. Intraspecific gametosomatic hybridization in *Petunia hybrida*. Plant Cell Rep. 7: 17–18.

Liscum III, E. & R.P. Hangarter, 1991. Manipulation of ploidy level in cultured haploid *Petunia* tissue by phytohormone treatments. J. Plant Physiol. 138: 33–38.

Magoon, M.L. & K.R. Khanna, 1963. Haploids. Caryologia 16: 191–235.

Magoon, M.L. & K.G. Shambulingappa, 1962. Cytological and morphological studies in some interspecific hybrids in Eu-Sorghums. Z. für. Vererbungslehre 193: 14.

Maheshwari, P. & N.S. Rangaswamy, 1965. Embryology in relation to physiology and genetics. In: R.D. Preston (Ed.), Advances in Botanical Research, Vol. 2, pp. 319–321. Academic Press, London.

Maheshwari, S.C., A. Rashid & A.K. Tyagi, 1982. Haploids from pollen grains – retrospect and prospect. Am. J. Bot. 69: 865–879.

Maheshwari, S.C., A.K. Tyagi, K. Malhotra & S.K. Sopory, 1980. Induction of haploidy from pollen grains in gymnosperms – the current status. Theor. Appl. Genet. 58: 193–206.

Maizonnier, D., 1976. Etude cytogénétique de variations chromosomiques naturelles ou induites chez *Petunia hybrida*. Ph.D. Thesis, Institut National de la Recherche Agronomique, Dijon.
Malhotra, K. & S.C. Maheshwari, 1977. Enhancement by cold treatment of pollen embryoid development in *Petunia hybrida*. Z. Pflanzenphysiol. 85: 177–180.
Martineau, B., M.R. Hanson & F.M. Ausubel, 1981. Effect of charcoal and hormones on anther culture of *Petunia* and *Nicotiana*. Z. Pflanzenphysiol. 102: 109–116.
Meyer, P., I. Heidmann, G. Forkmann & H. Saedler, 1987. A new *Petunia* flower colour generated by transformation of a mutant with a maize gene. Nature 330: 677–678.
Meyer, P., F. Linn, I. Heidmann, H. Meyer, I. Niedenhof & H. Saedler, 1992. Endogenous and environmental factors influence 35S promoter methylation of a maize A1 gene consruct in transgenic petunia and its colour phenotype. Mol. Gen. Genet. 231: 345–352.
Mitchell, A.Z., M.R. Hanson, R.C. Skvirsky & F.M. Ausubel, 1980. Anther culture of *Petunia*: genotypes with high frequency of callus, root or plantlet formation. Z. Pflanzenphysiol. 100: 131–146.
Murashige, T. & F. Skoog, 1962. A revised medium for rapid growth and bioassays with tobacco tissue cultures. Physiol Plant. 15: 473–497.
Nitsch, J.P., 1969. Experimental androgenesis in *Nicotiana*. Phytomorphology 19: 339–346.
Nitsch, J.P. & C. Nitsch, 1969. Haploid plants from pollen grains. Science 163: 85–87.
Nitzsche, W. & G. Wenzel, 1977. Haploids in plant breeding. Verlag Paul Parey, Berlin.
Pelletier, G. & M. Ilami, 1972. Les facteurs de l'androgenèse *in vitro* chez *Nicotiana tabacum*. Z. Pflanzenphysiol. 68: 97–114.
Raghavan, V., 1990. From microspore to embryoid: Faces of the angiosperm pollen grain. In: H.J.J. Nijkamp, L.H.W. van der Plas & J. van Aartrijk (Eds.), Progress in Plant Cellular and Molecular Biology, pp. 407–409. Kluwer Academic Publishers, Dordrecht.
Rao, P.S. & P. Suprasanna, 1996. Methods to double haploid chromosome numbers. In: S.M. Jain, S.K. Sopory & R.E. Veilleux (Eds.), *In Vitro* Haploid Production in Higher Plants, Vol. 1, Fundamental Aspects and Methods, pp. 317–339. Kluwer Academic Publishers, Dordrecht.
Raquin, C., A. Cornu, E. Farcy, D. Maizonnier, G. Pelletier & F. Vedel, 1989. Nucleus substitution between *Petunia* species using gamma ray-induced androgenesis. Theor. Appl. Genet. 78: 337–341.
Raquin, C. & M.V. Pilet, 1972. Production de plantules à partir d'anthères de *Petunias* cultivées *in vitro*. C.R. Acad. Sci. Paris. 274: 1019–1022.
Raquin, C., 1982. Genetic control of embryo production and embryo quality in anther culture of *Petunia*. Theor. Appl. Genet. 63: 151–154.
Raquin, C., 1985. Induction of haploid plants by *in vitro* culture of *Petunia* ovaries pollinated with irradiated pollen. Z. Pflazenzüchtg. 94: 166–169.
Raquin, C., 1990. Association de l'irradiation des gamétophytes et de la culture *in vitro* pour l'haploïdisation et les transfers nucléocytoplasmiques. Colloques de l'INRA, No. 51: 187–194.
Raquin, C., M. Amssa, Y. Henry, J. De Buyser & S. Essad, 1982. Origine de plantes polyploides obtenues par culture d'anthère. Analyse cytophotométrique *in situ* et *in vitro* des microspores de *Petunia* et de blé tendre. Z. Pflanzenzüchtg. 89: 265–277.
Raquin, C., 1983. Utilization of different sugars as carbon sources for *in vitro* cultures of *Petunia*. Z. Pflanzenphysiol. 111: 453–457.
Sangwan, R.S. & B. Norreel, 1975. Induction of plants from pollen grains of *Petunia* cultured *in vitro*. Nature 257: 222.
San Noeum, L.H., 1976. Haploïdes d'*Hordeum vulgare* L. par culture *in vitro* d'ovaires non fécondés. Ann. Amélior. Plantes. 26: 751–754.
San Noeum, L.H., 1979. *In vitro* induction of gynogenesis in higher plants. In: A.M. van Harten & A.C. Zeven (Eds.), Proc. Conf. Broadening Genetic Base of Crops, Wageningen, July 3–7, 1978, pp. 327–329. Pudoc, Wageningen.
Santos, R.F. & W. Handro, 1983. Morphological patterns in *Petunia hybrida* plants regenerated from tissue cultures and differing by their ploidy. Theor. Appl. Genet. 66: 55–60.

Santos, R.F., W. Handro & E.I.S. Floh, 1983. Ploidy in *Petunia hybrida* plants regenerated from tissue culture. Rev. Brasil Bot. 6: 33–39.

Sestili, S. & N. Ficcadenti, 1996. Irradiated pollen for haploid production. In: S.M. Jain, S.K. Sopory & R.E. Veilleux (Eds.), *In Vitro* Haploid production in Higher Plants, Vol. 1, Fundamental Aspects and Methods, pp. 263–274. Kluwer Academic Publishers, Dordrecht.

Sims, T., 1994. Molecular genetics of gametophytic self-incompatibility in *Petunia hybrida*. In: E.G. Williams, A.E. Clarke & R.B. Knox (Eds.), Genetic Control of Self-Incomatibility and reproductive Development in Flowering Plants, pp. 92–110. Kluwer Academic Publishers, Dordrecht.

Singh, I.S. & A. Cornu, 1976. Recherche de plantes haploïdes de *Petunia* à noyau androgenetique et a cytoplasme gynogenetique determinant de la stérilité pollinique. Ann. Amélior. Plantes 26: 565–568.

Sink, K.C., 1984. Monographs on Theoretical and Applied Genetics. Vol. 9: *Petunia*. Springer-Verlag, Berlin.

Sopory, S.K., 1972. Physiology of development of pollen embryoids in *Datura innoxia* Mill. Ph.D. Thesis submitted to University of Delhi, Delhi.

Sopory, S.K. & S.C. Maheshwari, 1973. *In vitro* production of plants from the anthers of *Petunia hybrida* Vilmour. Curr. Sci. 42: 247–248.

Steffen, A. & O. Schieder, 1984. Biochemical and genetical characterization of nitrate reductase deficient mutants of *Petunia*. Plant Cell Rep. 3: 134–137.

Straub, J., 1973. Die genetische Variabilität haploider Petunien. Z. Pflanzenzüchtg. 70: 265–274.

Sunderland, N., 1971. Anther culture: a progress report. Sci. Prog., London 59: 527–597.

Sunderland, N., 1974. Anther culture as a means of haploid induction. In: K.J. Kasha (Ed.), Haploids in Higher Plants: Advances and Potential, pp. 91–122. University of Guelph, Guelph, Ontario.

Swamy, R.D & E.K. Chacko, 1973. Induction of plantlets and callus from anthers of *Petunia axillaris* (LAM, B.S.P.) cultured *in vitro*. Hort. Res. 13: 41–44.

Van der Krol, A.R. & N.H. Chua, 1993. Flower development in *Petunia*. Plant Cell 5: 1195–1203.

Van Tunen, A.J., L.A. Mur, G.S. Brouns, J.D. Rienstra, R.E. Koes & J.N.M. Mol, 1990. Pollen and anther-specific chi promoters from *Petunia*: Tandem promoter regulation of the chi A gene. Plant Cell 2: 393–401.

Vasil, I.K. & C. Nitsch, 1975. Experimental production of pollen haploids and their uses. Z. Pflanzenphysiol. 76: 191–212.

Wagner, G. & D. Hess, 1974. Haploide, diploide und triploide pflanzen von *Petunia hybrida* aus Pollenkörnern. Z. Pflanzenphysiol. 73: 273–276.

Wenzel, G., U. Frei, A. Jahoor, A. Graner & B. Foroughi-Wehr, 1995. Haploids – an integral part of applied and basic research. In: M. Terzi, R. Cella & A. Falavigna (Eds.), Current Issues in Plant Molecular and Cellular Biology, pp. 127–135. Kluwer Academic Publishers, Dordrecht.

Yang, H. & C. Zhou, 1982. *In vitro* induction of haploid plants from unpollinated ovaries and ovules. Theor. Appl. Genet. 63: 97–104.

5. *In vitro* induction of haploid plants from the gametophytes of lily and tulip

R.W. VAN DEN BULK and J.M. VAN TUYL

Contents

1. General introduction

As a result of sporophytic cell division and proliferation of the male or female gametophytes in culture, embryos or calli may develop from which haploid plantlets may be regenerated. The first haploid plant from the *in vitro* culture of anthers was verified in 1964 for *Datura innoxia* (Guha & Maheshwari, 1964), whereas the first haploid plant from unpollinated ovaries was obtained in 1976 for *Hordeum vulgare* (San Noeum, 1976). The induction of haploid plants from male or female gamethophytes by *in vitro* culture is now possible for many species (Yang & Zhou, 1990; Sangwan & Sangwan-Norreel, 1990). *In vitro* culture of anthers has been the most efficient method for regenerating (doubled) haploid plants, especially for brassicaceous, graminaceous, or solanaceous species. In the last decade, microspore culture has become an important technique to obtain haploid or doubled haploid plants. The culture of isolated microspores has several advantages over culture of whole anthers. Regenerated plants will undoubtedly have been derived from the microspores and not from anther tissue. Also, microspores can be isolated in large numbers and are promising targets for *in vitro* manipulation and ontogenetic studies. Moreover, for some crops, e.g., *Hordeum vulgare* (Hoekstra *et al.*, 1992) and *Brassica napus* (Siebel & Pauls, 1989), a higher efficiency of haploid production can be reached with microspore culture. When anther or microspore culture cannot be applied, e.g., when microspores are not formed due to male-sterility, or when androgenesis is hampered by low response or albinism of regenerated plants, culture of

S.M. Jain, S.K. Sopory & R.E. Veilleux (eds.), In Vitro *Haploid Production in Higher Plants, Vol. 5*, 73–88.

unpollinated ovaries or ovules may be an alternative to haploid derivation (Yang & Zhou, 1982).

Breeding with doubled haploids is usually focused on the rapid production of inbred lines for F_1 hybrid production, as for *Brassica* species. For other crops, like the vegetatively propagated bulbous crops, a successful procedure to produce (doubled) haploid embryos and plants from gametophytes may lead to the improvement of agronomically important traits. Recessive traits are expressed and, because of their homozygous state, can be selected directly. The doubled haploids can be used in crossing programs, simplifying genetic analyses of the progeny and enabling a more directed transfer of specific traits. For the construction of genetic maps by molecular marker analysis, it is a prerequisite to utilize homozygous parents in crosses (Melchinger, 1990). Furthermore, the technique of regenerating plants from microspores can be employed for biotechnological applications to improve cultivars, like genetic modification through transformation or *in vitro* selection for specific traits in a microspore population.

2. Bulbous crops

The bulbous crops lily (*Lilium* spp. L.) and tulip (*Tulipa* spp. L.), which are grown for both bulb and cut flower production, represent an important economic enterprise in The Netherlands. Because of governmentally imposed restrictions on the use of chemical compounds for disease control, current cultivars urgently need improvement by introduction of resistance to viral and fungal diseases and insect pests. In the breeding of these vegetatively propagated crops, interspecific hybridization has played an important role. Therefore the cultivars, which are genetically very heterozygous, often originated from complex species crosses resulting in a broad range of flower forms and colours (Van Eijk *et al.*, 1991; Van Tuyl *et al.*, 1993). Genetic analyses are complex due to the heterozygous interspecific character. This is an important reason to initiate haploid production in lily and tulip. Breeding at the tetraploid level, using mitotic and meiotic polyploidization, is an important strategy in lily and tulip improvement. The reasons for inducing polyploidy include larger flower size, stronger stems (especially important for forcing under low light conditions during the winter) and, in the case of interspecific hybridization, restoration of F_1 sterility at the tetraploid level. Using haploid induction methods, the ploidy level can be reduced from tetraploid to diploid. This can hasten the breeding process, because breeding at the diploid level is more efficient than at the tetraploid level. Also, reduction of ploidy might improve the crossability of polyploid plants.

In this chapter, we review the literature on inducing (doubled) haploids from male and female gametophytes of lily and tulip and describe our experiences with the induction of haploids in microspore culture of these bulbous crops.

3. Anther and ovary culture

3.1. *Anther culture*

In vitro production of haploids from gametophytes of lily has been studied to some extent, but for tulip no information is available. Regeneration of, presumably haploid, plants via anther culture has been reported for *Lilium longiflorum* (Sharp *et al.*, 1971; Qu *et al.*, 1988) and *L. davidii* (Gu & Cheng, 1982). Sharp and coworkers (1971) described the formation of haploid callus from cultured anthers of *L. longiflorum*. Besides the haploid cell types, mixoploid populations also were present. The haploid cell lines were subcultured regularly and, after eight months in culture, bulblet formation was observed. The root tips of the resulting plantlets contained the haploid chromosome number. Unfortunately, these plantlets could not be maintained in culture and mature plants were not obtained (W.R. Sharp, personal communication).

The effects of 60 treatment combinations including sucrose concentration, growth regulators [2,4–dichlorophenoxyacetic acid (2,4–D) and benzyladenine (BA)], pre-treatment of flower buds at 7°C for 14 days and temperature treatment of anthers (25°C continuously vs. two days at 35°C before transfer to 25°C) on anther culture response of lily were studied by Qu *et al.* (1988). Moreover, 108 diploid genotypes of *L. longiflorum* were tested, assigned randomly to the different treatments. Both genotypic effect and the presence of 2,4–D were the most important factors for callus induction. Callus was derived from anthers of 28 of 108 genotypes, and gave rise to bulblets when transferred to regeneration medium. Mitotic analysis of root tip cells of plants derived from these bulblets showed a mixoploid or diploid chromosome number. The occurrence of a mixoploid population of cells has also been reported for shoot and root apices from regenerated plants derived from diploid callus (Bennici, 1979). Nevertheless, some plants showed predominantly haploid cells, up to 80%, at early stages, suggesting gametic origin. Consequently, it is possible that the diploid plants obtained actually were doubled haploids. However, since suitable biochemical markers were not available, this could not be proven (M.C. Mok, personal communication).

Successful anther culture for *L. davidii* was reported by Gu & Cheng (1982). Anthers with microspores at the late uninucleate stage were cultured on a modified MS (Murashige & Skoog, 1962) medium with zeatin or 2,4–D in combination with kinetin or BA; calli or embryos were produced after 30 days. The presence of callus with predominantly haploid ($2n = 1x = 12$) cells (about 70%) was proven. After transfer to regeneration medium, calli or embryos could develop into plantlets in approximately 2 1/2 months. Root tips of regenerated plantlets contained haploid, diploid and aneuploid cells, with haploid cells predominant (86%). These results strongly suggest that haploid plantlets had been derived from the male gametes. The results of this study emphasized that growth regulators were necessary for the induction of androgenesis in this *Lilium* species, and that anthers should be plated without the filament. If callus was produced on the filament, the formation of callus or embryos in the anther was suppressed.

3.2. *Ovary culture*

Anther culture was compared with ovary culture for an Oriental lily hybrid by Prakash & Giles (1986). Globular structures developed occasionally on anthers, but were found to be of anther wall origin after anatomical studies. Embryos derived from microspores were not obtained in their studies. However, they succeeded in the induction of gynogenic haploids through the culture of transverse sections, 2–3 mm thick, of unpollinated ovaries. From each responsive ovule, one embryo originated within 28 days and more than 70% of those germinated and developed into normal plantlets within three to four weeks. Unfortunately, the frequency of embryo formation was not reported. Sectioning the unpollinated ovaries in a suspension of activated charcoal, and culture of the sections on a medium containing a high (10%) sucrose concentration and glutamine was essential for the successful induction of gynogenesis. Anatomical studies of cultured ovules confirmed the "embryo sac" origin of the regenerants. Chromosome counts in root tip cells of some regenerants showed the presence of haploid and diploid cells, suggesting spontaneous doubling of the chromosomes in some cells of these regenerants.

For *Lilium davidii* a successful approach via ovary culture has also been reported (Gu & Cheng, 1983). Calli of about 6 to 12 mm in size were produced on cultured unpollinated young ovaries after 40 days, and could be regenerated into plantlets. Some ovaries produced embryos directly, which also were able to develop into plantlets. Microscopic analysis of the chromosome number in cells of regenerated plantlets showed 66% to be haploid and 34% to be diploid. Histological studies suggested that haploid plants originated from the megaspore tetrad, whereas diploid plants originated from somatic cells, probably of the nucellus. However, a definite proof of the haploid origin of plants was not given.

Notwithstanding the fact that the induction of (doubled) haploids has been reported, a reproducible and efficient protocol for haploid production has not yet become available for lily.

4. Microspore culture of lily

4.1. *Introduction*

With respect to the culture of microspores or pollen of lily *in vitro*, many studies have been conducted by Tanaka and coworkers in Japan. They studied the gametophytic development of explanted microspores of *Lilium longiflorum* (Tanaka & Ito, 1980, 1981b, 1982; Tanaka *et al.*, 1979, 1980). Occasionally, abnormal binucleate cells were observed. These aberrant cells had evenly stained nuclei, without normal nuclear differentiation, and were sometimes divided symmetrically by a septum (Tanaka & Ito, 1980; Tanaka *et al.*, 1980). It was suggested that these abnormal cells were possibly undergoing sporophytic development toward haploid embryo formation. However, subsequent divisions were never observed (I. Tanaka, personal communication).

Another approach for embryogenesis of gametophytes may be the use of pollen protoplasts as starting material for culture (Zhou & Wu, 1990). However, although pollen protoplasts of lily can be isolated and cultured, cellular divisions of these haploid protoplasts could not be induced (Tanaka *et al.*, 1987; Ueda *et al.*, 1990).

4.2. *Sporophytic development*

We have studied the induction of sporophytic development in microspore culture of lily, aimed at the production of haploid embryos and plants (Van den Bulk *et al.*, 1992). For that purpose we used microspores isolated from flower buds of the Asiatic hybrid "Whilito". Microspore populations with 80 to 90% viability could easily be obtained from this genotype. From each anther 35,000 to 80,000 microspores, depending on developmental stage of the anther, could be isolated. Microspores were cultured in NLN-medium with 13% sucrose and without growth regulators, as for microspore culture of *Brassica napus* (Lichter, 1982). After a heat shock treatment, i.e., culture for two days at 32°C and subsequently at 25°C, microspores which had undergone nuclear divisions were observed. These microspores showed predominantly three or four nuclei and were present in culture from day 4 onwards, reaching a maximum around day 10–12 (Fig. 1). Their subsequent decrease was probably due to degeneration. Generally, the viability of microspores in culture decreased rapidly from about 80% at the beginning to approximately 20% after one week and to 5–10% after two to three weeks of culture. Continuous culture at 25°C demonstrated that heat shock was unnecessary for induction of multinucleate microspores. Other stress conditions, e.g., starvation of microspores during the first four to six days or cold treatments to flower buds or anthers before isolation of microspores, did not have a positive effect on the formation of multinucleate structures.

Of the various developmental stages examined, late uninucleate microspores yielded the most multinucleate microspores, up to 5%, when microspores were cultured at a density of 10,000 per ml. Twenty-three other lily genotypes, belonging to *L. davidii*, *L. longiflorum*, the Asiatic hybrid group, and the Oriental hybrid group, were tested for their response using an identical protocol. However, none showed a percentage of multinucleate microspores exceeding 0.1% or other indication of putative sporophytic development.

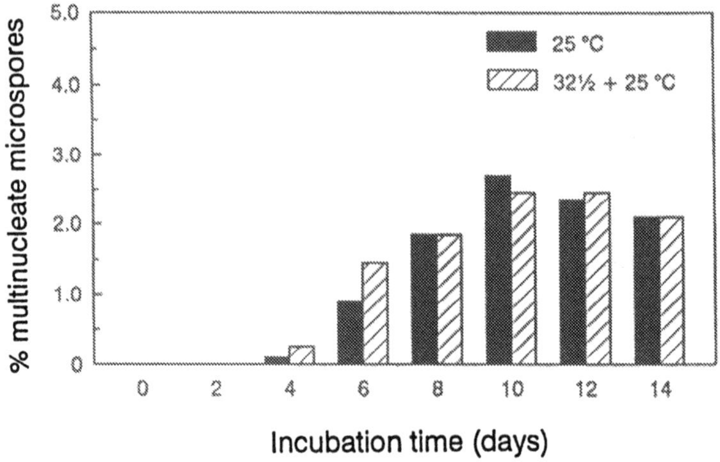

Figure 1. Percentage of multinucleate lily microspores observed over time in culture. Microspores of "Whilito" were incubated at 32°C for two days at the beginning of culture, before transfer to 25°C (heat shock treatment; hatched bars). Control cultures were incubated at 25°C continuously (filled bars).

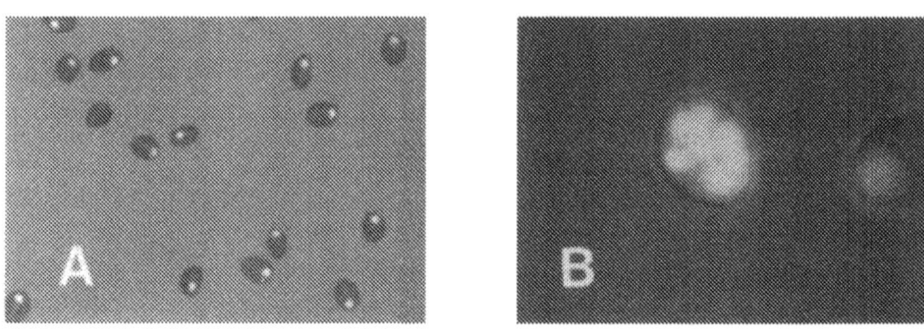

Figure 2. Lily microspores of "Whilito" in culture (nuclei stained with DAPI). (A) Population of late uninucleate microspores immediately after isolation. (B) Multinucleate microspore after 10 days of culture.

In cultures of "Whilito", microspores with three to eight nuclei have been observed (Fig. 2). Several factors, such as addition of antioxidants or growth regulators (combinations of 2,4–D and BA) to the medium, aeration, various regimes for refreshing culture media, and culturing fractions enriched with multinucleate microspores through Percoll gradient density centrifugation were tested, but did not result in further development of multinucleate microspores. Cell walls were not visible in these multinucleate microspores, even after staining with Calcofluor White. Multinucleate pollen grains, e.g., with 12 nuclei, have also been observed by Gu & Cheng (1982) in anther cultures of *L. davidii*, and in microspore or anther culture of other crops (Gaillard *et al.*, 1991; Idzikowska & Mlodzianowski, 1979). Because embryos could successfully be induced from microspores in these studies, the occurrence of multinucleate microspores is a strong indication of sporophytic development. So far, we have been unable to stimulate further nuclear and cellular divisions of multinucleate microspores leading to haploid embryos and plants of lily.

5. Microspore culture of tulip

5.1. *Introduction*

As with lily, Tanaka and coworkers also studied the response of explanted microspores of *Tulipa gesneriana*. The treatment of tulip microspores with demecolcine, a derivative of the spindle-disrupting agent colchicine, resulted in frequent formation of anomalous pollen with two equal nuclei, which did not undergo further nuclear division. Such pollen was capable of pollen tube development, indicating that symmetric division in microspores is not causally related to initiation of pollen embryogenesis (Tanaka & Ito, 1981a). Sporadically, anomalous binucleate pollen, derived from *in vitro* cultured uninucleate microspores, underwent further mitosis and produced tri- or tetra-nucleate pollen. Tanaka & Ito (1982) also reported that additional mitoses in explanted tulip microspores could be induced by bromodeoxyuridine. The nuclei in binucleate pollen did not show normal differentiation into vegetative and generative nuclei and were able to divide again. In bromodeoxyuridine-containing medium, the frequency of pollen with additional mitoses was about 30%, in contrast to 3% in control medium. Generally, pollen with four nuclei were produced this way, and cell walls between the four cells were frequently observed. The development of these cells usually ceased at the four to eight nucleate stage, but occasionally cell divisions continued, resulting in the formation of multicellular structures. Further development of these structures and regeneration into plantlets have not been reported.

5.2. *Starting material and culture conditions*

Microspore culture of tulip in our lab was carried out with *Tulipa gesneriana* cv. Leen van der Mark. Microspores were isolated from flower buds in bulbs stored at 17°C in the dark for three to four months after harvest and drying in the summer. During these months the development of the inflorescence inside the bulb occurred and anthers with young microspores had been formed. These bulbs could be used until about February, when the bulbs started to show symptoms of desiccation. Afterwards, bulbs stored at

Dissect the flower bud from the tulip bulb

↓

Disinfect the flower bud in 80% ethanol for 15 sec, rinse with tap water, and subsequently disinfect for eight min in a 2% sodium hypochlorite solution with 0.01% (v/v) 'Tween 20'

↓

Rinse with sterile water

↓

Excise the anthers and cut them transversely into thin segments of approximately 0.5 mm in isolation medium[1]

↓

Release microspores by gently agitating the medium with anther segments, filter the solution through a 150 μm nylon sieve and collect the microspores by centrifugation (2 min, 75xg)

↓

Resuspend the microspore pellet in fresh isolation medium and centrifuge again. Repeat this washing step and resuspend in culture medium[2]

↓

Adjust microspore density to 4×10^4 microspores/ml and culture 2.5 ml aliquots in 6-cm Petri dishes at 25°C in the dark

↓

Replace 1 ml of the culture medium with fresh medium at two-week intervals, starting three weeks after the beginning of the culture

[1] Composition of isolation medium: 1/4 MS-macro salts, full MS-micro salts, Nitsch & Nitsch vitamins, 100 mg/l myo-inositol, 7.5% mannitol, 3 mM 2-(N-morpholino)-ethanesulphonic acid, pH 5.8.

[2] Composition of culture medium: as for isolation medium, except mannitol was replaced by 13% sucrose.

Figure 3. Scheme for the isolation and culture of tulip microspores.

−2°C were used. These bulbs had been prepared after 3 1/2 months of storage at 17°C by rooting bulbs in a soil/sand mixture at 9°C for four weeks. Subsequently, bulbs were transferred to −2°C and stored for further use. The microspores were isolated and cultured according to the procedure described in Fig. 3. On average 800,000 microspores could be isolated from one flower bud. Three culture media, viz. NLN (Lichter, 1982), MS (Murashige & Skoog, 1962) and MS medium with quarter strength macro salts (1/4 MS), were initially compared. The 1/4 MS medium induced the best sporophytic response, i.e., development of microspores into multicellular structures, and was therefore adopted. Of several carbon sources tested, including maltose, sucrose, lactose, glucose, galactose and fructose, sucrose at a concentration of 13% was optimal. Additional organic components in the medium, e.g., amino acids or polyamines, had a negative effect on the sporophytic response.

5.3. *Development of multicellular structures*

Induced microspores with three or four nuclei were visualized by staining with the DNA specific fluorochrome 4′,6–diamidino-2–phenylindole (DAPI) after approximately 12 days of culture (Fig. 4A). Under optimal conditions, generally 0.5–1.0% of the plated microspores showed sporophytic development. Many induced microspores continued along the sporophytic pathway and the number of nuclei in these multinucleate microspores increased over time (Fig. 4B). After four to five weeks of culture, the first multicellular structures were observed (Fig. 4C). Based on the number of induced, multinucleate microspores, roughly 5–10% of these became multicellular.

Because the developmental stage of the microspores is critical for the induction of embryos in microspore culture, various stages of "Leen van der Mark" microspores were tested. Different stages could be obtained by isolating microspores from −2°C bulbs, which were incubated for various days in a controlled environment at 18°C under light. Under these conditions, sprouting of the bulbs and plant elongation occurred, resulting in progressive developmental stages of the microspores. The viability of the microspores/pollen, determined on the basis of the fluorochromatic reaction with fluorescein diacetate, depended on their developmental stage and varied from about 20% for early and mid-uninucleate populations to 60% for mature pollen. Results indicated that the highest number of multicellular structures, on average 131 structures per 100,000 plated microspores (0.13%), was obtained from a population consisting of early and mid-uninucleate microspores (Table 1). Because only about 20% of these plated microspores were viable, the response was greater by a factor of 5 when calculated for the viable fraction, i.e., about 0.7%.

5.4. *Response of various genotypes*

Microspores of 14 *T. gesneriana* cultivars, two *T. fosteriana* cultivars and 11 *Tulipa* species were isolated and cultured, in order to examine their sporophytic potential. For each genotype, one or two flower buds were used for microspore isolation. Microspores of approximately half the genotypes tested were able to divide sporophytically (Table 2). For *T. gesneriana* cvs. Leen van der Mark, Peach Blossom and Rosario, compact microcalli and embryo-like structures were obtained. Because microspore populations of "Leen van der Mark" occasionally did not respond in culture, it cannot be concluded that microspores of the non-responding genotypes are incapable of sporophytic development. The results suggest that induction of tulip microspores to sporophytic development may apply to many genotypes.

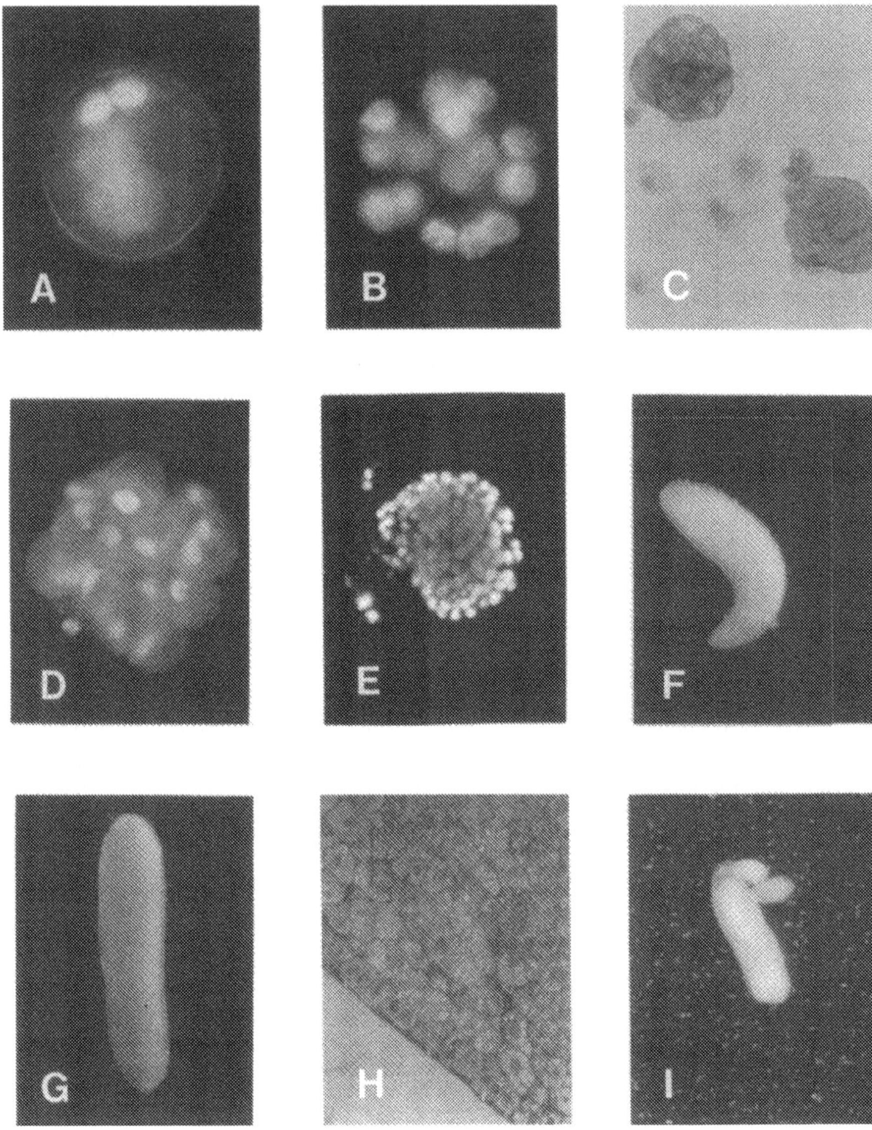

Figure 4. Various stages of embryo development in microspore cultures of tulip cv. Leen van der Mark. (A) Induced microspore with four nuclei, two larger vegetative and two smaller generative ones (DAPI staining). (B) Multinucleate microspore (DAPI staining). (C) Multicellular structures after 6 weeks of culture (ø ca. 0.3 mm). (D) Non-compact, multicellular structure (DAPI staining; ø ca. 0.25 mm). (E) Compact, globular structure with equally-sized cells (DAPI staining; ø ca. 0.3 mm). (F) Embryo-like structure, 2.5 mm in length after 12 weeks of culture. (G) Immature, zygotic embryo, 2.4 mm in length, isolated from an ovule 11 weeks after pollination. (H) Differentiation of cells in a microspore-derived embryo, showing the smooth, epidermal cell layer (8 μm thick longitudinal section stained with toluidine blue). (I) Formation of secondary embryos on a 12-week-old microspore-derived embryo.

Table 1. The effect of developmental stage of microspores of *Tulipa gesneriana* "Leen van der Mark" on sporophytic response, expressed as the number of multicellular structures scored after 6 weeks of culture. The microspores were isolated respectively from −2°C bulbs incubated for 1, 3, 6, 11 or 16 days at 18°C under light

Composition of microspore population (in %)				Number of multicellular structures[1]
Early uninucleate	Mid uninucleate	Late uninucleate	Binucleate	
40	58	2	0	131.0
40	49	11	0	14.0
18	56	26	0	5.6
3	9	22	66	0.2
0	0	2	98	0

[1] Figures represent mean values, expressed per 100,000 plated microspores, of four replicates, each consisting of eight Petri plates with 100,000 microspores.

5.5. *Embryo development*

The multicellular structures in cultures of "Leen van der Mark" microspores after six to eight weeks could be categorized into three distinct types. The first type was non-compact and irregular in structure, with cells of varying size (Fig. 4D). Most structures were of this type and usually stopped growing after the initial divisions. Two other types, accounting for 1–5% of the structures, showed organized growth. The second type was more compact, usually 0.2 to 0.8 mm diam after six to eight weeks of culture, and consisted of equally-sized cells (Fig. 4E). The third type was compact and elongated (Fig. 4F), resembling a zygotic embryo (Fig. 4G). The surface of these embryo-like structures was smooth and histological observations confirmed the presence of an epidermal layer (Fig. 4H). When multicellular structures were plated onto solidified culture medium with 8% sucrose and cultured at 17°C, most compact structures continued development. Subsequently, structures were transferred to solidified medium with 3% sucrose. For both "Leen van der Mark" and "Rosario", one to five elongated embryos, 3–8 mm in length, could be obtained per responsive flower bud. Several of the compact, non-elongated structures also had a smooth surface and may have been aberrant embryos. Occasionally, secondary embryogenesis occurred on a microspore-derived embryo (Fig. 4I).

5.6. *Germination of embryos*

Microspore-derived embryos of "Leen van der Mark" and "Rosario" did not regenerate on solidified medium with low sucrose at 25°C under light, as for *Brassica* spp. (Lichter, 1989; Duijs *et al.*, 1992). As for zygotic embryos

Table 2. Response in microspore culture of various tulip genotypes, determined after 6 weeks of culture. Experiments were limited to one or two bulbs for each genotype

T. gesneriana cultivars	Response[1]	Botanical species	Response[1]
Leen van der Mark	+++	*T. fosteriana* Madame Lefeber	+
Yokohama	-	*T. fosteriana* Princeps	-
Debutante	-	*T. eichleri*	+
Page Polka	-	*T. kaufmanniana*	+
Rosario	+++	*T. hungarica*	++
Lustige Witwe	+	*T. tubergeniana*	++
Monte Carlo	-	*T. lanata*	-
Lucky Strike	+	*T. marjoletti*	-
Gander	-	*T. neustruevae*	-
Dyanito	+	*T. polychroma*	-
Peach Blossom	+++	*T. turkestanica*	-
Garden Spot	-	*T. tarda*	-
White Dream	+	*T. orphanidea*	-
Couleur Cardinale	+		

[1] – = no sporophytic divisions; + = development of a few multicellular structures, less than 10 per Petri dish, that stopped growing after the initial divisions; ++ = development of several multicellular structures, at least 10 per Petri dish, of which some reached a size of 0.1–0.2 mm diameter; +++ = development of compact microcalli or embryo-like structures.

of tulip, microspore-derived embryos showed dormancy, which could be broken by a cold treatment. Incubation of the embryos at 4°C in the dark for two to three months was necessary to induce germination. Apparently, the induction of dormancy in microspore-derived embryos was part of the developmental program for embryogenesis.

So far, we have obtained several embryos of "Leen van der Mark" and "Rosario" that partially germinated, showing the formation of a primary root (Fig. 5A). Three "Rosario" embryos formed primary roots, green cotyledon-like structures and stolon-like droppers (Fig. 5B). This resembled the germination of some zygotic embryos cultured *in ovulo* (Custers *et al.*, 1992). At the end of the stolon, a new bulblet that may give rise to a haploid or doubled haploid plant may develop. Occasionally, shoots developed on partially germinated embryos, presumably originating from the cotyledonary

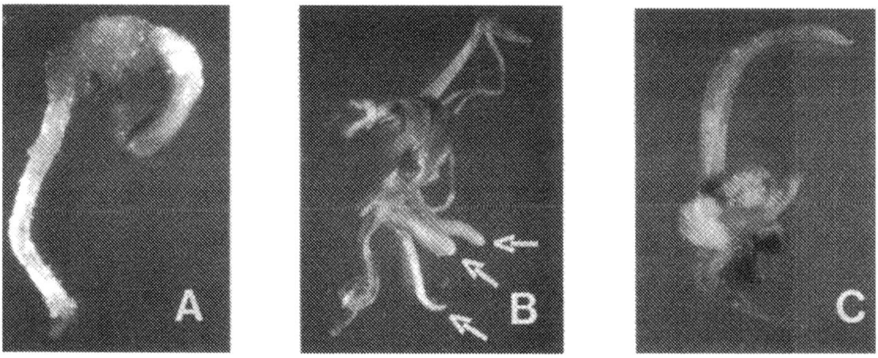

Figure 5. Germination and regeneration of microspore-derived tulip embryos. (A) Partial germination of an embryo of "Leen van der Mark" after a 3-month cold period and subsequent transfer to 17°C, resulting in the formation of a primary root. (B) Germination of a microspore-derived "Rosario" embryo after a cold treatment. Besides the primary root and cotyledon, stolon-like droppers have also been formed (indicated with arrows). (C) Regeneration of a shoot on a partially germinated embryo-like structure of "Leen van der Mark".

apex (Fig. 5C). Regeneration of leaf-like structures and shoots also took place on embryo-derived tissue of non-germinated "Rosario" embryos, six to twelve months after initiation of the culture.

The ploidy level of microspore-derived embryos was determined by flow cytometric analysis of nine "Leen van der Mark" and eight "Rosario" embryos. For "Leen van der Mark" two embryos were diploid, one was aneuploid, and six were haploid. The presence of haploid embryos confirmed that they originated from cultured microspores, suggesting that the diploid embryos were actually doubled haploids. Six embryos of "Rosario" were diploid, and two were tetraploid. "Rosario" embryos were more normal and vigorous than "Leen van der Mark" embryos. Thus, spontaneous doubling of the chromosome number of haploid embryos may be essential for their continued development.

6. Conclusions and future prospects

In contrast with expectations based on good regeneration potential of lily and on results in anther culture of lily, microspore culture of lily has not resulted in embryo development or regeneration thus far. Our results indicate better prospects for the production of haploid tulip plants. Embryos were obtained directly from microspores, without an intermediate callus phase. Thus, the chance for chromosomal aberrations to occur as a result of somaclonal variation may be limited. The first germinating embryos of "Rosario" developed the so-called droppers, at the end of which new bulblets typically form. These bulblets should give rise to (doubled) haploid tulip plants. Further studies are needed with regard to optimization of the method (culture conditions, media, maturation of embryos) for various cultivars to get a higher percentage of embryo formation, and with regard to germination of the microspore-derived embryos, including the formation of droppers and bulblets.

With respect to the ploidy level, the more vigorous "Rosario" embryos were predominantly diploid, whereas most of the weaker "Leen van der Mark" embryos were haploid. This phenomenon needs further study, because it may implicate that diploidization is required for successful production of functional embryos. The application of chromosome doubling agents such as colchicine or oryzaline (Van Tuyl *et al.*, 1992) at an early phase of microspore culture may enhance the formation of functional embryos (Hassawi & Liang, 1991). If diploidization plays a role, it also will be interesting to study the behavior of 2*n* gametes, resulting from meiotic diploidization, in microspore culture.

We have shown that embryogenesis in tulip microspores is not limited to just a few genotypes, but can be induced in several *Tulipa* species and cultivars. Implementation of this technique may improve efficiency in tulip breeding. Products of microspore embryogenesis can also be used to develop procedures for the production of somatic embryos as an alternative and rapid method of propagation. Also, microspore culture offers a possibility of *in vitro* selection. Thousands of microspores can be treated without the need of producing plants first. Time and labour can be saved and selections can be performed under standardized conditions, which can hardly be achieved when testing in the field. The use of heterozygous parent plants for microspore isolation guarantees a large amount of genetic variation through meiotic recombination and segregation. Microspores with an interesting trait can be selected and regenerated via embryogenesis into plants. With regard to disease resistance, this procedure may, for instance, be useful to obtain *Fusarium*-resistant plants by selecting for insensitivity towards the fungal toxin fusaric acid (Löffler *et al.*, 1990).

Furthermore, polyploidization plays an important role in the interspecific breeding of lilies and tulips. Reducing the level of ploidy using anther or microspore culture is essential for analytical plant breeding, with which efficiency of selection can be improved considerably.

7. Acknowledgements

Thanks are due to Dr. Raoul Bino, Dr. Hans Dons and Dr. Hein van Holsteijn for critical reading of the manuscript. Marleen de Vries, Bertus Meijer and Tobias Dümmen are greatly acknowledged for their assistance with the experiments. Part of the work was supported by the Urgency Programme for Research on Diseases and Breeding of Flower Bulbs.

8. References

Bennici, A., 1979. Cytological chimeras in plants regenerated from *Lilium longiflorum* tissues grown *in vitro*. Z. Pflanzenzüchtg. 82: 349–353.

Custers, J.B.M., W. Eikelboom, J.H.W. Bergervoet & J.P. van Eijk, 1992. In ovulo embryo culture of tulip (*Tulipa* L.); effects of culture conditions on seedling and bulblet formation. Scientia Hort. 51: 111–122.

Duijs, J.G., R.E. Voorrips, D.L. Visser & J.B.M. Custers, 1992. Microspore culture is successful in most crop types of *Brassica oleracea* L. Euphytica 60: 45–55.

Gaillard, A., P. Vergne & M. Beckert, 1991. Optimization of maize microspore isolation and culture conditions for reliable plant regeneration. Plant Cell Rep. 10: 55–58.

Gu, Z.-P. & K.-C. Cheng, 1982. Studies on induction of pollen plantlets from the anther cultures of lily. Acta Bot. Sin. 24: 28–32.

Gu, Z.-P. & K.-C. Cheng, 1983. *In vitro* induction of haploid plantlets from unpollinated young ovaries of lily and its embryological observations. Acta Bot. Sin. 25: 24–30.

Guha, S. & S.C. Maheshwari, 1964. *In vitro* production of embryos from anthers of *Datura*. Nature 204: 497.

Hassawi D.S. & G.H Liang, 1991. Antimitotic agents: effects of double haploid production in wheat. Crop Sci. 31: 723–726.

Hoekstra, S., M.H. van Zijderveld, J.D. Louwerse, F. Heidekamp & F. van der Mark, 1992. Anther and microspore culture of *Hordeum vulgare* L. cv. Igri. Plant Sci. 86: 89–96.

Idzikowska, K. & F. Mlodzianowski, 1979. Cell wall formation in multinucleate pollen grains of *Hordeum vulgare* anthers cultures *in vitro*. Acta Soc. Bot. Pol. 48: 377–380.

Lichter, R., 1982. Induction of haploid plants from isolated pollen of *Brassica napus*. Z. Pflanzenphysiol. 105: 427–434.

Lichter, R., 1989. Efficient yield of embryoids by culture of isolated microspores of different Brassicaceae species. Plant Breed. 103: 119–123.

Löffler, H.J.M., J.R. Mouris & M.J. van Harmelen, 1990. *In vitro* selection for resistance against *Fusarium oxysporum* in lily: prospects. The Lily Yearbook of the North Am. Lily Soc. 43: 56–60.

Melchinger, A.E., 1990. Use of molecular markers in breeding for oligogenic disease resistance. Plant Breed. 104: 1–19.

Murashige, T. & F. Skoog, 1962. A revised medium for rapid growth and bioassays with tobacco tissue cultures. Physiol. Plant. 15: 473–497.

Prakash, J. & K.L. Giles, 1986. Production of doubled haploids in oriental lilies. In: W. Horn, C.J. Jensen, W. Odenbach & O. Schieder (Eds.), Genetic Manipulation in Plant Breeding, pp. 335–337. Walter de Gruyter, Berlin.

Qu, Y., M.C. Mok, D.W.S. Mok & J.R. Stang, 1988. Phenotypic and cytological variation among plants derived from anther cultures of *Lilium longiflorum*. In Vitro Cellular and Devel. Biol. 24: 471–476.

Sangwan, R.S. & B.S. Sangwan-Norreel, 1990. Anther and pollen culture. In: S.S. Bhojwani (Ed.), Developments in Crop Science 19. Plant Tissue Culture: Applications and Limitations, pp. 220–241. Elsevier, Amsterdam.

San Noeum, L.H., 1976. Haploïdes d'*Hordeum vulgare* L. par culture *in vitro* non fécondés. Ann. Amélior. Plantes 26: 751–754.

Sharp, W.R., R.S. Raskin & H.E. Sommer, 1971. Haploidy in *Lilium*. Phytomorphology 21: 334–337.

Siebel, J. & K.P. Pauls, 1989. A comparison of anther and microspore culture as a breeding tool in *Brassica napus*. Theor. Appl. Genet. 78: 473–479.

Tanaka, I. & M. Ito, 1980. Induction of typical cell division in isolated microspores of *Lilium longiflorum* and *Tulipa gesneriana*. Plant Sci. Lett. 17: 279–285.

Tanaka, I. & M. Ito, 1981a. Control of division patterns in explanted microspores of *Tulipa gesneriana*. Protoplasma 108: 329–340.

Tanaka, I. & M. Ito, 1981b. Studies on microspore development in liliaceous plants III. Pollen tube development in lily pollen cultured from the uninucleate microspore stage. Plant Cell Physiol. 22: 149–153.

Tanaka, I. & M. Ito, 1982. Additional mitoses induced by bromodeoxyuridine in explanted tulip microspores. Plant Sci. Lett. 27: 37–42.

Tanaka, I., C. Kitazume & M. Ito, 1987. The isolation and culture of lily pollen protoplasts. Plant Sci. 50: 205–211.

Tanaka, I., T. Taguchi & M. Ito, 1979. Studies on microspore development in liliaceous plants. I. The duration of the cell cycle and developmental aspects in lily microspores. Bot. Mag. Tokyo 92: 291–298.

Tanaka, I., T. Taguchi & M. Ito, 1980. Studies on microspore development in liliaceous plants. II. The behavior of explanted microspores of the lily, *Lilium longiflorum*. Plant Cell Physiol. 21: 667–676.

Ueda, K., Y. Miyamoto & I. Tanaka, 1990. Fusion studies of pollen protoplasts and generative cell protoplasts in *Lilium longiflorum*. Plant Sci. 72: 259–266.

Van den Bulk, R.W., H.P.J. de Vries-van Hulten & J.J.M. Dons, 1992. Formation of multinucleate lily microspores in culture. Acta Hort. 325: 649–654.

Van Eijk, J.P., L.W.D. van Raamsdonk, W. Eikelboom & R.J. Bino, 1991. Interspecific crosses between *Tulipa gesneriana* cultivars and wild *Tulipa* species – a survey. Sex. Plant Rep. 4: 1–5.

Van Tuyl, J.M., H. Meijer & M.P. van Diën, 1992. The use of oryzalin as an alternative for colchicine in in-vitro chromosome doubling of *Lilium* and *Nerine*. Acta Hort. 325: 625–630.

Van Tuyl, J.M., M.C.M. van Creij, W. Eikelboom, D.M.F.J Kerckhoffs & B. Meijer, 1993. New genetic variation in the *Lilium* and *Tulipa* assortment by wide hybridization. In: T. Schiva & A. Mercuri (Eds.), Proceedings XVIIth Eucarpia Symposium, pp. 141–149. Istituto sperimentale per la Floricoltura, Sanremo.

Yang, H.Y. & C. Zhou, 1982. *In vitro* induction of haploid plants from unpollinated ovaries and ovules. Theor. Appl. Genet. 63: 97–104.

Yang, H.Y. & C. Zhou, 1990. *In vitro* gynogenesis. In: S.S. Bhojwani (Ed.), Developments in Crop Science 19. Plant Tissue Culture: Applications and Limitations, pp. 242–258. Elsevier, Amsterdam.

Zhou, C. & Y. Wu, 1990. Two pathways in pollen protoplast culture: cell divisions and tube growth. In: H.J.J. Nijkamp, L.H.W. van der Plas & J. van Aartrijk (Eds.), Progress in Plant Cellular and Molecular Biology, pp. 222–227. Kluwer Academic Publishers, Dordrecht.

6. Anther and microspore culture in *Camellia japonica*

M. CRISTINA PEDROSO and M. SALOMÉ PAIS

Contents

1. Introduction

Haploid plants have the gametophytic chromosome number. They are of great importance for the production of homozygous plants and for mutation studies. The use of anther and microspore cultures for the induction of pollen embryogenesis hastens the production of haploids and gametoclonal variants. These systems can be useful tools for cultivar improvement. *Camellia japonica* L. (Theaceae family) is one of the most important species in the genus *Camellia*. Economically, it is valuable as an ornamental woody species and has a potential to be used for oil and wood production. The available *Camellia* varieties are genetically highly heterozygous. In this genus, frost and pathogens cause serious damage that results in drastic annual economic losses. So far, *in vitro* culture of this species, and other *Camellia* species, has been mainly concerned with the establishment of protocols for micropropagation from juvenile and adult materials (Bennet, 1977, 1978; Bennet & Scheibert, 1982; Carlisi & Torres, 1986; Crezé, 1983; Kato, 1989a; Pedroso-Ubach, 1991; Samartin *et al.*, 1984, 1986; Samartin, 1989; Vieitez *et al.*, 1989a,b, 1992), and for mass propagation by somatic embryogenesis (Barciela & Vieitez, 1993; Kato, 1986, 1989b; Nakamura, 1988; Pedroso & Pais, 1993, 1994d,e; San-José & Vieitez, 1993; Vieitez & Barciela, 1990; Vieitez *et al.*, 1991). Plant improvement has been obtained in some species by field selection and artificial crossing techniques (Yamaguchi *et al.*, 1987). Anther culture, as a strategy for plant improvement, was reported for *C. sinensis* (Raina & Iyer, 1974; Chen & Liao, 1982, 1988) and *C. japonica* (Pedroso-Ubach, 1991). In *C. sinensis*, the production of haploid callus from anthers has been obtained in at least nine cultivars (Raina & Iyer, 1974; Chen & Liao, 1982, 1988) but haploid shoot regeneration was successful only in one of the cultivars (Chen & Liao, 1982, 1988). Embryo production has also been reported in this species by Shimokado *et al.* (1986) but apparently no plant

S.M. Jain, S.K. Sopory & R.E. Veilleux (eds.), In Vitro *Haploid Production in Higher Plants, Vol. 5*, 89–107.

regeneration was achieved. As far as we know, successful regeneration from anthers and microspore culture has been only reported in *C. japonica* (Pedroso-Ubach, 1991). Although pollen-derived plants have been obtained by anther culture in many species, plant regeneration from isolated microspores has only been successful in a few species (Bajaj, 1990). Microspore culture presents some potential advantages over anther culture, especially concerning *in vitro* selection strategies, genetic studies and genetic transformation. The culture of isolated microspores has become a valuable system for studying *in vitro* embryogenesis (Taylor *et al.*, 1990). Complementing earlier work (Pedroso & Pais, 1992, 1993, 1994c,d), direct microspore embryogenesis in *C. japonica* would be valuable for studying gene expression during microspore embryogenesis and the factors that divert microspore development onto an embryogenic pathway.

This chapter presents the major results obtained on plant regeneration from anthers and on the induction of the first stages of microspore embryogenesis in *Camellia japonica* L. cv. Elegans.

2. Anther culture and plant regeneration

Two adult plants (>50 years old) of *Camellia japonica* L. cvs. Elegans and Ville de Nantes were used as sources of unopened floral buds. The floral buds were surface-sterilised for 20 min in full strength domestic bleach (5% active chlorine), washed 4 times in sterile distilled water and immediately dissected under aseptic conditions for anther isolation. Optical and scanning microscopy studies were performed as earlier described (Pedroso & Pais, 1994a,b). The results showed that an increase in bud size was directly related to anther and microspore development. Anthers isolated from flower buds 9 to 10 mm diam were small (0.5 to 1 mm length) and demonstrated developing locules and a large petaloid anther (Fig. 1a). The microspores from these anthers were at the tetrad and early uninucleate microspore stages (Fig. 2). Anthers isolated from 9 to 12 mm diam flower buds were at intermediate ontogenic stages. Microscope observations showed that the appearance of bilocular morphology and the partial or total reduction of petaloid development occurred parallel with microspore maturation and with the increase of bud size (Fig. 2). Mature anthers (1.8–2.3 mm length) contained only binucleate microspores and, generally, absence of petaloid anthers.

Anthers at different stages of microsporogenesis were inoculated horizontally on 17 culture media (Table 1), preferentially with their locules upwards. Inoculum density was 5 anthers per ml of 0.4% (w/v) agarose-solidified medium (15 anthers/50 mm glass Petri dish sealed with parafilm). The cultures were incubated in growth chambers at 24 ± 1°C, in darkness or under 16-h photoperiod (26 $\mu mol \cdot m^{-2} \cdot s^{-1}$). Anther browning was common either in darkness or under light. However, only anther culture under light enabled

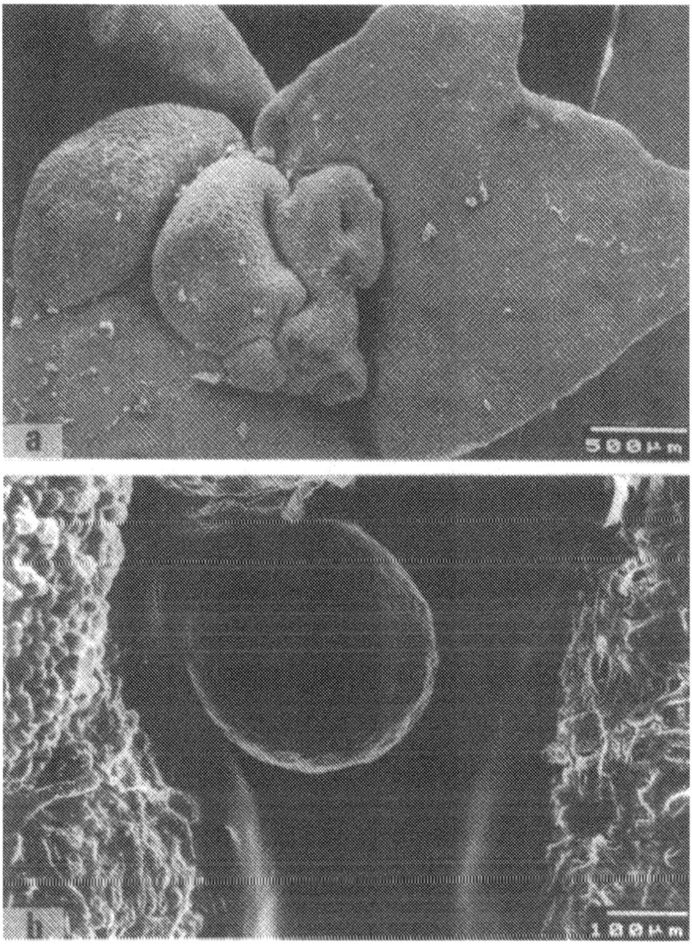

Figure 1. Scanning electron microscopy images of anther embryogenesis in *Camellia japonica* L. (a) Anther at tetrad microspore stage with large petaloid anthers at the explantation time. (b) Globular embryo and the embryogenic region of anther locule are covered by a layer of material of smooth texture, 8 weeks after culture initiation (Pedroso-Ubach, 1991).

embryo and callus formation. For this reason, only anther cultures under light conditions were used for further assays.

Anthers from both cultivars formed callus and embryos on MS10 medium (Table 2). Thirty to 40 days after culture initiation callus from the internal tissues of the anthers of both cultivars emerged after rupture of the browned anther wall. Three types of calluses were observed: a) friable and hyaline; b) opaque almond colour; and c) compact, yellow-greenish. Anther-derived callus and embryo production in cv. Ville de Nantes were significantly higher

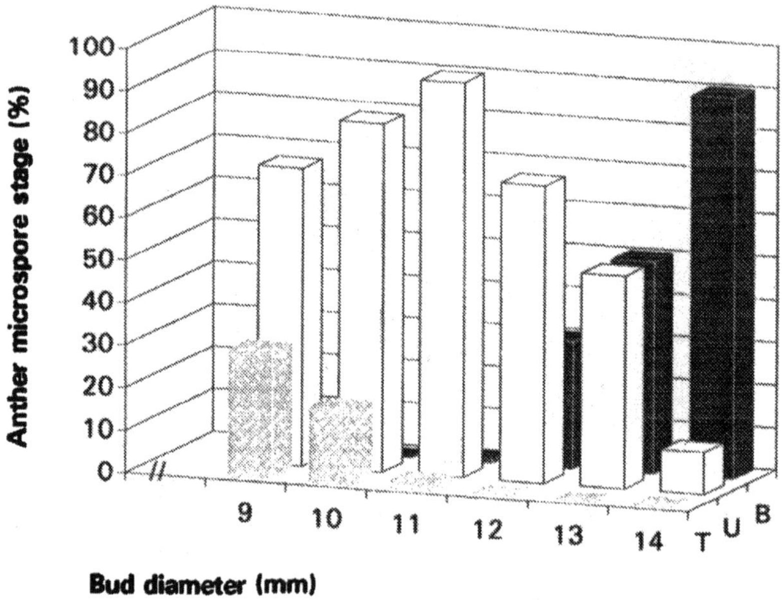

Figure 2. Anther microspore stage (% of isolated anthers at tetrad stage, T, at uninucleate stage, U, and at binucleate stage, B, as a function of flower bud diameter (mm) in *Camellia japonica* L. cv. Elegans.

than in cv. Elegans (Table 2). Petaloid anthers were more responsive than anthers. Globular embryos were observed 50 to 60 days after culture initiation on MS10 medium (Fig. 3a). Embryo formation in anthers and petaloid anthers occurred by direct embryogenesis in both cultivars. In 92% of the cultured embryogenic anthers, embryo and callus formation was concomitant. Embryogenesis was not synchronous and occurred only in anthers forming the opaque almond coloured callus. The first heart shaped embryo was observed at day 75 of culture (Fig. 3b). Scanning electron microscopy images showed that only anther-derived globular embryos were covered by a layer of unknown composition with smooth appearance (Fig. 1b). Petaloid-derived embryos were covered by a thin and discontinuous layer of the same material (image not shown). As shown on Fig. 4, embryogenesis occurred only in anthers cultured at the tetrad and early uninucleate stages. Although it is clear that the embryos originated from the inner part of the anther locule and were frequently observed surrounded by microspores, their origin could not be undoubtedly determined in all cases. Callus formation was highest when anthers at these stages were used, and decreased to 0% with an increase in the number of binucleate microspores (Fig. 4). Only 1.0% of the anthers cultured at mid to late uninucleate stage (6/594) formed callus (Fig. 3c), that is 0.23% (6/2574) of the total number of cultured anthers. With the exception

Table 1. Culture media tested for anther and microspore culture of *Camellia japonica* L.

Media	Suc.	Gluc.	m-Inos	Gln	Ser	2,4-D	Kin	IBA	BA	AC
	(g/liter)		(mg/liter)							g/liter
MS1	30	----	1000	----	----	1.0	0.1	-----	----	-------
MS2	30	----	1000	----	----	------	----	4.0	2.0	-------
MS3	30	----	1000	----	----	------	----	----	----	------
MS4	20	----	100	----	----	1.0	0.1	----	----	------
MS5	20	----	100	----	----	1.0	0.1	----	----	2.0
MS6	20	----	100	800	200	1.0	0.1	----	----	------
MS7	----	25	100	----	----	1.0	0.1	----	----	------
MS8	----	25	100	----	----	1.0	0.1	----	----	2.0
MS9	----	25	100	800	200	1.0	0.1	----	----	------
MS10	20	----	100	800	200	------	----	----	0.5	------
N1	100	----	100	----	----	1.0	0.1	----	----	------
N2	20	----	100	----	----	1.0	0.1	----	----	------
N3	20	----	100	800	200	1.0	0.1	----	----	------
N4	100	----	500	800	200	1.5	0.5	----	----	------
N5	100	----	100	----	----	------	----	----	----	------
N6	100	----	100	----	----	1.5	0.5	----	----	------
N7	----	25	100	----	----	1.0	0.1	----	----	------

MS = inorganic salts and vitamins of modified Murashige and Skoog medium (Murashige & Skoog, 1962). N = inorganic salts and vitamins of modified Chu medium (N6) (Chu, 1978). Suc. = sucrose; Gluc. = glucose; m-Inos = myo-inositol; Gln = glutamine; Ser = L-serine; 2,4-D = 2,4-dichlorophenoxyacetic acid; Kin = kinetin; IBA = indol-3-butyric acid; BA = 6-benzyl-adenine; AC = activated charcoal. (Adapted from Pedroso & Pais, 1994b.)

of petaloid response recorded for cv. Elegans (Table 3), no significant differences were recorded between both cultivars in terms of embryogenic anthers and embryo production/anther. Total embryogenic efficiency (embryos formed/total number of anthers or petaloid anthers) for cv. Elegans, at week 9 of culture, was significantly higher than that for cv. Ville de Nantes. In both cultivars, petaloid-derived embryos aborted on the 10th week of culture.

Twelve weeks after culture initiation in MS6, MS9 or MS10 medium, cultures were transferred to 0.7% (w/v) agar-solidified regeneration medium for plant regeneration via: a) embryogenesis, and b) organogenesis. The

Table 2. Callus and direct embryo formation from anthers and petaloid anthers[a] (%) of *C. japonica*, cv. Elegans and cv. Ville de Nantes, 9 weeks after culture initiation

Cultivar	Media	Anthers		Petaloids	
		Callus	Embryo	Callus	Embryo
Elegans	MS6	0	0	5.4c	0
	MS9	0	0	20.2d	0
	MS10	37.2a	24.8d	58.3e	31.7a
Ville de Nantes	MS6	0	0	72.7b	0
	MS9	0	0	75.0b	0
	MS10	73.5b	33.4a	100f	38.5a

[a] Cultured on 0.4% agarose-solidified media, under 16-h photoperiod, independently of microspore developmental stage. Values are the mean of three experiments, with five replications each (125 anthers). Values were analysed by Tukey HSD test before conversion into %; values followed by different letters are significantly different at $\alpha = 0.05$ (Pedroso & Pais, 1994a.)

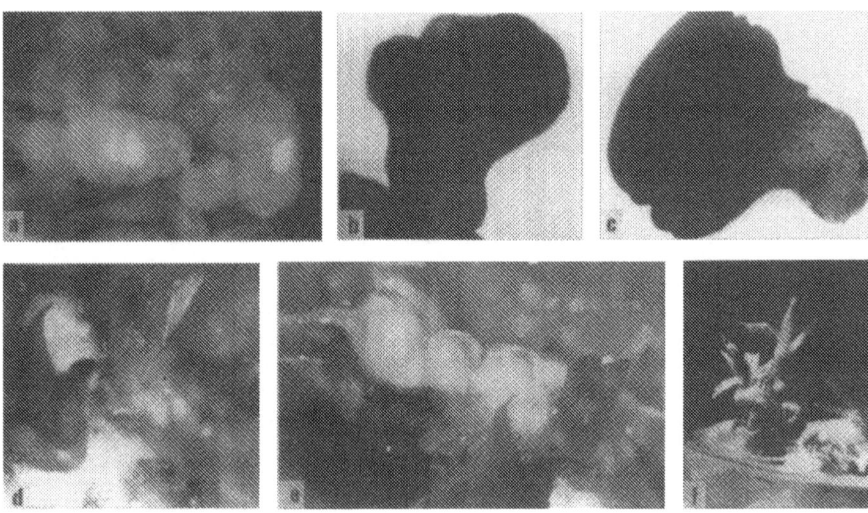

Figure 3. Regeneration from anthers of *Camellia japonica* L. cvs. Elegans and Ville de Nantes. (a) Direct embryo formation from anther locules and concomitant callus formation from petaloid in anthers of cv. Ville de Nantes, 8 weeks after culture initiation on MS10 medium (× 38). (b and c) observations under an inverted light microscope; (b) Anther-derived heart shaped embryo developed in cv. Ville de Nantes after 10 weeks of culture on MS10 medium (×21). (c) Callus formation on an anther at mid-late uninucleate stage in cv. Elegans, 7 weeks after culture initiation on MS10 medium (×18). (d) Conversion of anther-derived embryos (×6). (e) Shoot regeneration from anther-derived callus in cv. Elegans (×6). (f) Shoots obtained from anther-derived embryos of *C. japonica* L. cv. Ville de Nantes (Pedroso-Ubach, 1991).

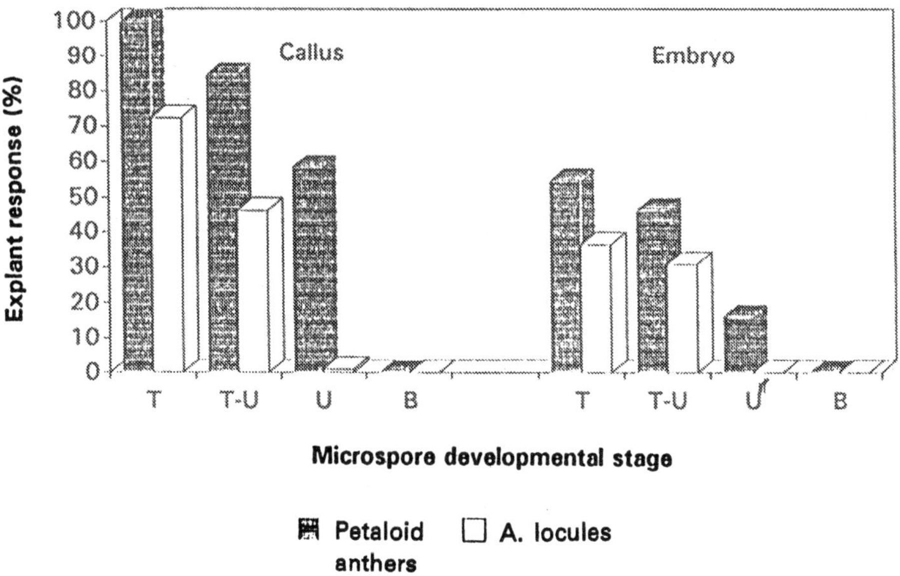

Figure 4. Explant response (%, callus and embryo formation in anthers and petaloid anthers) as a function of microspore developmental stage in isolated anthers. T, tetrad stage; T-U, tetrad and early uninucleate stage; U, uninucleate stage; B, binucleate stage.

cultures were kept under similar temperature and light conditions as for anther culture.

a) Anther-derived embryos were transferred to MS/2–25 medium (half strength basal MS medium supplemented with 25 g/l D-glucose) for maturation, and later, to MS28 medium (MS9 medium containing growth regulators replaced by 1 mg/l N^6-benzyladenine (BA) and 0.1 mg/l indol-3-butyric acid (IBA)) for embryo conversion and further plantlet development. Maturation of anther-derived-embryos occurred 2 to 3 months after transfer to MS/2–25 medium. The embryos detached easily. Regeneration efficiency through embryogenesis was 60% for cv. Ville de Nantes and 97.5% for cv. Elegans (Fig. 3d), as recorded after 12 weeks of transfer to MS28 medium. Plantlet production by embryogenesis was 0.9 (±0.3) plantlets/responsive anther. Callus growth on MS28 medium was drastically reduced and no callus-derived embryos were obtained.

b) Anther-derived calluses were directly transferred to MS28 medium for shoot induction. Bud formation was observed on 6.8% of the transferred calluses, 8 to 11 weeks later (Fig. 3e). Regeneration efficiency via organogenesis was identical for both cultivars and shoot production was 2.8 (±1.0) shoots/anther-derived callus, 12 weeks after transfer to MS28 medium. Anther-derived shoots were regenerated in 180 to 210 days (Fig. 3f) (Table 4). The shoots were rooted by immersion in 1g/l IBA solution for 7 min, followed by culture on solid half-strength MS (Murashige &

Table 3. Embryogenic response in anthers (An) and petaloid anthers (Pt) in two cultivars of *Camellia japonica*

	cv. Elegans			cv. Ville de Nantes		
	AT	AT-U	AU	AT	AT-U	AU [a]
Embryogenic An (%)[b]	36.4a	30.8a	0	38.5a	36.4a	0
Embryogenic Pt	54.5b	46.2b	16.1c	38.5a	38.5a	0
Embryo Production (no.)						
/ anther	1.4e	1.3e	0	1.5e	1.2e	0
/ petaloid anther	3.0f	3.5f	1.5e	2.5f	2.2f	0
Embryogenic efficiency (%)						
of anthers	5.3f	3.0f	0	4.1f	3.0f	0
of petaloid anthers	16.7c	10.7c	5.3g	8.9g	6.5g	0
Total (An+Pt)[c]	25.6h	13.6c	5.3g	13.0c	8.9g	0

[a] Anthers at tetrad stage (AT), tetrad and early uninucleate stage (AT-U) and at uninucleate stage (AU). 198 anthers/stage were cultured for 9 weeks on 0.4% agarose-solidified MS10 medium (Table 1) under 16-h photoperiod. Values are the mean of 8 to 12 replicates (one replicate = one Petri dish). Each experiment was repeated three times.
[b] Values before conversion into % were analysed by Tukey HSD test. Values followed by the same letter are significantly different at $\alpha = 0.05$.
[c] Embryogenic efficiency/inoculated explant (Pedroso & Pais, 1994a).

Skoog, 1962) basal medium supplemented with 20 g/l sucrose. The cultures were maintained in darkness for 11 days and then transferred to light (26 $\mu mol \cdot m^{-2} \cdot s^{-1}$) (Pedroso-Ubach, 1991). After 6 weeks, 77.9% (±4.5%) of the regenerated shoots had rooted. Plantlets were transferred to the acclimatisation chamber but died due to fungi infection.

3. Microspore isolation and culture in *C. japonica* cv. Elegans

As for anther culture, all culture media listed in Table 1 were assayed for microspore culture. The highest sporophytic induction was obtained using MS6, MS9 and MS10 media.

Microscopic observations during the isolation and culture of microspores indicated that microspores could be divided in three groups: group 1, isolated microspores at tetrad, early uninucleate and late binucleate stages, showing 0 to 90% volume increase, poor acetocarmine staining, and a breakdown

Table 4. Duration of each culture stage for plant regeneration from anthers of adult *Camellia japonica*

Stages for plant regeneration from anthers	Time (days)
Dedifferentiation	15
Callus formation	38 - 50
Proembryos	50 - 60
Globular embryos	70
Heart-shaped embryos	75
Vascular nodules	120 - 130
Buds	150 - 175
Shoots / plantlets	180 - 210

of gametophytic development after culture initiation; group 2, microspores increased in volume by 100 to 150% and demonstrated normal gametophytic development until germination of pollen grains; and group 3, comprised of microspores exhibiting an increased volume of 200 to 400% and a sporophytic pathway of development (Fig. 5a,b,c). These last two groups comprised microspores isolated at the uni- and binucleate stages, and 0.98% (±3.6) of them (uninucleate microspores with peripheral nucleus and a dense cytoplasm) were highly stained by acetocarmine. These observations indicated that cell volume increase during the first 2 weeks of culture could be used as an indicator of microspore-response. Two types of sporophytic response, embryo and callus formation, were observed in group 3 microspores. DAPI staining demonstrated that embryo and callus originated from uninucleate microspores. The first symmetric mitotic division gave rise to proembryos (Fig. 5d) whereas asymmetric divisions (Fig. 5f) formed callus (Fig. 5g). After transfer to MS10 medium (embryo maturation medium), proembryos slowly formed milky-white globular embryos. Their development was much slower than that of somatic embryos. Three of the embryos reached a torpedo-like shape but their further development ceased (Fig. 5e). Microscopic observation of hand sections showed that these embryos had two poles but protoderm seemed aberrant (image not shown) compared to the protoderm of somatic embryos (Pedroso & Pais, 1994e).

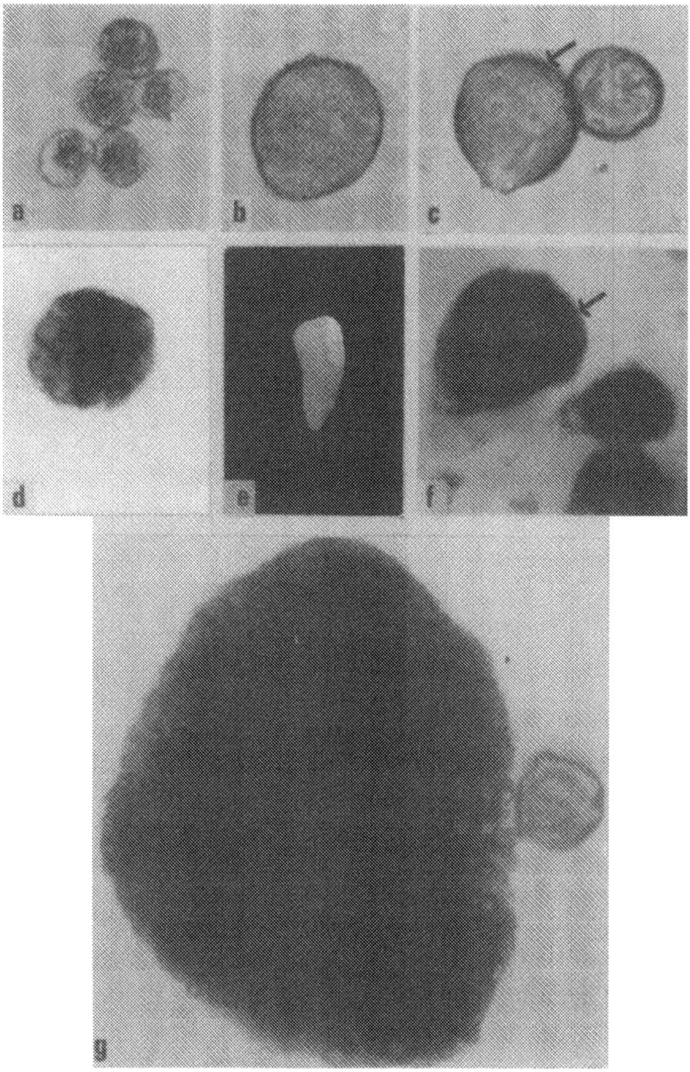

Figure 5. Microspore culture in *Camellia japonica* L. cv. Elegans, under 16-h photoperiod. (a, b, c, f, g) immobilised in 0.4% agarose. (a) Microspores at the uninucleate stage (×183). (b) Microspore volume increase compared to Fig. 5a (×220). (c) Microspore undergoing sporophytic development (arrow) and a microspore following the gametophytic pathway (×217). (d) Proembryo derived from a microspore isolated by method II and cultured at the uninucleate stage (×75). (e) Microspore-derived embryo, 4 weeks after transfer to MS10 medium (maturation stage) (×3). (f) Asymmetric sporophytic division (arrow) (×212). (g) Microspore-derived-microcallus in MS10 medium, originated by first asymmetric division (×145).

Three methods were employed for microspore isolation and culture: methods I and II, immediately after anther isolation; method III, after anther incubation.

Method I (spontaneous release). The microspores were spontaneously released through natural dehiscence of anthers cultured in 20 ml of liquid MS6, MS9 or MS10 media (Table 1) in 100 ml Erlenmeyer flasks at a density of 4 anthers per ml. The anther dehiscence in suspension culture occurred during the first 3 weeks of culture. The mean released microspores ranged from 8.9×10^2 to 2.3×10^3. The microspores were at the uninucleate stage with peripheral nucleus or at the binucleate stage and had 94.3% (±4.2) viability. Two months after culture initiation, anthers were removed from the medium and the microspores kept in suspension culture. Microspore suspension culture enabled the induction of embryogenesis. However, proembryo development was only obtained when microspore suspension cultures were plated over solid MS10 medium.

Method II (isolation through homogenisation and filtration). Anthers (8–10) were crushed in 5 ml of MS3 or N_5 medium (media without growth regulators, washing solution) for microspore release. The microspore suspensions were purified as earlier described (Pedroso & Pais, 1994b), re-suspended in MS6, MS9 or MS10 medium (Table 1), and then plated over identical solid medium and cultured under light (26 μmol $\cdot$ $m^{-2} \cdot s^{-1}$). The isolation of microspores by this method immediately released 2.3×10^3 to 2.4×10^4 microspores per anther with 99.4% (±0.6) viability. The isolated microspores were initially at the uninucleate stage (92–100%). Purification of microspore suspensions did not significantly affect microspore viability (Pedroso-Ubach, 1991). The development of microspores *in vitro* was followed by periodic observation of cultures immobilised in 0.4% agarose medium. The observations were recorded at the 6th week of culture. Values are the mean of at least three experiments with six to ten replicates each (one replicate = 5 Petri dishes; pollen from 20 anthers).

The periodic observation of cultures under an inverted light microscope showed that the microspores cultured in MS4, MS6 to MS10, N5, and N6 media underwent sporophytic divisions (Fig. 5c,d). However, successful induction of globular embryo and/or callus formation was only achieved in MS6, MS9, and MS10 media. The highest value (1.3 microspores undergoing division per 1000 microspores) was recorded for microspores cultured in MS9 medium. First sporophytic divisions were observed after 2 weeks of culture in 0.56% of the cultured microspores and continued to increase until 4 weeks. The culture of microspores on solid medium induced the first sporophytic divisions, callogenesis and proembryo development. However, proembryo formation occurred only in cultures grown under light. After 6 weeks of culture, the number of microspores undergoing division per 100 anthers under light was 3.6 fold higher than that of cultures in the dark (380 vs. 105). For microspores cultured in the dark, embryo and callus induction stopped after the first or second mitotic division within 6 weeks of culture.

Table 5. Effect of the percentage of microspores cultured at the uninucleate stage and of inoculum density on the number of microspores undergoing sporophytic divisions in *Camellia japonica* cv. Elegans[a]

Assay	Microspores at the uninucleate stage (%)	Inoculum density[b]	Number of microspores undergoing sporophytic divisions
1	100	8.0×10^3	0
2	9.5	2.0×10^4	0
3	34	2.9×10^4	40
4	60	5.6×10^4	78
5	65	1.5×10^5	750
6	70	3.7×10^5	1.0×10^4
7	45	5.7×10^5	2.9×10^3
8	24	1.0×10^6	0

[a] Microspores were isolated following method II. Values are the mean of 3 experiments with 6 replicates each. Results were recorded after 6 weeks in culture in liquid over solid MS9 medium (Table 1) under 16-h photoperiod.
[b] Results are expressed in number of microspores per ml of culture medium.

Twenty different inoculum densities were tested ranging from 2.0×10^3 to 1.1×10^6 microspores per ml. The optimal values of inoculum density varied from $3.7–5.7 \times 10^5$ microspores per ml (Table 5, assays 6, 7). Lower microspore densities (2.9×10^4 to 5.6×10^4) stopped sporophytic development after 6 weeks in culture (Table 5, assays 3, 4). Inoculum densities lower than 2.9×10^4 and higher than 5.7×10^5 inhibited sporophytic induction (Table 5, assays 1, 2, 8). The inoculation of 9.5 to 60% of uninucleate microspores did not result in successful sporophytic induction (Table 5, assays 2, 3, 4). Identical results were obtained when a low inoculum density (8×10^3) consisting of 100% uninucleate microspores was used (Table 5, assay 1). Only microspores at the uninucleate stage ($>$ 45% of total population) inoculated at optimal densities enabled high sporophytic induction (Table 5, assays 6, 7).

Method III (microspore isolation after anther incubation). Anthers (6 per ml) were incubated at 24 ± 1°C for 1 week on 0.4% agarose-solidified MS6, MS9 and MS10 media under light (26 $\mu mol \cdot m^{-2} \cdot s^{-1}$). After anther incubation, microspores were isolated following method II and plated over

Table 6. Induction of sporophytic divisions, production of globular embryos and callus, and embryogenic and callogenic efficiency (%) in microspore cultures of *Camellia japonica* L. cv. Elegans, 12 weeks after culture initiation[a]

Method	Media	Sporophytic Induction		Medium	Production				Efficiency (%)	
					Globular Embryos		Callus		Embryogenic	Callogenic
		/100an	/1000mic		/100an	/1000mic	/100an	/1000mic		
I	MS6	603	0.602	MS10	4.92	(0.0005)	1.96	0.006	0.08	0.96
	MS9	200	0.186	MS10	0.86	(0.0001)	0	0	0.04	0
	MS10	0	0	MS10	0	0	0	0	0	0
II	MS6	380	1.430	MS10	14.28	0.053	2.33	0.013	3.71	0.91
	MS9	629	1.300	MS10	6.45	0.013	6.45	0.013	1.00	1.00
	MS10	63.4	0.090	MS10	0	0	2.86	0.004	0	4.44
III	MS6	2484	2.170	MS10	2.86	0.003	0	0	0.14	0
	MS9	1250	4.500	MS10	0.93	0.007	1.85	0.006	0.14	0.13
	MS10	1666	2.940	MS10	0	0	6.67	0.001	0	0.03

[a] Results are expressed in dividing microspores, globular embryos and callus per 100 anthers (an) and per 1000 microspores (mic) for the three methods of isolation and culture tested (I, II, and III). The media MS6, MS9, and MS10 (Table 1) were used for induction stage, and MS10 medium for the production stage. Values were recorded after 6 weeks in induction medium, and after 12 weeks in the production medium. Embryogenic and callogenic efficiency was determined 12 weeks after culture initiation. (Adapted from Pedroso & Pais, 1994b.)

solid MS6, MS9 and MS10 media and cultured under light (26 μmol $\cdot$ $m^{-2} \cdot s^{-1}$). Microspore culture, initiated on the same day without previous anther incubation (method II), was also performed for comparison purposes.

We considered it as "sporophytic induction" (opposed to gametophytic) when the pathway of development lead to embryo or callus formation. Embryogenic and callogenic efficiency (percentage of embryos and calluses produced per total dividing microspores) of the three methods was determined for uninucleate microspores plated at a density of 1.6×10^5 microspores per ml over solid MS6, MS9, and MS10 media, under light conditions. Microspore culture over solid media was performed by adding, every 4 weeks, 2 ml of fresh liquid medium per dish. Twelve weeks after culture initiation, microspore cultures both in the liquid and solid media were transferred to MS10 medium (production stage).

The isolation and culture of microspores with method III enhanced the sporophytic induction (Table 6) compared with direct microspore culture using method II (Table 6). Microspores underwent the first sporophytic division after 2 weeks of culture and reached the maximal values recorded for sporophytic induction 4 weeks after culture initiation. Thus, the use of method III induced earlier sporophytic divisions in all the induced microspores. Sporophytic induction was synchronous in MS10 medium but not in MS6 and MS9 media. However, in MS9 medium, all the induced microspores (4.5 microspores undergoing division per 1000 microspores) had divided 4 weeks after culture initiation, and no increase in sporophytic induction was recorded since that time (Table 6). Independently of the methods, globular embryo formation occurred only when the induction was performed in MS6 or MS9 medium (Table 6). Method III was the most efficient for sporophytic induction (proembryo and callus formation) but not for embryo development and callus proliferation. Globular embryo and callus production (production stage) were higher using method II and MS6 medium for the induction stage (Table 6). Induction of microspores in MS10 medium inhibited proembryo development during the production stage. Embryogenic efficiency (Table 6) was higher in MS6 medium using method II, while callogenic efficiency was highest when MS10 medium and method II were used.

4. Discussion and conclusions

4.1. *Anther culture*

Flower bud-size can be used as an indicator of the anther and microspore stage of development (Harada *et al.*, 1988). *Camellia* flower buds 9 to 10 mm in diameter were the most appropriate for anther culture. In *C. japonica* the induction of callus and embryo formation were dependent on light treatment, although at low intensity. This is in agreement with the results obtained for leaf-derived morphogenic cultures of *C. japonica* (Pedroso-Ubach, 1991; Pedroso & Pais, 1993). An increase of anther culture efficiency was also reported in *Malus domestica* Borkh. cv. Jonathan under light conditions (Milewska-Pawliczuk & Kubicki, 1977).

The sensitivity of *C. japonica* anthers to growth regulators as well as their callogenic and embryogenic efficiencies were dependent on the developmental stage of microsporogenesis and on the cultivar used. Embryo formation occurred only when anthers were cultured at the tetrad or early uninucleate stage on MS10 medium containing BA. Callus formation appeared to be independent of stage of microsporogenesis and of the cytokinin in the medium. The requirements of *C. japonica* anthers for callus formation were similar to those reported for other species, including *C. sinensis* (Bajaj, 1990; Raghavan & Nagmani, 1989). In *C. sinensis*, haploid callus formation was obtained on Nitsch culture medium (1969) supplemented with 15% coconut milk (Raina & Iyer, 1974) or on Nitsch medium supplemented with 800 mg/l L-glutamine, 100 mg/l serine, 0.5 mg/l 2,4–dichlorophenoxyacetic acid and 2 mg/l kinetin (Chen & Liao, 1982, 1988; Shimokado *et al.*, 1986). Similarly to *C. japonica*, no embryo formation was obtained in *C. sinensis* anther cultures in the presence of 2,4–dichlorophenoxyacetic acid and kinetin. The results suggest that a culture media containing cytokinin will probably induce embryogenesis from *C. sinensis* anthers. Although media composition differed, the presence of L-glutamine and serine was common in the media used for both species, suggesting that these amino acids are necessary for successful anther culture in both *C. sinensis* and *C. japonica*.

The abortion of petaloid-derived globular embryos by the 10th week of culture was possibly caused by callus proliferation around the embryo or by abnormal embryo formation. SEM images showed that the morphology of anther-derived globular embryos was identical to stem (Pedroso-Ubach, 1991) and leaf-derived embryos (Pedroso & Pais, 1992) but different from petaloid-derived embryos. The presence of a layer of material covering the embryogenic regions and globular embryos of *C. japonica* seems necessary for true embryo formation in this species (Pedroso & Pais, 1992). Histochemical analysis showed that this layer is cutin or a substance close to it (Pedroso & Pais, 1994f).

In *C. sinensis*, shoot regeneration from haploid calluses was obtained only in one of the nine cultivars used, and on Nitsch medium containing 2 mg/l zeatin, 20 mg/l adenine and 10 mg/l lactoalbumin hydrolysate (Chen & Liao, 1988). Plant regeneration from *C. japonica* anthers was more efficient by embryogenesis than by organogenesis. However, as reported for *C. sinensis*, the application of the protocols here described to ten other cultivars of *C. japonica* have not been successful for all of them (unpublished results), suggesting that culture conditions need to be adjusted to each genotype to obtain similar *in vitro* responses.

Compared with axillary bud culture from mature field-grown material of *C. japonica* (Pedroso-Ubach, 1991), the establishment of shoot cultures from anthers prevented initial contamination problems and reduced explant browning. Embryo production/anther in this culture system can possibly be further increased by synchronising embryo formation and preventing the formation of callus on embryogenic anthers. Additionally, both callus and embryos obtained are a continuous source of plant material to study somaclonal and gametaclonal variation in this species.

4.2. *Microspore culture*

The best culture conditions for microspore-derived embryo production were the isolation of microspores by method II, and culture of $1.5–5.7 \times 10^5$ uninucleate microspores per ml, under light ($26\ \mu mol \cdot m^{-2} \cdot s^{-1}$), in MS6 medium for the induction stage, and in MS10 medium for the production stage.

Anther incubation prior to microspore isolation (method III) was advantageous for the induction of sporophytic divisions but less efficient for embryo maturation than method II. Microspore isolation from incubated anthers seems more traumatic than isolation from freshly collected anthers, probably due to the deficiencies of the homogenisation medium.

Embryo and callus induction from microspores at the uninucleate stage was highest in MS6, MS9, and MS10 media, also the best media for successful anther culture. However, the presence of BA in MS10 medium, the best for embryo formation from anthers (Pedroso-Ubach, 1991; Pedroso & Pais, 1994a), enabled only the formation of microspore-derived callus. According to Raghavan & Nagmani (1989), the presence of cytokinins alone in the culture medium can negatively affect pollen embryogenesis, inducing only callus formation and stopping the development of callus-derived embryos. Although no study was performed concerning the effects of cytokinins on microspore embryogenesis in *C. japonica*, an inhibitory effect caused by the presence of BA cannot be ruled out. The apparent contradictory growth regulator requirements to promote embryogenesis from anthers and from microspores is probably a consequence of the explant used (tissues vs single cells) and not of tissue ploidy (somatic vs. gametophytic). The presence of endogenous growth regulators, nutritional tissue gradients and cell-to-cell interactions are three of many factors that must be considered in anther culture and can be responsible for a different growth regulator requirement, compared to that needed to promote an identical response on a single cell culture system. Microspore embryogenesis occurred when microspores were cultured in the presence of both 2,4–D and kinetin (MS6 and MS9 media) followed by transfer to MS10 medium for further development. In *C. japonica*, culture medium requirements to induce embryogenesis from microspores were identical to those used in the induction of embryogenesis from single somatic cells (Pedroso & Pais, 1994d). These results may indicate that growth regulator requirements to promote embryogenesis are not significantly affected by tissue ploidy. The occurrence of symmetric and asymmetric divisions in responding microspores originating, respectively, embryos and callus, seems to indicate that sporophytic response of microspores in *C. japonica* cv. Elegans follows at least two pathways of development, identical to those reported in *Hevea brasiliensis* Muell.-Arg. (Chen, 1990), and corresponding, respectively, to pathways B and A described by Sunderland & Dunwell (1974). According to Chen (1990), the majority of the microspores follow pathway B of development in the presence of 2,4–D and kinetin, which is in agreement with the results obtained for *C. japonica*. The death of microspore-derived embryos has also been reported for other woody species (Milewska-Pawliczuk & Kubicki, 1977; Milewska-Pawliczuk, 1990; Zhang, *et al.*, 1990) and is frequently associated with an inadequate selection of the culture medium. Although it could be the case, our results suggest that abnormal protoderm formation was the cause of embryo death. The results described above on anther-derived embryo formation and earlier results on embryogenesis from leaves (Pedroso & Pais, 1992), suggest that the presence of a normal protoderm covered by a layer of cutin, or a substance close to it, is essential for successful embryo development (Pedroso & Pais, 1994f).

The differences in densities of microspores isolated from the same number of crushed anthers lead to a contradiction when results from embryo and callus induction and maturation were expressed per 100 crushed anthers and per 1000 microspores. We opted to present both forms of quantification for comparison with results reported for other woody species. In *C. japonica*, embryo and callus induction per 1000 microspores and embryo maturation per 100 anthers were, respectively, 3- and 357-fold higher than those reported for *Malus domestica* Borkh (Milewska-Pawliczuk, 1990). The highest value obtained for callus formation in *C. japonica* was 6.7 calluses per 100 anthers, while for *Populus* spp. was 2.25–9.92 calluses per 100 anthers (Wang *et al.*, 1975) and for *Litchi chinensis* Sonn was 10 calluses per 100 anthers (Lianfang, 1990). The optimization of anther culture in *Hevea brasiliensis* Muell.-Arg. (Chen *et al.*, 1978) led to the production of 44 embryos per 100 anthers and 34 embryos per 1000 microspores, respectively, ca. 3– and 7.5-fold higher than those recorded for *C. japonica*.

In conclusion, microspore-derived embryos in *C. japonica* can be produced through direct microspore culture without previous anther incubation or floral bud cold-pretreatment. This method enables the inoculation of microspores with precise control of the optimal stage of microsporogenesis and direct observation of cultures and the stages of microspore embryogenesis. Compared to anther culture, microspore culture seems less dependent on genotype (unpublished results). This culture system, when optimised, can be used as an experimental system to study, at the cellular and molecular levels, the induction of microspore embryogenesis in this woody angiosperm.

5. Acknowledgements

We thank Eng. Oliveira (Administração Forestal de Sintra, Portugal) for his permission to collect plant material in Parque de Monserrate and to Mr. Chaveiro for his technical assistance in scanning electron microscope. This research was supported by I.N.I.C (Centro de Engenharia Biológica) and by J.N.I.C.T, Ph.D. grant BD/239/90–IF.

6. References

Bajaj, Y.P.S. (Ed.), 1990. Haploids in Crop Improvement I. Biotechnology in Agriculture and Forestry, Vol. 12, pp. 101–541. Springer-Verlag, Heidelberg.

Barciela, J. & A.M. Vieitez, 1993. Anatomical sequence and morphometric analysis during embryogenesis on cultured cotyledon explants of *Camellia japonica* L.. Ann. Bot. 71: 395–404.

Bennet, W.Y., 1977. Tissue culture for Camellias? II American American Camellia Yearbook, pp. 188–190.

Bennet, W.Y., 1978. Tissue culture for Camellias. II American American Camellia Yearbook, pp. 102–104.

Bennet, W.Y. & P. Scheibert, 1982. *In vitro* generation of callus and plantlets from cotyledons of *C. japonica*. Camellia J. 37: 12–15.

Carlisi, J.A. & K.C. Torres, 1986. *In vitro* shoot proliferation in camellia Purple Down. HortScience 21: 314.

Chen, Z., 1990. Rubber (*Hevea brasiliensis* Muell. Arg.): *in vitro* production of haploids. In: Y.P.S. Bajaj (Ed.), Biotechnology in Agriculture and Forestry, Vol. 12, pp. 215–236. Springer-Verlag, Heidelberg.

Chen, Z. & H.H. Liao, 1982. Obtaining plantlets by anther culture of tea plants. J. Chinese Tea 4: 6–7.

Chen, Z. & H.H. Liao, 1988. Study on the induction of haploid plants from tea anther culture. J. Fujian Agric. College 17(3): 185–190.

Chen, Z., F. Chen, C. Chien, C-H. Wang, S-C. Chang, H-C. Hsu, H-H. Ou, Y-T. Ho & T-M. Lu, 1978. A process of obtaining pollen plants of *Hevea brasiliensis* Muell.-Arg. Sci. Sin. 22: 81–90.

Chu, C.C., 1978. The N6 medium and its applications to anther culture of cereal crops. In: Proc. Symp. Plant Tissue Culture Beijing, pp. 43–50. Science Press, Peking.

Crézé, J., 1983. The graft of *Camellia* meristem *in vitro*. Camellia Rev. 42: 11–12.

Harada, H., M. Kyo & J. Imamura, 1988. The Induction of Embryogenesis in *Nicotiana* Immature Pollen Culture. Applications of Plant Cell and Tissue Culture 137. Ciba Foundation Symposium, pp. 59–74. Wiley, Chichester.

Kato, M., 1986. Micropropagation through cotyledon culture in *Camellia japonica* L. and *C. sinensis* L. Jpn. J. Breed. 36: 31–38.

Kato, M., 1989a. *Camellia sinensis* L. (Tea): *In vitro* regeneration. In: Y.P.S. Bajaj (Ed.), Biotechnology in Agriculture and Forestry, Vol. 7: Medicinal and Aromatic Plants II, pp. 82–98. Springer-Verlag, Berlin.

Kato, M., 1989b. Polyploids of *Camellia* through culture of somatic embryos. HortScience 24: 1023–1025.

Lianfang, F., 1990. Litchi (*Litchi chinensis* Sonn.): *In vitro* production of haploids. In: Y.P.S. Bajaj (Ed.), Biotechnology in Agriculture and Forestry, Vol. 12, pp. 265–274. Springer-Verlag, Heidelberg.

Milewska-Pawliczuk, E., 1990. Apple (*Malus domestica* Borkh.): *In vitro* induction of androgenesis. In: Y.P.S. Bajaj (Ed.), Biotechnology in Agriculture and Forestry, Vol. 12, pp. 250–265. Springer-Verlag, Heidelberg.

Milewska-Pawliczuk, E. & B. Kubicki, 1977. Induction of androgenesis *in vitro* in *Malus domestica*. Acta Hort. 78: 271–276.

Murashige, T. & F. Skoog, 1962. A revised medium for rapid growth and bioassays with tobacco tissue cultures. Physiol. Plant. 15: 473–497.

Nakamura, Y., 1988. Efficient differentiation of adventitious embryos from cotyledons culture of *Camellia sinensis* and other *Camellia* species. Tea Res. J. 67: 1–12.

Nitsch, J.P., 1969. Experimental androgenesis in *Nicotiana*. Phytomorphology 19: 389–404.

Pedroso-Ubach, M.C., 1991. Contribuição para a preservação e o melhoramento de *Camellia japonica* L., pp. 23–50; 89–103 (English abstract). Master's Thesis, Faculdade de Ciências da Universidade de Lisboa, Lisboa.

Pedroso, M.C. & M.S. Pais, 1992. A SEM and X-ray microanalysis study during induction of morphogenesis in *Camellia japonica* L. Plant Sci. 87: 99–108.

Pedroso, M.C. & M.S. Pais, 1993. Direct embryo formation in leaves of *Camellia japonica* L. Plant Cell Rep. 12: 639–643.

Pedroso, M.C. & M.S. Pais, 1994a. Regeneration from anthers of adult *Camellia japonica* L. In Vitro Cell. Dev. Biol. Plant 29P: 155–159.

Pedroso, M.C. & M.S. Pais, 1994b. Induction of microspore embryogenesis in *Camellia japonica* L., cv. Elegans. Plant Cell Tiss. Organ Cult. 33: 339–345.

Pedroso, M.C. & M.S. Pais, 1994c. Early detection of embryogenic competence and of polarity in *Camellia japonica*. Plant Sci. 96: 189–201.

Pedroso, M.C. & M.S. Pais, 1994d. Plant regeneration from embryogenic suspension cultures of *Camellia japonica*. In Vitro Cell. Dev. Biol. Plant 31: 31–35.

Pedroso, M. C. & M.S. Pais, 1994e. Explant region-specific embryogenic competence and plant recovery in *Camellia japonica*. In Vitro Cell. Dev. Biol. Plant 31: 8–14.

Pedroso, M.C. & M.S. Pais, 1994f. Early marker of embryogenic competence in *Camellia japonica*. In: Abstracts of the 4th International Congress of Plant Molecular Biology, organized by The International Society for Plant Molecular Biology, Amsterdam, June 19–24, no. 660.

Raina, S.K. & R.D. Iyer, 1974. A study of response to anther culture. Indian J. Genet. Plant Breed. 34: 283–286.

Raghavan, V. & R. Nagmani, 1989. Cytokinin effects on pollen embryogenesis in cultured anthers of *Hyoscyamus niger*. Can. J. Bot. 67: 247–257.

Samartin, A., A.M., Vieitez, E. Vieitez, 1984. *In vitro* propagation of *Camellia japonica* seedlings. HortScience 19: 225–226.

Samartin, A., A.M. Vieitez, & E. Vieitez, 1986. Rooting of tissue cultured *Camellias*. J. Hort. Sci. 61: 113–120.

Samartin, A., 1989. A comparative study of effects of nutrient media and cultural conditions on shoot multiplication of in-vitro cultures of *Camellia japonica* explants. J. Hort. Sci. 64: 73–79.

San-José, M.C. & A.M. Vieitez, 1993. Regeneration of *Camellia* plantlets from leaf explants by embryogenesis and caulogenesis. Sci. Hort. 54: 303–315.

Shimokado, T., T. Murata & Y. Miyaji, 1986. Formation of embryoid by anther culture of tea. Jpn. J. Breed., Suppl. 2: 282–283.

Sunderland, N. & J.M. Dunwell, 1974. Pathways in pollen embryogenesis. In: H.E. Street (Ed.), Tissue Culture and Plant Science, pp. 141–167. Academic Press, New York.

Taylor, D.C., N. Weber, E.W. Underhill, M.K. Pomery, W.A. Keller, W.R. Scowcroft, R.W. Wilen, M.M. Moloney & L.A. Molbrook, 1990. Storage-protein regulation and lipid accumulation in microspore embryos of *Brassica napus* L. Planta 181: 18–26.

Vieitez, A.M., J. Barciela & A. Ballester, 1989a. Propagation of *Camellia japonica* cv. Alba Plena by tissue culture. J. Hort. Sci. 64: 177–182.

Vieitez, A.M., M.C. San-José & A. Ballester, 1989b. Progress towards clonal propagation of *Camellia japonica* cv. Alba Plena by tissue culture techniques. J. Hort. Sci. 64: 605–610.

Vieitez, A.M. & J. Barciela, 1990. Somatic embryogenesis and plant regeneration from embryonic tissues of *Camellia japonica* L. Plant Cell Tiss. Organ Cult. 21: 267–274.

Vieitez, A.M., M.C. San-José, F.J. Vieitez & A. Ballester, 1991. Somatic embryogenesis from roots of *Camellia japonica* plantlets cultured *in vitro*. J. Am. Soc. Hort. Sci. 116: 753–757.

Vieitez, A.M., M.L. Vieitez, A. Ballester & E. Vieitez, 1992. Micropropagation of *Camellia* spp. In: Y.P.S. Bajaj (Ed.), Biotechnology in Agriculture and Forestry High-Tech and Micropropagation III, Vol. 19, pp. 361 387. Springer-Verlag, Berlin.

Wang, C., Z. Chu & C. Sun, 1975. The induction of *Populus* pollen plants. Acta Bot. Sin. 17: 56–59.

Yamaguchi, S., T. Kunitake & S. Hisatomi, 1987. Interspecific hybrid between *Camellia japonica* cv. Chochidori and *C. chrysantha* produced by embryo culture. Jpn. J. Breed. 37: 203–206.

Zhang, Y.X., Y. Lespinasse & E. Chevreau, 1990. *In vitro* androgenesis in apple (*Malus domestica* Borkh.). In: Abstracts VIIth International Congress on Plant Tissue and Cell Culture, Amsterdam, June 24–29, A6–10, p. 185.

7. Haploidy in asparagus by anther culture

H.-S. TSAY

Contents

1. Introduction

Asparagus (*Asparagus officinalis* L., $2n = 2x = 20$) is a seed-propagated herbaceous perennial vegetable grown for its tender spears that are harvested primarily in the spring in temperate climates. Due to dioecy, it is obligately cross-pollinating except for occasional self-pollination of perfect flowers that occur on andromonoecious plants (Ellison, 1986). Hybrid cultivars have been developed through double cross breeding schemes or by single crosses after tissue culture methods became available for rapid multiplication of parental clones. Propagation of hybrid asparagus by seed results in almost equal numbers of male and female plants and individual plants vary in yield ability (Ito & Currence, 1965; Tsay *et al.*, 1980). Sex expression is under simple genetic control with maleness dominant to femaleness. Because male plants tend to be more productive than female plants, the potential of all-male hybrid cultivars has attracted the attention of many asparagus breeders. Haploids and doubled haploids can be efficiently utilized to realize such breeding schemes. Anther culture allows the rapid production of haploids and subsequent chromosome doubling can lead to completely homozygous diploid male (YY) or female (XX) plants. Homogeneous all male F_1 hybrids can then be obtained by crossing homozygous female and supermale parents (Cateland & Lambert, 1973; Corriols, 1979, 1985; Corriols *et al.*, 1990; Doré, 1974, 1979a,b; Falavigna, 1979; Falavigna *et al.*, 1990; Hondelmann & Wilberg, 1973; Pelletier *et al.*, 1972; Rameau, 1990; Raquin, 1973; Reuther, 1984; Price & Baughan, 1990; Sink & Padmanabhan, 1977; Tsay *et al.*, 1980, 1982a,b; Tsay & Hsu, 1985; Wricke, 1979; Yakuwa *et al.*, 1972). The first hybrid cultivar resulting from a single cross between two selected doubled

S.M. Jain, S.K. Sopory & R.E. Veilleux (eds.), In Vitro *Haploid Production in Higher Plants, Vol. 5,* 109–134.

haploids, one derived by anther culture and the other derived parthenogenically, was released recently (Corriols *et al.*, 1990).

Traditionally asparagus has been grown in temperate climates. Therefore, most asparagus cultivars are adapted to temperate climates. Harvesting of spears is restricted to the spring and early summer and the plants are dormant from late fall through the winter. However, in 1956, asparagus was introduced to the tropical and subtropical climates of Taiwan where it grows throughout the year without ever entering a completely dormant period. Through development of the so-called "mother-stalk culture method" for commercial cultivation of asparagus (Huang, 1979; Tu, 1985) in conjunction with research on breeding, cultural improvement, fertilization, pest control and postharvest treatment, Taiwan has become the primary producer of processed asparagus in the world (Huang, 1985).

In Taiwan, asparagus grows, flowers and sets fruit in any season as long as fertilizer and water are supplied. Experience with cultivar trials has demonstrated the importance of local adaptation of asparagus cultivars (Ellison, 1986). The development of cultivars suitable for production in the alien climate of Taiwan may be hastened by application of haploid breeding methods. This chapter summarizes the methods that have been used and developed in the laboratory of the Taiwan Agricultural Research Institute (TARI).

2. Establishment of aseptic anther culture

Flower buds with pollen at different developmental stages (Table 1) were collected from the field, sterilized in 0.5% sodium hypochlorite for 10 min, and rinsed three times with sterile water. The anthers were removed aseptically from the flower buds, and cultured on a nutrient medium. During excision of anthers, care must be taken to avoid injuring them. The basal medium contained the following components (in mg l^{-1}): 1/2 strength of the inorganic salts described by Murashige & Skoog (1962) (except Na-Fe-EDTA); thiamine HCl, 0.4; myo-inositol, 100; nicotinic acid, 0.5; pyridoxine HCl, 0.5; glycine, 1.0; sucrose, 60,000; Difco-agar, 8,000. Different plant growth regulators in varying concentrations have been added to the medium (Table 2). The medium was adjusted to pH 5.7 with either 1N NaOH or HCl, and autoclaved at 121°C at 15 lbs in^{-2} for 15 min. Each culture tube contained 10 ml medium. Cultures were maintained in darkness at 26°C.

Table 1. Influence of illumination on callus formation from asparagus anthers cultured at different developmental stages of the microspore (cv. 309)

Developmental stage of microspore	Treatment	No. of anthers cultured	Anthers forming callus		Anthers turning brown	
			No.	%	No.	%
Uninucleate	Light	118	69	58.5	118	100.0
Mitosis	Light	116	22	19.0	112	96.5
Binucleate	Light	183	22	12.0	85	46.0
Uninucleate	Dark	106	70	66.0	106	100.0
Mitosis	Dark	103	55	53.4	100	97.1
Binucleate	Dark	110	11	10.0	40	36.4

Basal medium: 1/2 MS salts supplemented with 2 mg l^{-1} NAA, 1 mg l^{-1} BA and 6% sucrose.

3. Haploid plant production

3.1. *Callus initiation*

The highest frequencies of callus induction were obtained from anthers with uninucleate microspores cultured on 1/2 strength MS inorganic salts (except Na-Fe-EDTA), supplemented with full strength organic substances, 2 mg l^{-1} naphthaleneacetic acid (NAA), 1 mg l^{-1} benzyladenine (BA) and 6% sucrose (Tables 1 and 2). Most reports on asparagus confirm that uninucleate microspores are most responsive to anther culture (Falavigna, 1979; Feng & Wolyn, 1991; Yakuwa *et al.*, 1972). Callus could be induced under either light (100 μE/m^{2}s or 2500 lux) or dark conditions, although better results were obtained when anthers with microspores at the mitotic stage were cultured in darkness.

Conditions conducive to callus development from anthers of asparagus have been defined by many scientists. The concentrations of cytokinins and auxins has been varied from 0.1–2 mg l^{-1} kinetin (kin) or BA and from 0.1–2 mg l^{-1} NAA, indolebutyric acid (IBA) or 2,4–dichlorophenoxyacetic acid (2,4–D) (Doré, 1973; Feng & Wolyn, 1991; Pelletier *et al.*, 1972; Yakuwa *et al.*, 1972; Yan *et al.*, 1991). Ziauddin *et al.* (1993) reported that an auxin transport inhibitor, 2,3,5–triiodobenzoic acid (TIBA), improved the response of anthers from plants stressed by cold or insects.

The success of callus induction from asparagus anther culture has been greatly affected by genetic and environmental factors. Falavigna *et al.* (1981) reported maximal callus formation under the following conditions: a) length

Table 2. Effect of plant growth regulators and sucrose on callus formation from asparagus anthers cultured at late uninucleate pollen stage (cv. UC 309)*

Sucrose concentration	2,4-D	NAA	BA	Kin	No. of anthers cultured	% of anthers forming callus
		- mg l^{-1} -				
3%	0	0	0	0	196	0.0 (0- 2)**
3%	0	2	0	2	72	16.7 (9-27)
3%	1	0	0	0	57	5.3 (1-13)
6%	0	2	0	4	116	12.9 (7-21)
6%	0	2	2	1	104	21.1 (14-30)
6%	2	0	0	2	63	17.5 (8-28)
6%	0	4	0	2	108	0.0 (0- 4)
6%	0	2	0.5	0	109	15.6 (9-24)
6%	0	2	1.0	0	115	42.6 (33-53)
6%	0	2	2.0	0	110	13.6 (8-22)
6%	0	2	4.0	0	110	26.4 (18-36)
6%	0	2	8.0	0	115	16.5 (10-26)

* Basal medium: 1/2 MS salts.
** Data in parentheses are the 95% confidence limits.

of selected flower buds from 1.6–2.1 mm corresponding to the uninucleate stage, b) cold pretreatment of flower buds at 7°C for 10 days or 6°C for 6 days (Peters & Augustin, 1989), c) solid medium containing 4% sucrose, and d) anthers taken from field-grown plants in the summer. Yan *et al.* (1991) found that anthers with uninucleate stage microspores collected from yellow-green buds from April to May had the highest callus induction frequency after a pretreatment for 6–8 days at 6°C. Usually, two kinds of callus have been induced: one is friable with few haploid cells and little potential for differentiation and the other is compact and more easily regenerable.

Ellison (1985) reported considerable variation for yield and quality of spears among plants within a cultivar of asparagus. In open-pollinated populations, ten-fold differences in spear yield between the highest and lowest individuals could be found. Such differences can be expected, given the genetic diversity of an F_1 population developed from non-inbred parents. Similar variation for anther culture competence could also be expected.

Table 3. Plant-to-plant variability in the breeding efficiency of asparagus anther culture*

Cultivar and plant no.		No. of anthers cultured	% of anthers producing callus (A)	% of callus forming green plants in		Breeding efficiency A X (B+C) - % -
				M medium**	MS medium**	
UC309	-17	1,430	29.9	4.0	8.0	3.6
	-64	76	36.8	0.0	10.5	3.9
UC500	- 5	943	25.6	4.2	19.0	5.9
	-12	306	34.3	6.7	12.1	6.4
UC711	- 3	178	26.4	0.0	9.1	2.4
	-11	108	25.0	0.0	11.1	2.8
UC New500	-6	96	35.4	3.7	4.2	2.8
	-20	114	25.4	3.4	4.8	2.1
	-27	106	34.0	0.0	14.7	5.0
UC500W	-20	127	63.0	9.3	13.9	14.6
	-23	226	47.8	0.0	6.4	3.1
UC Dwarf5	- 6	225	36.4	15.2	5.1	7.4
	- 8	108	30.6	3.0	13.3	5.0

* Data on plants with breeding efficiency over 2.0% are presented. Callus induction medium: 1/2 MS salts supplemented with 2 mg l^{-1} NAA, 1 mg l^{-1} BA and 6% sucrose.
** M medium: Murashige *et al.* (1972) basic salts with 1 mg l^{-1} NAA, 0.5 mg l^{-1} BA and 2.5% sucrose. MS medium: MS salts with 3% sucrose.

Anthers from 80 plants comprising 11 cultivars were used to test the callus induction and culture efficiency of anthers. Callus formation ability differed among cultivars as well as among plants within cultivars (Tables 3 and 4). Within cultivar UC500W, the callus induction percentage of plants 20 and 23 was 63% and 48%, respectively (Table 3) whereas other plants of this cultivar did not respond at all (data not shown). Likewise, Doré (1973) and Falavigna (1979) reported that the frequency of anthers producing callus was genotype dependent for asparagus.

Indirect production of haploid plants via callus formation in asparagus anther culture (Figs. 1 and 2) has been described in several reports (Doré, 1973; Falavigna, 1979; Melle Dore, 1973; Pelletier *et al.*, 1972; Tsay *et al.*, 1980; Yakuwa *et al.*, 1972). However, direct embryogenesis of microspores after 1–2 months of culture has also occurred for all cultivars examined in our study (Fig. 3). The percentage of anthers that formed embryos ranged from 0.1 to 1.7% (Table 5). Further differentiation of embryos into shoots and roots has been achieved by transfer to MS basal medium without growth regulators (Fig. 4). Most of the embryo-derived plants were haploid. Em-

Table 4. The breeding efficiency of asparagus anther culture among cultivars*

Cultivars	No. of plants used	No. of anthers cultured	% of anthers forming callus	% of callus forming green plants in M** medium (B)	% of callus forming green plants in MS** medium (C)	Breeding efficiency A X (B+C) - % -
UV157	8	3,191	22.8	0.4	3.3	0.8
UC309	6	2,269	27.5	2.8	7.1	2.7
UC500	4	3,363	18.7	3.3	12.4	2.9
UC711	11	2,522	11.5	1.6	5.2	0.8
UC800	1	185	11.9	0.0	4.6	0.5
UC New 500	9	1,641	16.8	0.8	6.7	1.3
UC500W	2	353	53.3	4.0	9.4	7.1
UC Dwarf 5	6	794	11.3	6.8	6.0	1.4
UC High 5	6	900	17.1	2.1	10.9	2.2
500009	15	2,443	21.6	0.2	4.0	0.9
Mary Washington	12	2,494	16.8	1.0	5.5	1.1
Total & Average	80	20,155	20.8	1.9	6.6	2.0

* Callus induction medium: Same as Table 3.
** Same as Table 3.

bryogenic callus was obtained by Feng & Wolyn (1991) from anthers with microspores at the late-uninucleate stage cultured at 32°C in the dark for 3–4 weeks after which it was grown at 25°C under a 16 h photoperiod for an additional 3–4 weeks. Medium containing 0.65 mg l^{-1} ancymidol, 0.1 mg l^{-1} NAA and 0.1 mg l^{-1} kin was optimal for embryo maturation. Half of the embryos regenerated on MS medium containing 1 mg l^{-1} GA_3. Production of haploid plants through embryogenesis may be more efficient; however, the role of plant growth regulators in the process remains to be clarified.

3.2. *Organ initiation*

Once anther callus has been obtained, that derived from filaments and anther wall must be discarded leaving only that derived from split anthers to be

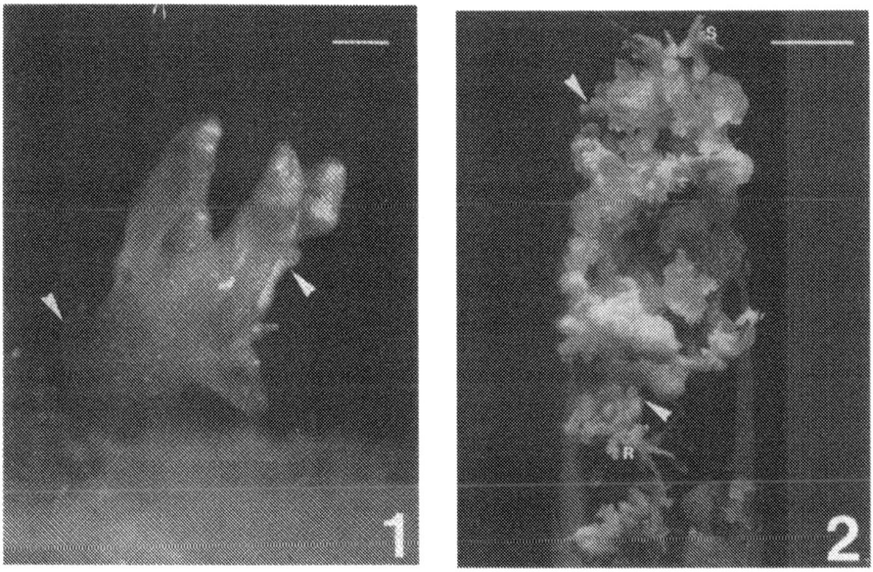

Figures 1–2. Indirect production of a haploid plant via callus formation in asparagus anther culture. The haploid plants (arrow) were produced by organogenesis of haploid callus (Fig. 1, arrow) and an embyro (Fig. 2, arrow) initiated from haploid callus; differentiation of shoots (S) and roots (R) can be observed from these embyros. (Bars equal 2 mm in Fig. 1 and 1 cm in Fig. 2).

used for organ initiation. In a study of the effect of plant growth regulators on morphogenesis of asparagus anther callus, it was found that a combination of 1 mg l^{-1} NAA and 0.5 mg l^{-1} BA was most effective for promoting differentiation of shoots and roots. The cultures were given 16 h daily illumination under 1,500 lux fluorescent light. In some cases, embryos were induced from callus cultured on basal medium with 0.3 mg l^{-1} NAA and 0.1 mg l^{-1} kin. Differentiation of shoots and roots was observed from these embryos (Fig. 2). Organs did not differentiate from callus grown continuously on a callus-inducing medium. Most of the callus remained undifferentiated even after two months of culture on medium containing 1 mg l^{-1} NAA and 0.5 mg l^{-1} BA. A second subculture to either fresh medium of the same composition or to basal medium without plant growth regulators was made. After two months, both shoot and root formation occurred on basal medium (Tables 3 and 4), i.e., after the second subculture, the presence of NAA and BA had no positive effect on organ differentiation in anther callus. The induction of callus mostly occurred 30–60 days after the anthers were cultured. The greatest differentiation ability was obtained from 50 day old callus (Table 6) (Lai *et al.*, 1991).

Varietal differences and plant-to-plant variability have also been observed

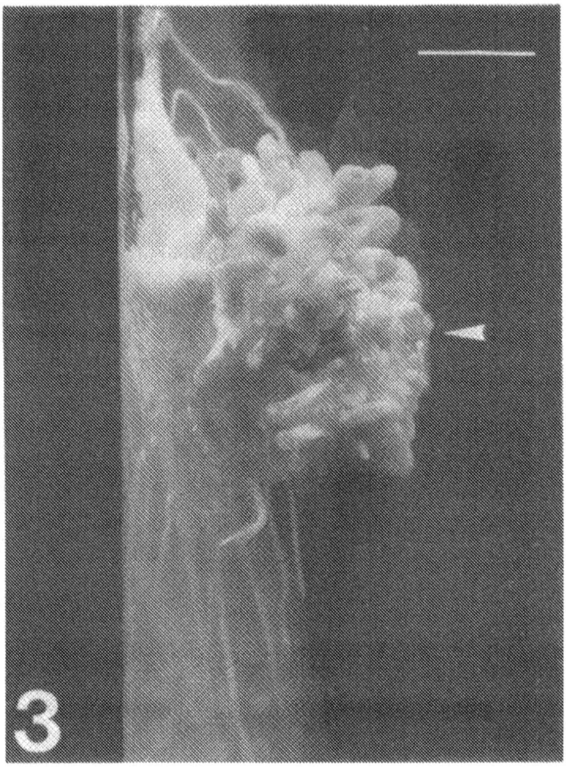

Figure 3. Direct production of haploid embyros from asparagus anther culture (Bar equal 1 cm).

for plant regeneration from anther-derived callus (Table 3). Of 80 plants tested, only 13 were found to have anther breeding efficiency (number of green plants produced per 100 cultured anthers) greater than 2%; UC500W-20 exhibited the highest anther breeding efficiency (14.6%, Table 3) for an individual plant whereas UC500W was the cultivar with the highest anther breeding efficiency (7.1%; Table 4).

The influence of genotype on the response of cultured anthers was also reported by Lai *et al.* (1991). Qiao & Falavigna (1990) reported that the percentage of responding anthers varied from 0.7 to 12.8 according to genotype of the anther donor. However, environmental factors were also important. Falavigna *et al.* (1981) obtained better results from vigorous plants grown in the field than from plants grown in the greenhouse. Furthermore, they reported that the size of flowers was strongly influenced by the genotype and position on the plants. Routine microscopic observations were necessary to identify buds which correlated with the highest frequency of microspores at the uninucleate stage (Falavigna *et al.*, 1985).

Table 5. Percentages of callus and embyro formation in asparagus anther culture

Cultivar	No. of anthers cultured	Callus-forming anthers		Embryo-forming anthers	
		No.	- % -	No.	-%-
UC309-4	2,334	891	38.2 (35-41)	4	0.17
UC500-5	2,003	562	28.1 (25-31)	23	1.15
UC500-12	2,423	1,422	58.7 (56-62)	40	1.65
UC500W-20	2,003	326	16.3 (14-18)	3	0.15
UC New500-27	4,841	469	9.7 (8-12)	1	0.02
UC Dwarf 5-6	2,065	527	25.5 (23-28)	3	0.15
UC Dwarf 5-8	1,367	265	19.4 (17-22)	1	0.07
500009-29	1,905	655	34.4 (31-37)	12	0.63

Data in parentheses are the 95% confidence limits.

3.3. *Anther browning*

Many reports have emphasized the importance of the condition of the anther wall for anther culture (Lai *et al.*, 1980; Mii, 1976; Pelletier & Ilami, 1972; Sunderland, 1974). In rice and tobacco, e.g., early anther browning was associated with low callus induction. However, the appearance of brown color at four weeks of culture coincided with vigorous callus or embryo differentiation in anther culture (Tsay, 1981). Pelletier & Ilami (1972) reported that it was necessary for anthers to remain green during the first few days of culture in order for microspore embryogenesis to occur. After this critical stage, the browning of anthers was not harmful.

In our studies of asparagus anther culture, the color of cultured anthers turned from green to brown or white after three weeks of culture and callus induction coincided with the color change. The percentage of anthers forming callus and turning brown was investigated at 30 day intervals over a two-month period. Of 389 anthers with a brown color, 281 (72.3%) produced callus, whereas callus was rarely induced from anthers which turned white (10 of 77 or 13.0%). By pooling the results of 27 experiments on different cultivars, a positive correlation (0.80) was found between anther browning and callus formation in asparagus (Tsay *et al.*, 1982a). The regression line between these two factors is shown in Fig. 5.

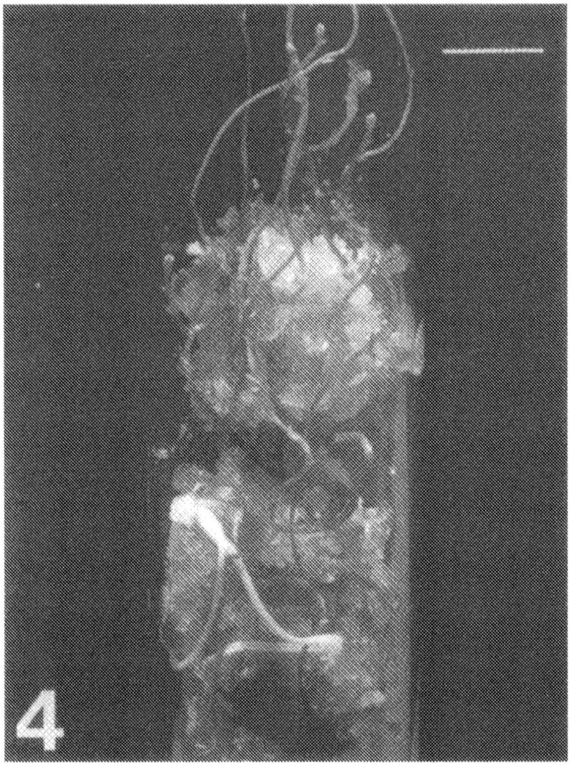

Figure 4. Differentiation of embyros into shoots and roots was achieved by transferring them to MS basal medium without plant growth regulators (Bar equal 1 cm).

3.4. *Chromosome variation*

Cytological analysis of anther callus and mitotic cells in root tips of anther-derived asparagus plants (Fig. 6) revealed variation in chromosome number ranging from haploid (8.2%), diploid (60.6%), triploid (5.3%), to tetraploid (24.5%) and mixoploid (1.3%) (Table 7; Fig. 7). Such variation in callus tissue of asparagus has also been reported by Cateland & Lambert (1973), Doré (1974), Falavigna *et al.* (1990) and Yan *et al.* (1991). The incidence of polyploids may be partially due to the use of plant growth regulators in the culture medium and the long period that callus remains on the culture medium before regeneration occurs. For producing haploid plants, the presence of polyploid and aneuploid callus is undesirable as they reduce the proportion of haploid cells. Increasing the haploid cell population can be achieved by applying chemicals, such as para-fluorophenylalanine, in the culture medium to inhibit the growth of polyploid cells or by shortening the

Table 6. Influence of callus age on differentiation ability in asparagus anther culture (cv. UC500–5)

Days after callus emergence	Differen-tiation medium used*	No. of calli cultured	% of calli forming green plants in 1st medium (B)	2nd medium (C)	Total	Breeding efficiency A**x(B+C) - % -
10	I	268	4.5	10.7	15.2	7.4
	II	179	1.1	0.8	1.9	0.9
20	I	309	9.7	20.9	31.6	14.8
	II	250	4.0	1.6	5.6	2.7
30	I	282	9.9	21.1	31.0	15.0
	II	275	2.9	0.4	3.3	1.6
40	I	255	12.2	24.6	36.8	17.8
	II	276	5.1	0.4	5.5	2.7
50	I	23	30.4	28.6	59.0	28.6
	II	26	23.1	6.3	29.4	14.4
60	I	39	15.4	22.9	38.3	18.6
	II	39	20.5	0.0	20.5	9.9
70	I	66	19.7	15.7	35.4	17.2
	II	52	13.5	0.0	13.5	6.5
80	I	13	23.1	20.0	43.1	20.9
	II	25	12.0	4.6	16.6	8.0
90-100	I	5	0.0	40.0	40.0	19.4
	II	5	0.0	0.0	0.0	0.0
Total and average	I	1260	13.9	22.7	36.6	17.7
	II	1127	10.5	1.6	12.1	5.2

* I: Anther-derived callus was first transferred onto M medium for organ induction; two months later, non-regenerative callus was transferred onto MS medium (2nd medium) for another two months. M & MS medium: same as Table 3. II: Anther-derived callus was first transferred onto MS medium for organ induction; two months later, non-regenerative callus was transferred onto fresh MS medium (2nd medium) for another two months.
** A: Percentage of callus induction = 48.5%.

culture duration of anther-derived callus (Gupta & Carlson, 1972; Matthews & Vasil, 1976; Tomes & Collins, 1976). Direct regeneration of haploid asparagus plants from cultured anthers is rare and therefore the homozygosity of regenerated diploids must be confirmed by test crosses.

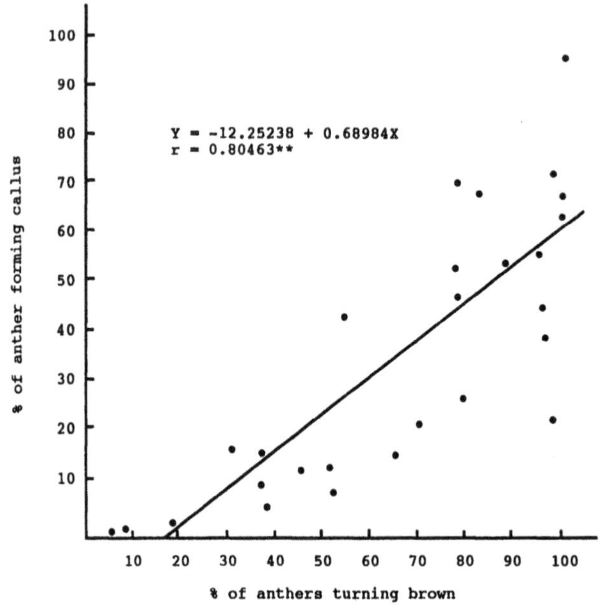

Figure 5. Regression line between anther browning and callus formation in asparagus anther culture.

3.5. *Phenotypic appearance of anther-derived plants*

Plant type, internode length, leaf color, phyllocladia length and flower size were compared among haploid, diploid and tetraploid plants. Plant height, internode length and phyllocladia length of haploid plants were the shortest among the three ploidy levels. Dense phyllocladia, light green leaf color and small flowers (bud length × width was only 3.2 × 1.7 mm at the uninucleate stage) were typical for haploid plants. On the contrary, tetraploid plants were taller, with longer internodes, with phyllocladia distributed more sparsely, darker green leaf color, and larger flowers (6.7 × 3.3 mm). The phenotypic appearance of diploid plants was intermediate between haploid and tetraploid (Figs. 8 and 9).

It has been possible to distinguish ploidy levels of anther-derived plants by their stem diameter, leaf color and internode length *in vitro*. Haploid plants possessed thin stems, dark green leaf color and indistinct internodes. On the contrary, tetraploid plants possessed thicker stems, light green leaf color and more obvious internodes. The stem diameter, leaf color and internode length of diploid plants were intermediate (Figs. 10–12). Albino plants were observed in some cases (Fig. 13).

Early phenotypic evaluation, either *in vitro* or *in vivo*, is valuable because cytological examination of root tips frequently causes serious damage to the

Figure 6. Anther-derived plants were acclimatized in a growth chamber for 2–3 weeks then transferred to a greenhouse. At that time their ploidy level was checked by counting chromosomes in meristematic cells of root tips.

anther-derived plants and thus lowers their survival on transplanting. Early phenotypic evaluation can also save the time and labor required for cytological examination. Qiao & Falavigna (1990) reported that doubled-haploid calli regenerated normal shoots and rooted quickly, whereas polyploid calli gave distorted shoots and large roots.

A positive correlation between the multiplication rate of callus on regeneration medium and vigor of the plants in the field has been observed. A better understanding of these correlations will allow us, in the future, to apply *in vitro* selection and reduce the labor required for rooting and field evaluation of inferior doubled haploid clones.

4. Chromosome doubling of haploid plants

The chromosome number of anther-derived haploid plants needs to be doubled to produce homozygous lines useful in a breeding program. Thevenin & Doré (1976) described *in vitro* culture of asparagus meristems on a medium containing colchicine for doubling the chromosome number of

Table 7. Ploidy distribution of anther-derived plants of asparagus

Cultivars	Number of plants					Total
	Haploid	Diploid	Triploid	Tetraploid	Mixoploid	
UC157	3	12	1	4	0	20
UC309	14	21	5	14	4	58
UC500	22	212	18	84	0	336
C711	0	8	0	3	0	11
UC New 500	0	1	1	3	0	5
UC500W	0	8	0	0	0	8
UC New Dwarf-5	0	6	0	2	0	8
500009	0	6	0	0	1	7
UC New High	0	6	0	1	1	8
Mary Washington	0	4	0	4	0	8
Total	39	284	25	115	6	469
Percentage	8.3	60.6	5.3	24.5	1.3	100

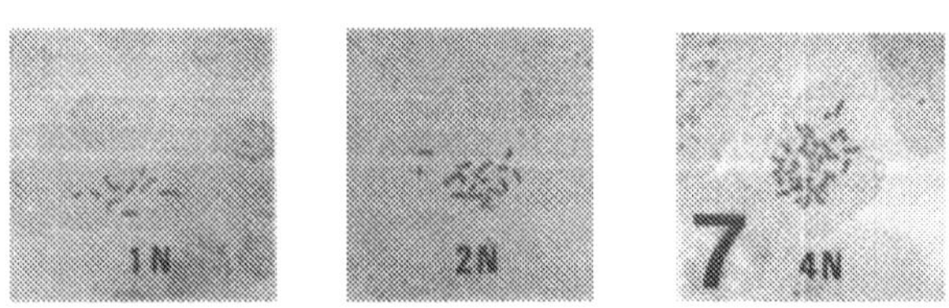

Figure 7. Cytological examination of meristematic cells from root tips of anther-derived plants revealed haploid (N), diploid (2N) and tetraploid (4N) mitoses.

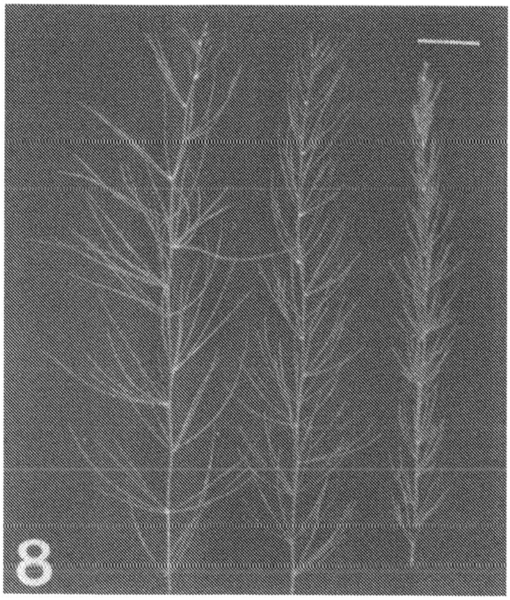

Figure 8. Phenotypic appearance (phyllocladia) of anther-derived haploid, diploid and tetraploid plants (from right to left) of asparagus (Bar equal 1 cm).

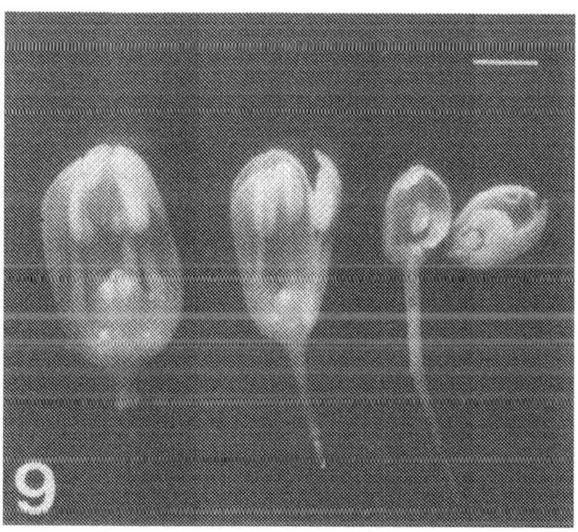

Figure 9. Flowers and anthers of anther-derived haploid, diploid and tetraploid plants (from right to left) of asparagus (Bar equals 2 mm).

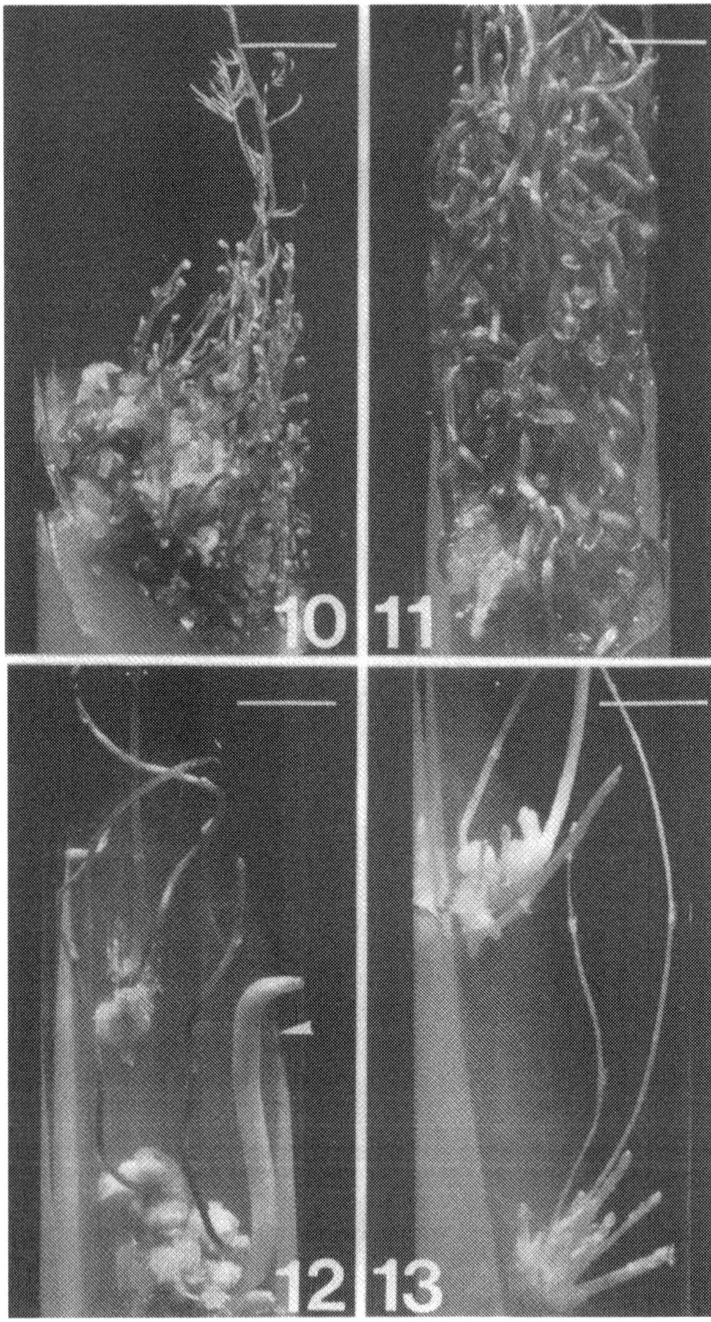

Figures 10–13. The phenotypic appearance of anther-derived haploid (Fig. 10), diploid (Fig. 11), tetraploid (Fig. 12, arrow) and albino plants (Fig. 13) *in vitro* (Bars equal 1 cm in all figures).

genotypes for which *in vivo* treatments had failed. This *in vitro* procedure was considered efficient and inexpensive and could be applied for the purpose of haploid chromosome doubling. Broertjes (1974) reported the application of colchicine solution to *Saintpaulia* petioles to obtain polyploid plantlets. The plantlets proved to be sensitive to the chemical, and lethality was observed under high concentrations. According to Broertjes, an increase in concentration and treatment time was directly related both to an increase of polyploids and to the low production of plantlets. On the other hand, Espino & Vazquez (1981) found that in *Saintpaulia*, the time of exposure to the colchicine did not influence the percentage of polyploid plantlets obtained. These opposing results reveal that the technique of chromosome doubling by colchicine treatment may be unpredictable. In addition to colchicine concentration and treatment duration, the treatment was also strongly influenced by temperature and the sterilization method of colchicine. Griesbach (1981) reported that autoclaving colchicine at 20 lbs/in^2 for 20 min considerably reduced its activity since colchicine is thermolabile and degrades upon heating. Thus, filter-sterilization of colchicine is recommended for *in vitro* treatments (Hsu *et al.*, 1991).

Although there are not many references on the use of colchicine available for asparagus, it is known that the addition of colchicine to tissue or cell culture media has a polyploidizing effect (Bayliss, 1976; Chen & Kallemeyn, 1979; Heinz & Mee, 1970; Perry & Lyrene, 1984; Thevenin & Doré, 1976). Haploid shoot tips or callus were treated with different colchicine concentrations for various durations to determine the best method for chromosome doubling. Supermale and homozygous female plants have been successfully obtained through the following techniques:

1. Spontaneous doubling: Haploids were spontaneously doubled at a rate of 6.0% through subculturing shoot tip meristems five times (Hsu *et al.*, 1991).
2. Colchicine treatment: Doubling ratio of 21–97% could be obtained from colchicine treatment of shoot tips of seedlings *in vivo* by applying 1.2% colchicine in lanolin (Figs. 14–15). For *in vitro* applications, both colchicine-containing solid medium and colchicine soaks resulted in higher doubling rates than colchicine suspension method. Among the many methods tested, haploid plantlets cultured on 2,000 ppm colchicine solid medium or soaked in 2,000 ppm liquid medium for 4 h or 4,000 ppm medium for 2 h produced the best result for increasing the stem diameter as a visual indication of chromosome doubling. However, the chromosome doubling effect was not reliable. In all treatments, mixoploids, polyploids, aneuploids, cytochimeras, morphological abnormalities, necrosis and growth inhibition have been observed. In general, higher colchicine concentrations and longer treatment periods were associated with lower survival rates (Hsu *et al.*, 1991).
3. Regeneration from haploid shoot tip-derived callus: Shoot tips of haploids were cultured on 2,4–D-containing medium for callus induction. The

Figures 14–15. Chromosome doubling was achieved (arrows) by colchicine treatment of shoot tips of haploid plantlets by spotting (Fig. 14) and lanolin method (Fig. 15) *in vivo* (Bars equal 0.5 cm in both figures).

induced calli were then transferred to differentiation medium for plant regeneration. Some plants derived from shoot tip-callus were homozygous diploids at a proportion of 10–50%. The doubling effect was more stable than that from colchicine treatment; moreover, rooting of these plantlets was much easier (Hsu *et al.*, 1991).

5. Rooting of supermales

Supermale plantlets could be regenerated from chromosome-doubled shoot-tips of haploids cultured on MS medium containing kin and NAA. However, these cultures often developed excessive callus and produced thin, weak roots, or no roots at all. Successful transfer of these supermales into soil has been dependent upon formation of thick vigorous roots. Chin (1982) reported that 0.13 mg l^{-1} (0.5 μM) ancymidol added to MS medium containing NAA and kin enhanced the rate of root and shoot development from single node explants of asparagus. Roots and shoots were also stronger when the antigibberellin was present. Callus growth from the explants was suppressed, and the time required for the production of transplantable plantlets was reduced from 20 to 8 (or even 3) weeks. Various combinations of auxins and cytokinins did not improve the plantlets. Incorporation of a number of growth retardants, e.g., B-995, phosforn, Amo 1618, cycocel and paclobutrazol, promoted growth of stronger shoots and roots. The effectiveness of the growth retardants varied, with ancymidol being most effective and cycocel least effective (Khunachak *et al.*, 1987). Ancymidol also enhanced embryo maturation and germination of anther-derived callus and was more critical in the solid than liquid medium. The best response was observed when ancymidol concentrations were 0.1 and 0.5 mg l^{-1} in liquid and solid medium, respectively (Feng & Wolyn, 1993). Chang & Peng (1993) reported that phloroglucinol or tryptone significantly increased shoot number and root length of asparagus nodal sections cultured *in vitro*, whereas ancymidol reduced shoot length, but significantly increased diameter of shoots and roots and the number of multiple shoots.

Sucrose concentration may also play an important role on root formation of asparagus *in vitro*. Desjardins & Tiessen (1985) described that high levels of sucrose (over 5%) in conjunction with 5 μM ancymidol produced the best rooting in the shortest time. The shoot growth was good in media containing 2, 3, 5, and 7% sucrose. They suggested that ancymidol is required for the initiation of a crown. Once the crown has been initiated, roots can then be induced to form. The triggering mechanism for root production is the high concentration of sucrose in the medium.

The rooting process of supermales in our laboratory has been: chromosome-doubled shoot tips obtained from colchicine lanolin treatment were cultured on a rooting medium containing 1/2 MS basal salts supplemented with 5% sucrose, 0.6% agar, 0.1 mg l^{-1} NAA, 0.1 mg l^{-1} kin and 0.13 mg l^{-1} ancymidol for one month (Fig. 16); transfer to ancymidol-free liquid medium for another month for faster growth (Fig. 17).

6. Progeny test

A morphological evaluation in the field of doubled haploid androgenic clones is a fundamental step of the asparagus all-male breeding program. The parents of the all-male hybrids must be selected following at least three years of evaluation. The main characteristics of the selected clones should be vigor, fertility, spear size, disease resistance, survival, and capacity for *in vitro* multiplication and rooting (Falavigna *et al.*, 1990).

More than 100 supermale plants (Fig. 18) were selected during the period of 1980–1985 in Taiwan Agricultural Research Institute for progeny test. These supermales were evaluated in the open field for morphological and physiological traits, resistance to disease and plant survival. Numerous F_1 hybrids were obtained by pollinating homozygous female plants with pollen from homozygous supermale clones.

In order to check the androgenic origin of the more vigorous male clones,

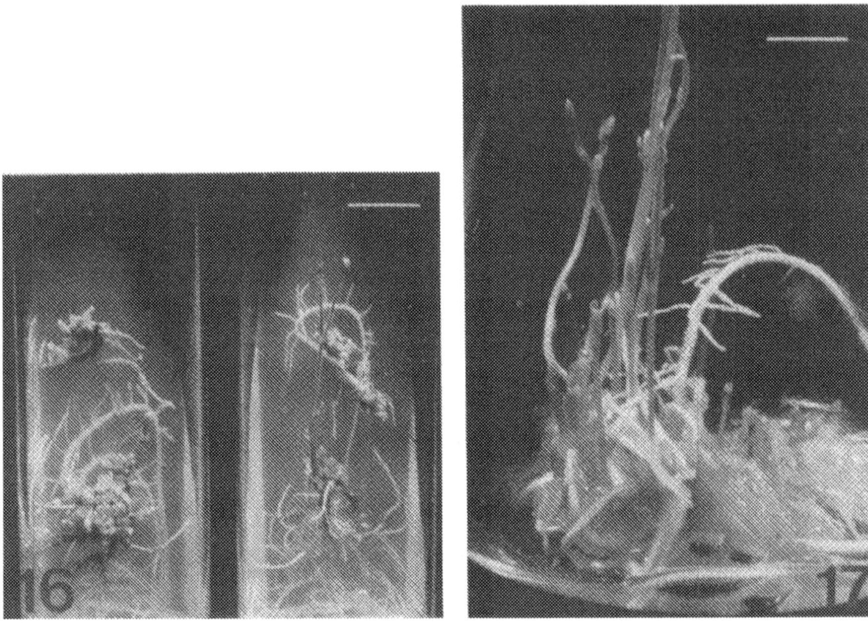

Figures 16–17. Chromosome-doubled shoot tips obtained from colchicine lanolin treatment on rooting medium (1/2 MS + 5% sucrose, 0.6% agar, 0.1 mg l^{-1} NAA, 0.1 mg l^{-1} kin and 0.13 mg l^{-1} ancymidol) in the first month (Fig. 16), then after transfer to ancymidol free liquid medium (Fig. 17) (Bars equal 1 cm in both figures).

sex expression of the F_1 progeny was examined. If these progenies segregated 1 male:1 female, their male parent was considered to be derived from somatic cells rather than from a microspore. F_1 hybrids were tested for their combining ability in order to observe heterosis. By using a polycross system, the optimal combination of female and male genotypes could be identified (Reuther, 1984). Figures 19 and 20 show seedlings and field evaluation of the hybrids. It was found that 153 of 157 combinations generated all-male hybrids. Only one of 20 putative supermales was heterozygous.

7. Conclusion

Using the technique of anther culture, completely homogeneous F_1 asparagus hybrids can be obtained by crossing two homozygous parents. These hybrids can be all male if one of the progenitors is a supermale. Since no selection can be made before this step, field tests need to be undertaken to select the best cross combinations. All-male F_1 hybrids differ with respect to yield, quality and disease resistance (Reuther, 1984). Corriols (1985) reported that

Figure 18. Super-male plants obtained from asparagus anther culture (Bar equal 6 cm).

Figure 19. Numerous F_1 hybrids obtained after pollinating female plants with pollen of super-males.

Figure 20. Field evaluation of all-male progenies.

unselected all-male hybrids were not necessarily higher yielding than good mixed varieties (composed of males and females). If all-male hybrids were not selected carefully enough, they could actually express a reduction in spear diameter. It is necessary to cross a genotype that has good combining ability for spear number with another that has good combining ability for spear diameter (Rameau, 1990). The limiting factors for large-scale F_1 hybrid seed production are inbreeding depression of supermale parents and failure of vegetative propagation by conventional methods. Therefore, cloning of parent plants by *in vitro* culture is important for setting up a hybrid production field for generation of large amounts of F_1 hybrid seed in a short time (Reuther, 1984).

A more efficient method of haploid plant or callus formation in asparagus is necessary. Microspore or ovary culture may provide alternatives. Unfortunately, these methods have not been reported for asparagus. As for genetic engineering or mutation breeding, suspension and protoplast culture are prerequisite techniques that need to be established (Duc *et al.*, 1975; Hsu *et al.*, 1989). Successful regeneration of plants from protoplasts can provide a model system for studying genetic modification through transformation and protoplast fusion.

8. Acknowledgements

The author would like to thank Prof. Richard E. Veilleux, Department of Horticulture, Virginia Tech, Virginia, U.S.A. and Dr. Dah-Jiang Liu, Department of Agronomy, Taiwan Agricultural Research Institute, for their kindness in reading and correcting the manuscript and making valuable suggestions.

9. References

Bayliss, M.W., 1976. Variation of cell cycle duration within suspension cultures of *Daucus carota* and its consequence for the induction of ploidy changes with colchicine. Protoplasma 88: 279–285.

Broertjes, C., 1974. Production of polyploids by the adventitious bud technique. In: Polyploid and Induced Mutations in Plant Breeding, pp. 29–34. I.A.E.A., Vienna, -PL-503/6.

Cateland, B. & A. Lambert, 1973. L'androgenèse *in vitro* par culture d'anthères d'*Asparagus officinalis*. II. Les problèmes de polyploïdie. In: L. Thevenin (Ed.), 4ème Réunion sur la Section de l'Asperge, pp. 183–189. Eucarpia Sect. Horticole, Versailles.

Chang, D.C.N. & K.H. Peng, 1993. Ancymidol, phloroglucinol and tryptone enhanced *in vitro* rooting of asparagus nodal sections. In: 8th Intl. Symp. Abstract, Palmerston North, Nov. 21–26, 1993, p. 59.

Chen, C.H. & Y.G. Kallemeyn, 1979. *In vitro* induction of tetraploid plants from colchicine-treated diploid daylily callus. Euphytica 28: 705–709.

Chin, C.K., 1982. Promotion of shoot and root formation in asparagus *in vitro* by ancymidol. HortScience 17: 590–591.

Corriols, T.L., 1979. Different methods in asparagus breeding. In: G. Reuther (Ed.), Proc. 5th Intl. Asparagus Symposium, pp. 8–22. Eucarpia Section Vegetables, Geisinheim Forschungsanstalt.

Corriols, T.L., 1985. Behaviour of several hybrid schemes towards asparagus breeding tests. In: E.C. Lougheed & H. Tiessen (Eds.), Proc. 6th Intl' Asparagus Symp., Univ. of Guelph, Guelph, pp. 8–21.

Corriols, T.L., C. Doré & C. Rameau, 1990. 1990: Commercial release in France of Andreas, the first asparagus all-male F_1 hybrid. Acta Hort. 271: 249–252.

Doré, C., 1973. Androgenèse *in vitro* par culture d'anthères d'*Asparagus officinalis*. I. État actuel des recherches. In: L. Thevenin (Ed.), 4ème Réunion sur la Sélection de l'Asperge, pp. 173–181. Eucarpia Sect. Horticole, Versailles.

Doré, C., 1974. Production de plantes homozygotes mâles et fémelles à partir d'anthères d'asperge, cultivées *in vitro* (*Asparagus officinalis* L.). C.R. Acad Sci. 278: 2135–2138.

Doré, C., 1979a. Different utilizations of *in vitro* cloning techniques. In: G. Reuther (Ed.), Proc. 5th Asparagus Symposium, pp. 146–149. Eucarpia Section Vegetables, Geisinheim Forschungsanstalt, Geisinheim.

Doré, C., 1979b. Anther culture as a way towards F_1 hybrids. In: G. Reuther (Ed.), Proc. 5th Asparagus Symposium, pp. 87–90. Eucarpia Section Vegetables, Geisinheim Forschungsanstalt, Geisinheim.

Desjardins, Y. & H. Tiessen, 1985. The effect of sucrose concentration and ancymidol on the rooting of asparagus *in vitro*. In: E.C. Lougheed & H. Tiessen (Eds.), Proc. 6th Intl. Asparagus Symp. Univ. of Guelph, Guelph, p. 384.

Duc, B.D.H., B. Noreel & A. Masset, 1975. Regeneration of *Asparagus officinalis* L. through callus cultures derived from protoplasts. J. Exp. Bot. 26: 263–270.

Ellison, J.H., 1985. Breeding true varieties of asparagus. In: Proc. 6th Intl. Asparagus Symp., Univ. of Guelph, Guelph, p. 7.
Ellison, J.H., 1986. Asparagus breeding. In: M.J. Bassett (Ed.), Breeding Vegetable Crops, pp. 521–569. Avi Publishing Co., Westport, Connecticut.
Espino, F.J. & A.M. Vazquez, 1981. Chromosome numbers of *Saintpaulia ionantha* plantlets regenerated from leaves cultured *in vitro* with caffeine and colchicine. Euphytica 30: 847–853.
Falavigna, A., 1979. Pure lines of *Asparagus officinalis* L. obtained by *in vitro* culture in Italy. In: E.C. Lougheed & H. Tiessen (Eds.), Proc. 5th Asparagus Symp., pp. 91–99. Eucarpia Sect. Vegetables, Geisinheim Forschungsanstalt, Geisinheim.
Falavigna, A., E. Chiapparini & G.P. Soressi, 1981. Genetic and environmental factors effecting *in vitro* androgenesis of asparagus (*Asparagus officinalis* L.). In: Proc. 25th Annual meeting of the S.L.G.A. Research Institute for Vegetabale Crops, Rome, pp. 59–61.
Falavigna, A., M.G. Tacconi & P.E. Casali, 1985. Aspagagus breeding through anther *in vitro* culture: eight years of activity in Italy. In: E.C. Lougheed & H. Tiessen (Eds.), Proc. 6th Intl. Asparagus Symp., pp. 31–40. Univ. of Guelph, Guelph.
Falavigna, A., P.E. Casali & M.G. Tacconi, 1990. Potential of *in vitro* anther culture technique for asparagus breeding in Italy. Acta Hort. 271: 39–46.
Feng, X.R. & D.J. Wolyn, 1991. High frequency production of haploid embryos in asparagus anther culture. Plant Cell Rep. 10: 574–578.
Feng, X.R. & D.J. Wolyn, 1993. Development of haploid asparagus embryos from liquid cultures of anther-derived calli is enhanced by ancymidol. Plant Cell Rep. 12: 281–285.
Griesbach, R.J., 1981. Colchicine-induced polyploidy in *Phalaenopsis* orchids. Plant Cell Tiss. Organ Cult. 1: 103–107.
Gupta, N. & P.S. Carlson, 1972. Preferential growth of haploid plant cells *in vitro*. Nature 239: 86.
Heinz, D.J. & G.W.P. Mee, 1970. Colchicine induced polyploids from cell suspension cultures of sugar cane. Crop Sci. 10: 696–699.
Hondelmann, W. & B. Wilberg, 1973. Breeding all male varieties of asparagus by utilization of anther- and tissue-culture. Z. Pflanzenzüchtg. 69: 19–24.
Hsu, J.Y., C.C. Yeh, T.P. Yang, W.C. Lin & H.S. Tsay, 1989. Initiation of cell suspension cultures and plant regeneration from protoplast of asparagus. Acta Hort. 271: 135–143.
Hsu, J.Y., P.C. Lai, C.C. Yeh & H.S. Tsay, 1991. Breeding for all-male plants through anther culture in asparagus II. Chromosome doubling techniques of anther-derived haploid plants. J. Agric. Res. China 40: 102–114.
Huang, C.H., 1979. Asparagus research in Taiwan. In: G. Reuther (Ed.), Proc. 5th Asparagus Symposium, pp. 240–254. Eucarpia Section Vegetables, Geisinheim Forschungsanstalt, Geisinheim.
Huang, C.H. 1985. Current asparagus research in Taiwan. In: E.C. Lougheed & H. Tiessen (Eds.), Proc. 6th Intl. Asparagus Symp., pp. 345–354. Univ. of Guelph, Guelph.
Ito, P.J. & T.M. Currence, 1965. Inbreeding and heterosis in asparagus. Proc. Amer. Soc. Hort. Sci. 86: 338–346.
Khunachak, A., C.K. Chin, T. Le & T. Gianfagna, 1987. Promotion of asparagus shoot and root growth by growth retardants. Plant Cell Tiss. Organ Cult. 11: 97–110.
Lai, P.C., H.S. Tsay & N.C. Chi, 1980. The relationship between anther browning and callus formation in rice anther culture. J. Agric. Res. China 29: 137–146.
Lai, P.C., J.Y. Hsu, C.C. Yeh & H.S. Tsay, 1991. Breeding for all-male plants through anther culture in asparagus I. The anther culture efficiency among different varieties and plants. J. Agric. Res. China 40: 94–101.
Matthews, P.S. & I.K. Vasil, 1976. The dynamics of cell proliferation in haploid and diploid tissues of *Nicotiana tabacum*. Z. Pflanzenphysiol. 77: 222–236.
Melle Dore, C. 1973. Androgènes *in vitro* par culture d'anthères d'*Asparagus officinalis* I. État actuel des recherches. In: L. Thevenin (Ed.), 4ème Réunion sur la Section de l'Asperge, pp. 173–182. Eucarpia Sect. Horticole, Versailles.

Mii, M., 1976. Relationships between anther browning and plantlet formation in anther culture of *Nicotiana tabacum* L. Z. Pflanzenphysiol. 80: 206–214.

Murashige, T. & F. Skoog, 1962. A revised medium for rapid growth and bioassays with tobacco tissue cultures. Physiol. Plant. 15: 473–497.

Murashige, T., M.N. Shabed, P.M. Hasegawa, F.H. Takatori & J.B. Jones, 1972. Propagation of asparagus through shoot apex culture. I. Nutrient medium for formation of plantlets. J. Amer. Soc. Hort. Sci. 97: 158–161.

Pelletier, G. & M. Ilami, 1972. Les facteurs de l'androgenèse *in vitro* chez *Nicotiana tabacum*. Z. Pfanzenphysiol. 68: 97–114.

Pelletier, G., C. Raquin & G. Simon, 1972. La culture *in vitro* d'anthères d'asperge (*Asparagus officinalis*). C.R. Acad. Sci. 274D: 848–851.

Perry, J.L. & P.M. Lyrene, 1984. *In vitro* induction of tetraploidy in *Vaccinium darrowi*, *V. elliottii*, and *V. darrowi* × *V. elliottii* with colchicine treatment. J. Am. Soc. Hort. Sci. 109: 4–6.

Peters, J.A. & E. Augustin, 1989. Anther culture in asparagus. In: 7th Intl. Asparagus Symp. Abstracts, June 19–23, 1989, Ferrara, p. 27

Price, H.C. & R.A. Baughan, 1990. A six year summary of yields with New Jersey hybrids in Michigan. Acta Hort. 271: 159–162.

Qiao, Y.M. & A. Falavigna, 1990. An improved *in vitro* anther culture method for obtaining double-haploid clones of asparagus. Acta Hort. 271: 145–147.

Rameau, C., 1990. Fifteen years of experiments on asparagus F_1 hybrids: synthesis, evaluation of homozygous parents and application to the French breeding programme. Acta Hort. 271: 47–54.

Raquin, C., 1973. Étude de l'androgenèse *in vitro* chez *Petunia hybrida* et *Asparagus officinalis*. Soc. Bot. Fr. Mem. 1973: 269–274.

Reuther, G., 1984. Asparagus. In: W.R. Sharp, D.A. Evans, P.V. Ammirato & Y. Yamada (Eds.), Handbook of Plant Cell Culture. Vol. 2, Crop Species, pp. 211–242. Macmillan Publishing Co., New York.

Sink, K.C., Jr. & V. Padmanabhan, 1977. Anther and pollen culture to produce haploids: progress and application for the plant breeder. HortScience 12: 143–147.

Sunderland, N., 1974. Anther culture as a means of haploid induction. In: K.J. Kasha (Ed.), Haploids in Higher Plants: Advances and Potential, pp. 91–122. Guelph Univ. Press, Guelph.

Thevenin, L. & C. Doré, 1976. L'amélioration de l'asperge (*Asparagus officinalis* L.) et son atout majeur, la culture *in vitro*. Ann. Amélior. Plantes 26: 633–674.

Tomes, D.T. & G.B. Collins, 1976. Factors affecting haploid plant production from *in vitro* anther cultures of *Nicotiana* species. Crop Sci. 16: 837–840.

Tsay, H.S., 1981. The relationship between anther browning and callus formation in tobacco, rice and asparagus anther culture. J. Agric. Res. China 30: 121–131.

Tsay, H.S. & J.Y. Hsu, 1985. The breeding of all male asparagus plants by means of anther culture. In: E.C. Lougheed & H. Tiessen (Eds.), Proc. 6th Intl. Asparagus Symp., pp. 22–30. Univ. of Guelph, Guelph.

Tsay, H.S., P.C. Lai & N.C. Chi, 1980. Studies on anther culture and haploid plant regeneration of asparagus. J. Agric. Res. China 29: 309–319.

Tsay, H.S., P.C. Lai & N.C. Chi, 1982a. The development of haploid plants of *Asparagus officinalis* L. Proc. 4th Intl. Symp. SABRAO, Kuala Lumpur, pp. 313–325.

Tsay, H.S., P.C. Lai & L.J. Chen, 1982b. Breeding for all male plants through tissue culture in asparagus. In: S.C. Hsieh & D.J. Liu (Eds.), Proc. Symp. Plant Breeding, pp. 210–218, Agric. Assoc. Chian & Regional Society of SABRAO (ROC).

Tu, C.C., 1985. Major diseases of asparagus and their control in Taiwan. In: E.C. Lougheed & H. Tiessen (Eds.), Proc. 6th Intl. Asparagus Symp., pp. 168–180. Univ. of Guelph, Guelph.

Wricke, G., 1979. Breeding research in *Asparagus officinalis* introductory remarks. In: G. Reuther (Ed.), 5th Asparagus Symposium, pp. 1–7. Eucarpia Sect. Vegetables, Geisinheim Forschungsanstalt, Geisinheim.

Yakuwa, T., T. Harada, N. Inagaki & Y. Shiga, 1972. Studies on the anther culture of horticultural crops. I. Callus and organ formation in anther culture of asparagus. J. Jpn. Soc. Hort. Sci. 41: 272–280.

Yan, R.L., L. Zhang, C.J. Zhang and G.R. Liu, 1991. A study on the *in vitro* anther culture of asparagus and haploid plant regeneration. Act Agric. Boreali-Sinica 7: 75–82.

Ziauddin, A., X.R. Feng & D.J. Wolyn, 1993. Advances in asparagus anther culture. In: 8th Intl. Asparagus Symp. Abstract, Nov. 21–26, 1993, Palmerston North, p. 33.

8. Haploidy in *Nicotiana* spp.

ALFREDO CERSOSIMO and CIRO SORRENTINO

Contents

1. Introduction

The genus *Nicotiana* belongs to the tribe of *Nicotianeae*, of the family Solanaceae. It was classified by Goodspeed (1954) into three subgenera (*Rustica*, *Tabacum* and *Petunioides*), 14 sections and 60 species. Forty-five are indigenous to North and South America and 15, the *Suaveolentes* section of the subgenus *Petunioides*, to Australia. Five new species (*N. umbratica*, *N. cavicola*, *N. amplexicaulis*, *N. hesperis* and *N. simulans*) were added to the Australian group by Burbidge (1960) who also changed the name of *N. stenocarpa* to *N. rosulata*. In the same year, Wells (1960), after finding a continuous intergradation of *N. palmeri* with *N. trigonophylla*, removed the validity of species rank to *N. palmeri*. A complete list of the 64 species was reported by Smith (1979). Subsequently, two other species with uncertain taxonomic position were described. The first, *N. africana* (Merxmüller & Buttler, 1975), was found on the mountains in the middle of Namib (Southwest Africa) and although some characteristics were reminiscent of the subgenera *Rustica* and *Tabacum*, it was placed in the subgenus *Petunioides*. The authors state that: "*N. africana* can be considered as an endemic relict of considerable age. It may form a parallel to the *Suaveolentes* section (endemic to Australia and the South Pacific) with which it agrees in being

S.M. Jain, S.K. Sopory & R.E. Veilleux (eds.), In Vitro *Haploid Production in Higher Plants, Vol. 5*, 135–163.

diploid ($2n = 2x = 46$)". So far, *N. africana* is the only *Nicotiana* known from Africa. The second new species, *N. kawakamii* (Ohaski, 1976) was found in the Andes of South America and although some cytological features are unique to this species and it is day neutral, it seems to fit in the *Tomentosae* section of the subgenus *Tabacum* (Smith, 1979).

Except for *N. rustica* and *N. tabacum*, both with $2n = 4x = 48$ chromosomes, all species of the subgenera *Rustica* and *Tabacum* have $2n = 2x = 24$ chromosomes. The subgenus *Petunioides* presents a wide range of chromosomal variation including species with $2n = 18$, 20, 24, 32, 36, 38, 40, 42, 44, 46, and 48 chromosomes.

Although it has worldwide distribution, *Nicotiana* can be described as a New World genus because 75% naturally occurs in the Americas and 25% in Australia and the South Pacific. Of the 66 species of *Nicotiana*, 36 (55%) occur in South America (Argentina: 18 spp., Peru: 12 spp., Chile and Bolivia: 11 spp., Brazil: 6 spp. Paraguay: 4 spp. Uruguay: 3 spp., Ecuador: 2 spp.), 9 (14%) in North America, 20 (30%) in Australia and the South Pacific and one in Africa (Goodspeed, 1954; Burbidge, 1960; Wells, 1960; Merxmüller & Buttler, 1975; Ohaski, 1976).

Most *Nicotiana* species are not cultivated. *N. sylvestris* and *N. alata* are grown as ornamental plants. *N. tabacum* and *N. rustica* are cultivated for industrial purpose and are called "tobacco". Nevertheless, we will use this term to indicate exclusively *N. tabacum*. *N. rustica* is grown more or less extensively in Russia (C.S.I.) and in some of the Asiatic and North African countries, such as India, Pakistan, Libya, Algeria, Tunisia and Morocco, where it is still of local importance. *N. tabacum* is grown worldwide, from 69° N to 45° S latitude, and from sea level to 4,200 meters altitude (South America), throughout a wide range of climatic conditions (Goodspeed, 1954).

The world production of tobacco has been estimated at 7–8 million metric tons annually, and China is the largest producer and consumer of tobacco in the world (35–38%). The major tobacco producing countries are listed in Table 1. The trend in world tobacco production has been increasing in the lesser developed countries and decreasing in the developed ones. Approximately 60% of worldwide tobacco production is flue-cured tobacco (Virginia Bright), 80% of which is cultivated in the four major tobacco-producing countries: China (60%), USA (10%), Brazil (7%) and India (3%).

Although *N. tabacum* is the most widely grown species of *Nicotiana*, it does not exist in the wild state because it is an interspecific hybrid (amphidiploid) between *N. sylvestris* and *N. tomentosiformis*, both with $2n = 2x = 24$ chromosomes (Gerstel, 1961; Sheen, 1972; Gray *et al.*, 1974). The origin of *N. rustica* is also amphidiploid, probably between *N. undulata* and *N. paniculata* (Akehurst, 1981) both with $2n = 2x = 24$ chromosomes. *N. tabacum* is a polymorphic species in which it is possible to distinguish many botanical varieties such as: *macrophylla*, *havaniensis*, *brasiliensis*, *virginica*, *fruticosa*, *lancifolia*, *attenuata*, *albiflora*, *rubriflora*, and others (Goodspeed, 1954). *N.*

Table 1. The major tobacco producing countries (×1000 metric tons)

Country	1970	1980	1990	1993
China	806	920	2,646	3,050
USA	865	810	738	733
Brazil	244	405	445	663
India	337	439	552	581
Turkey	150	228	296	334
Zimbabwe	55	125	130	205
Greece	95	117	136	168
Italy	78	126	215	145
World total	4,645	5,250	7,048	7,898

(F.A.O. – S.O.F.A. 94).

rustica is also a polymorphic species, although less so than *N. tabacum*, in which it is possible to distinguish some varieties including: *pavonii*, *pumila*, and *brasilia* (Goodspeed, 1954).

Most *Nicotiana* species are day neutral and flower in any season, but others will only flower under specific conditions of day-length and temperature (Steinberg, 1957, 1959; Burk, 1967). Chaplin & Burk (1979) reported the requirements of day-length and temperature for floral induction and seed production of *Nicotiana* species. Although *N. tabacum* is considered a day-neutral species, some accessions, such as the Mammoth type, will only flower under short day conditions (Chaplin & Burk, 1979).

One peculiarity of *Nicotiana* species is the presence of alkaloids. Many species contain primarily one of three identified alkaloids: nicotine (formed in the roots and translocated apically through the xylem), nornicotine (a demethylation product of nicotine formed in the shoots) and anabasine (formed in both roots and shoots) (Mothes, 1955; Tso, 1972).

N. tabacum is generally self-pollinated but can also be easily cross-pollinated. Its fruits are capsules that may contain from 2,500 to 5,000 seeds each (in species of the subgenus *Tabacum* and in *N. sylvestris*), from 900 to 1,300 seeds (in the woody species of section *Paniculatae* and in *N. longiflora* and *N. plumbaginifolia*) or from 100 to 700 seeds (in the remaining species of subgenus *Petunioides*) (Goodspeed, 1954).

The genus *Nicotiana* can be considered a useful research tool. It has been widely used in genetic research because of numerous advantages, such as the simplicity of controlled pollination, the abundant production of seeds and progeny, the great variation and evolutionary stages of its numerous species, the rich reservoir of genetic resources available, the numerous distinctive hereditary characters which provide useful markers for scientific investigation, the aptitude to regenerate whole plants from single cells (microspores, protoplasts, etc.) and callus tissues, the possibility to produce haploid plants and dihaploid lines, and many others.

A major objective of plant breeders has been the production of near-homozygous lines. Pure line selection can approach homozygosity but some heterozygosity is always retained. With traditional methods of selection, seven or more generations are required to develop pure lines. This chapter deals with different techniques for obtaining haploid plants of *Nicotiana* and factors that influence these techniques. Furthermore, problems regarding the doubling of chromosomes (diploidization), breeding of dihaploids and their field performance are discussed.

2. Methods for haploid production

Haploid plants are the quickest and most reliable means of obtaining completely homozygous lines after chromosome doubling. Methods for haploid production can be divided into three groups: 1) by seed, through interspecific hybridization, 2) by gametic embryogenesis, through *in vitro* culture of anthers, uninucleate microspores (pollen grains), or unfertilized ovules, and 3) by adventitious bud formation from *in vitro* pollen-derived callus, leaf protoplast-derived callus from haploid plants, etc.

2.1. *Interspecific hybridization*

The occurrence of haploid plants in nature is infrequent in *Nicotiana*. However, they can be induced by a variety of treatments including interspecific hybridization (Kostoff, 1929, 1943; Clausen & Lammerts, 1929; Kehr, 1951), chromosome alteration resulting from high-frequency radiation (Goodspeed, 1954), and colchicine treatment (Smith, 1943). Haploidy in F_1 interspecific hybrid progeny may be induced by germination of foreign incompatible pollen on the stigma to stimulate parthenogenetic development of seeds. Haploid plants by interspecific hybridization may have either paternal or maternal characters.

2.1.1. *Haploids showing paternal characters*

Clausen and Lammerts (1929) first produced natural haploid plants showing paternal characters by pollinating flowers of *N. digluta*, a synthetic amphidiploid ($2n = 3x = 36$), with pollen of a white-flowered mutant of *N. tabacum* ($2n = 4x = 48$). Among the progeny of the F_1 hybrid, they found a haploid plant showing characters of *N. tabacum*. Similar results were also obtained by Kostoff (1929, 1943) who pollinated flowers of *N. tabacum* var. *macrophylla* ($2n = 6x = 72$) with pollen of *N. langsdorffii* ($2n = 2x = 18$) and by Kehr (1951) who made a cross between *N. glutinosa* ($2n = 2x = 24$) and *N. repanda* ($2n = 4x = 48$). The few seedlings produced were thought to originate from single male gametes through a parthenogenetic process called "androgenesis". Haploids obtained in this way were likely due to genetic incompatibility of chromosome synapsis between species and to the dominance of the pollen nucleus over the egg nucleus which regressed (Cersosimo, 1996). Such haploid plants would retain the effects of extrachromosomal inheritance of the egg cell (cytoplasmic inheritance).

2.1.2. *Haploids showing maternal characters*
The second method of obtaining haploid plants *in vivo*, with transmission of maternal characters only, was described by Burk *et al.* (1979), using interspecific hybridization with *N. africana*. They found that when *N. tabacum* was pollinated with *N. africana*, most of the resulting seeds germinated but the seedlings died before they developed true leaves. However, 0.25 to 1.42% of seedlings grew vigorously and were readily identified as either F_1 hybrids or maternal haploids. Hybrid plants were distinguished from haploids by morphological differences in leaves, trichomes and glands. Furthermore, haploids flowered earlier than the hybrids, had an abundance of flowers of reduced size, aborted pollen, and anthers subordinate to the stigma. *N. africana* pollen was applied to stigmas of monosomics of *N. tabacum* (Clausen & Cameron, 1944), the disomic line Red Russia, used to maintain the monosomics, a burley hybrid VT-9 and an inbred strain of cv. NC95. Monosomics of *N. tabacum* yielded fewer seeds than disomics. An exception was the cross haplo "H" × *N. africana* which yielded many seeds that germinated well but resulted in viable hybrid plants. The disomics Red Russian, VT-9 and NC 95, pollinated with *N. africana*, yielded maternal haploids in the percentages of 15, 4 and 16, respectively, of the seeds that germinated normally. These percentages were comparable to those obtained from reciprocal intraspecific crosses of *N. tabacum* in which one of the parents was marked with a recessive seedling trait. The authors suggested that the frequencies obtained in their experiment probably reflected a normal rate of spontaneous gynogenic haploidy.

The low number and frequency of naturally occurring haploids makes it difficult to use them effectively in breeding programmes. This led researchers to look for alternative methods for producing haploid plants. The use of seed-derived haploid plants has been confined largely to genetic investigations.

2.2. *Gametic embryogenesis*

A new method for producing haploid plants efficiently was based on the possibility that immature gametic cells, not yet differentiated as gametes (uninucleate microspores), when subjected to suitable inductive stimuli, can change their natural gametic development and acquire an embryogenic capacity that should be named "ephebogenesis[1]". This can be obtained either through *in vitro* culture of anthers, isolated microspores (pollen grains) or unfertilized ovules.

[1] Term of Greek origin for which the botanical meaning refers to the parthenogenetic development of a plant from a male gamete. The process is similar to androgenesis, and this term is currently used to indicate both phenomena. There is, however, a difference: whereas in ephebogenesis only the male gametic cell becomes an embryo, in androgenesis it is the egg cell that, under the stimulus of the male gametic nucleus during division (its own regressed), starts a segmentation process and originates a haploid embryo. In this case, the embryo obviously acquires the whole cytoplasmatic inheritance of the egg cell.

2.2.1. *Anther and isolated microspore culture*

The pioneering work of Guha & Maheshwari (1964, 1966) on *Datura innoxia* and of Bourgin & Nitsch (1967) on *N. tabacum*, demonstrated that haploid plants could be obtained efficiently in a short time through *in vitro* culture of anthers. Kameya & Hinata (1970), using a similar technique, observed cell cluster formation by mature pollen grains of *Brassica*. However, they did not clarify whether the observed clusters were derived from a single grain or were only agglomerates of pollen grains. Subsequently, Sharp *et al.* (1972), with *Lycopersicon esculentum*, and Pelletier (1973), with *N. tabacum*, obtained colony formation from isolated microspores with a nurse-culture technique using *Petunia* callus tissue. However, whole plants could only be obtained in this way from *Nicotiana*.

These events started a new era of biological research. During the last decades, further development in this area was made possible by extensive work on *Nicotiana* species and a few other solanaceous plants, used for numerous studies of androgenesis with both anther and isolated microspore cultures. These species were used as model plants for the optimization of androgenic techniques and their application to recalcitrant but agronomically important species (Maheshwari *et al.*, 1980, 1982).

The *in vitro* culture of anthers or isolated microspores of tobacco has generally been accomplished according to the following techniques: closed floral buds containing immature pollen grains are excised from the mother plant, sterilized, cold pretreated and aseptically cultured on a suitable agar-solidified medium (Plate 1(4)) and under suitable incubation conditions. For isolated microspore cultures, pollen grains are extracted by crushing cultured anthers after induction has occurred (2–3 days of culture) and cultured in a liquid medium at a suitable density (10^4–10^5 grains per ml) (Nitsch, 1977). Embryos and plantlets of *N. tabacum* cv. Samsun and *N. rustica* cv. Rustica were obtained by Imamura *et al.* (1982) through direct pollen culture, without prior anther culture. Isolated microspores were cultured in a medium without sucrose, then transferred into Nitsch's H medium containing 2% sucrose and 5 mM glutamine. The optimal medium for the initial culture was simply water (sugar starvation), whereas six days was the optimal period of culture. A method giving a high rate of induction of cell division and embryogenesis was developed by Kyo & Harada (1985). Binucleate microspores were fractionated by 35/45% Percoll density gradient centrifugation and cultured in 0.4 M mannitol at 30°C (first culture). After three days, microspores were recollected by a second Percoll fractionation (0/30%) and cultured in MS (Murashige & Skoog, 1962) macro-elements medium, 0.4 M mannitol, 40 mM galactose, 3 mM glutamine, and 5 μM ABA (abscisic acid) for ten days (second culture). The dividing pollen (approximately 80%) was transferred to MS medium containing 0.4 M mannitol, 3 mM glutamine and no growth regulators (third culture). In this way, about 40% of dividing pollen developed into embryos or embryogenic calli.

Androgenic plants, originated via embryogenesis from microspores or pollen grains, have generally been euploid. In *N. tabacum*, most of the androgenic plants have been haploid. Normally, haploid embryos originate from the vegetative cells of binucleate microspores after repeated divisions (McComb & McComb, 1977; Sunderland & Dunwell, 1977). Microspores may also undergo an equal division without any difference in size or chromatin structure between the two derivative nuclei (Sunderland & Dunwell, 1977). In this case, either cell or both may contribute to the embryo. When differentiation occurs, the generative cell generally does not contribute to the embryo (D'Amato, 1985). An exception is *H. niger* where the generative cell may undergo repeated divisions and originate a haploid embryo (Raghavan, 1976).

2.2.2. *Unfertilized ovule culture*

Haploid plants have also been obtained by *in vitro* culture of unfertilized ovules (Ran, 1980; Zhu *et al.*, 1980; Wu & Cheng, 1982) but with considerably more difficulty than by androgenesis. The main problem has been the abundant proliferation of undifferentiated callus tissue, both of somatic and gametic origin, which makes their separation extremely difficult. The somatic callus, moreover, has an inhibiting effect on the gametic one (Maheshwari, 1961; Hsu & Stewart, 1976; Mullins & Srinivasan, 1976). The production of haploid embryos has been accompanied by somatic embryos (diploids) originating from the ovary wall or from the nucellar tissue (Kuo, 1982). In this type of culture, it has also been important to regulate the ratio of growth regulators (auxins/cytokinins) to be more favourable to the production of callus from the embryo sac. Excised ovaries have been cultured on an agar-solidified medium (Nitsch & Nitsch, 1969) enriched with 1 mg l^{-1} IAA (3-indoleacetic acid), 0.2 mg l^{-1} BA (benzyladenine), 4% sucrose and 0.8% agar. After a ten-day incubation, ovules were aseptically extracted from the ovaries and cultured on the same fresh medium, without BA and with half the sucrose concentration (2%). After about 20 days, the ovules, by now enlarged, were sectioned along the longitudinal axis and transferred to MS medium containing 1 mg l^{-1} IAA and 0.4 mg l^{-1} kinetin (6-furfurylaminopurine) to facilitate callus production. To regenerate shoots (organogenesis), callus was cultured on agar-solidified medium (Nitsch & Nitsch, 1969) amended with 0.2 mg l^{-1} BA. Regenerated shoots were transferred to rooting medium supplemented with 0.2 mg l^{-1} IAA.

Once ovules were separated from ovaries and sectioned, after about 20 days in culture, they produced ovular and stigmatoid excrescences, both originated from the somatic tissue. Stigmatoid excrescences came from the primine (outer integument) whereas ovular excrescences came from the secondine (inner integument). Besides these formations, ovule culture may give rise to callus tissue of gametic origin from which it is possible to re-

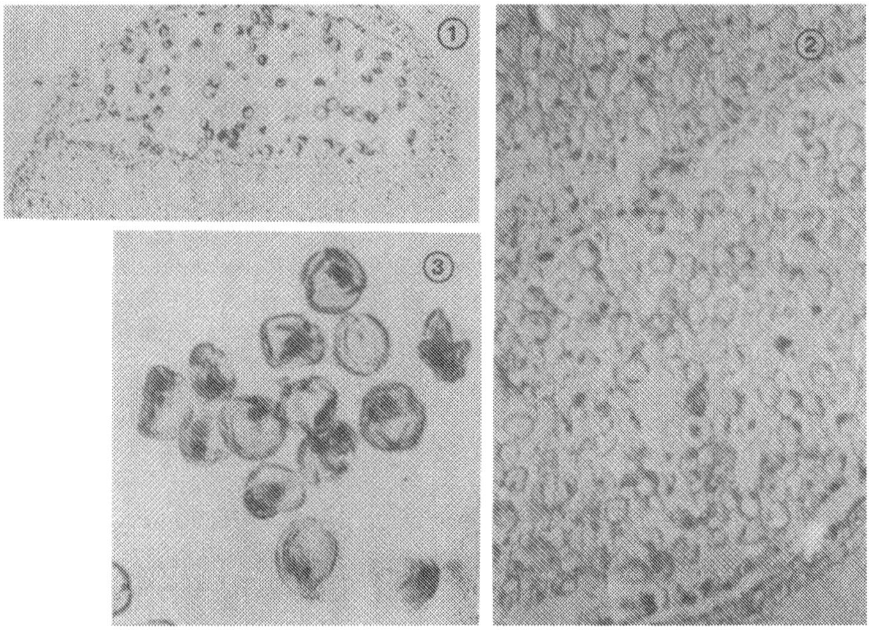

Plate 1. 1) Anther cross section (180 ×) from a plant grown *in vitro* (there are few microspores inside the locule but they are more sensitive to embryogenic induction). 2) Anther cross section (180×) from a plant grown in a greenhouse (there are many microspores inside the loculus but they are not very sensitive to embryogenic induction). 3) Same as 2) (288×). 4) Petri dish with cultured anthers on solid medium. 5) Embryogenesis and organogenesis emerging from anthers after 4–5 weeks of culture. 6) Haploid plantlets emerging from anthers after 4–5 weeks of culture.

generate whole plants through organogenesis. Regenerated plants often show various levels of ploidy, including triploids and mixoploids, which are probably due to incomplete maturation of synergids and antipodal cells (Sorrentino *et al.*, 1987). Zhu *et al.* (1983) reported haploid plantlet regeneration from unpollinated ovaries of *N. tabacum* derived directly from ovules (Plate 2). These plants were shorter and weaker than diploids and produced small flowers and sterile seeds. Plantlets derived from embryo sacs differed from those derived from pollen grains in morphology and electrophoretic banding pattern. Haploid plantlets were also obtained from immature embryo sacs by culturing unpollinated ovaries of haploid plants derived from ovary culture. In a study on the ploidy level and characteristics of plantlets regenerated from *in vitro* cultures of unpollinated tobacco ovaries, Wu & Zhu (1988) found that regenerated plants of the sterile and maintainer lines of "NC2326" and "Copus Yeusuheku 4" (CopusY4) were regenerated directly from embryos while those of "Daye-Huang" (DayeH) were regenerated from ovary-derived calli. The percentages of haploid, diploid and tetraploid regenerants

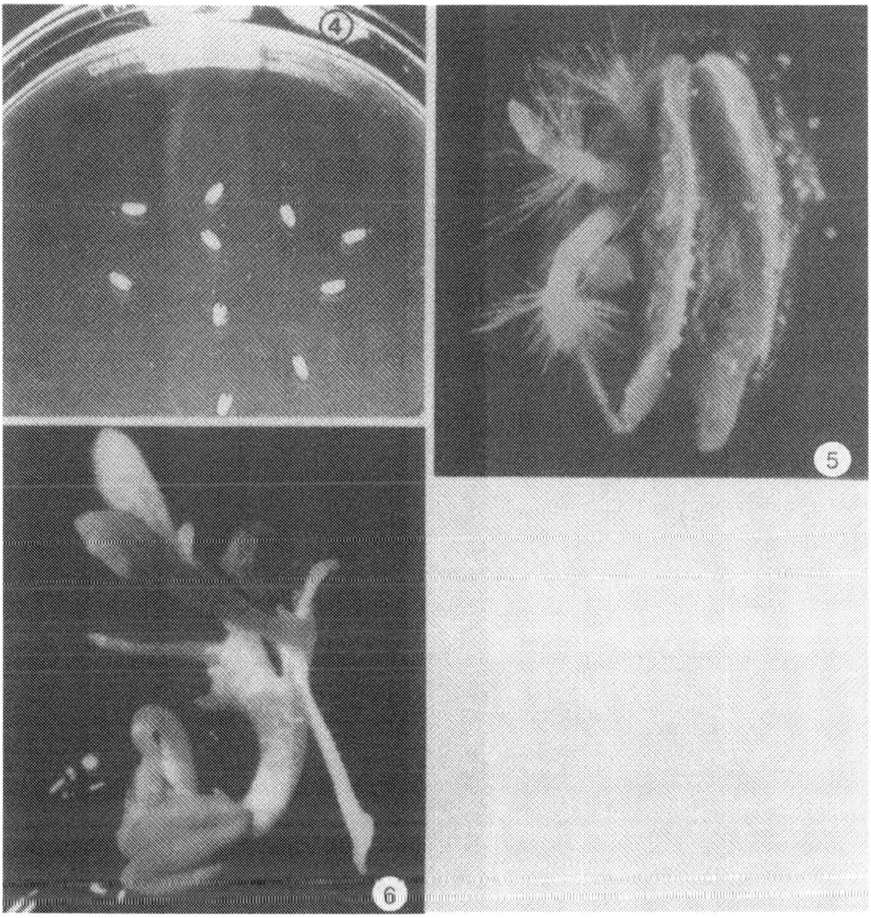

Plate 1. Continued.

were 77, 23 and 0% in CopusY4, 49, 50 and 0.6% in DayeH, and 28, 68 and 4% in haploid CopusY4. The regenerated plants of NC 2326 were mostly haploid.

2.3. *Pollen-derived callus culture*

Androgenic plants can originate by adventitious bud formation on pollen-derived callus (Durr & Fleck, 1980). In this case, aneuploids and mixoploids have occurred among the regenerated plants. This was explained by variation for chromosome number during the induction and development of callus and because of the multicellular origin of the adventitious buds that may originate

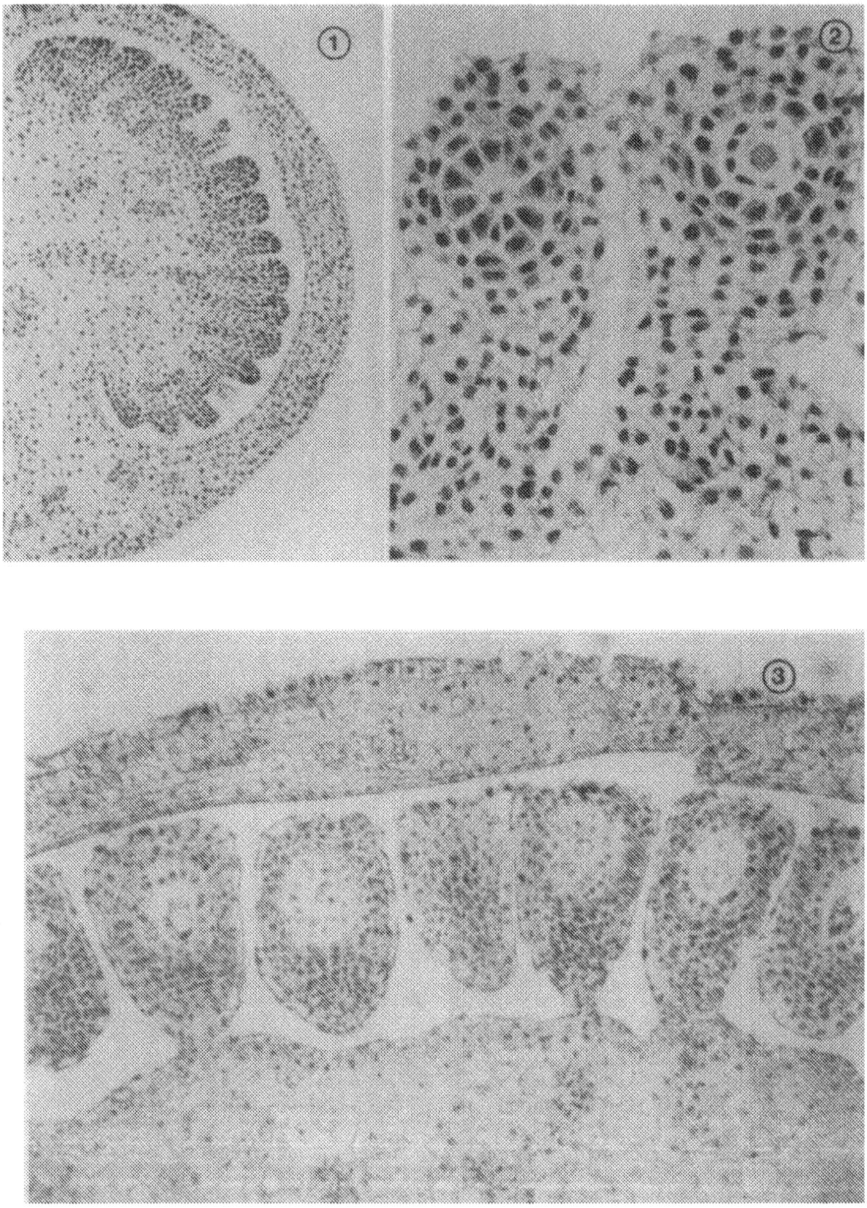

Plate 2. 1) Ovary cross section (117×) from flowers grown *in vitro*. 2) Longitudinal section of ovules (312×) from flowers grown *in vitro*, in an early phase of embryogenic development. 3) Longitudinal section of ovules (195×) as in 2), but in a subsequent stage of embryogenic development.

from a genetically heterogeneous cell population (D'Amato, 1985). In *N. tabacum* there is also a tendency to form callus by anthers cultured after pollen mitosis. Formation of callus from microspore or pollen culture has been observed in cultured anthers of a number of species, including *Solanum melongena* (Raina & Iyer, 1973), *Solanum tuberosum* (Sopory & Tan, 1979), *Lycopersicon esculentum* (Zagorska *et al.*, 1982), and others. In these *in vitro* cultures, anther somatic tissue may grow faster than and eventually submerge callus from haploid microspores. The long-term presence of anthers in culture medium containing growth regulators promotes the formation of haploid callus but inhibits the development of newly-formed embryos. Plants regenerated from this type of callus always display numerous chromosome alterations and various abnormal chromosome numbers (Nitsch, 1977).

2.4. *Culture of protoplasts from haploid plants*

Once haploid plants have been obtained, haploid somatic cells (protoplasts) can be produced by enzymatic extraction (0.5–1.0% cellulase R-10 Onozuka; 0.1–0.5% macerozyme, in 0.5 M mannitol). Plants regenerated from protoplasts are useful for cell genetics, haploid breeding, somatic hybridization, and induction of mutations by chemical, physical and biological agents. However, protoplast-derived callus has shown genetic instability (Bajaj *et al.*, 1978; Barfield *et al.*, 1985) that often results in mixoploidy. In cultures of haploid and diploid protoplasts of *N. tabacum* and *A. belladonna*, Bajaj *et al.* (1978) observed that protoplast-derived callus of *A. belladonna* exhibited three models of morphogenesis: 1) embryogenesis, 2) shoot-bud differentiation and 3) rhizogenesis. The regenerated plants ranged from haploid to diploid, triploid and hexaploid, along with various aneuploids. In *Nicotiana*, the regenerated plants were predominantly haploid (91%) with occasional formation of diploids.

In culture of protoplasts isolated from a predominantly haploid cell suspension culture of *N. plumbaginifolia*, in which the ploidy had remained stable for nearly four years in culture, Barfield *et al.* (1985) regenerated several plants. Of 16 regenerated plants examined in detail, three were haploid and many others were sterile diploids exhibiting various degrees of abnormality. Eight regenerated fertile diploids were selfed and, among the progeny, most plants were uniform within a group, suggesting that the regenerated diploid plants were doubled haploids (DHs). Genetic instability and diversity in protoplast-regenerated plants have been reported in many species, including *Nicotiana* (Melchers, 1974), *Datura* (Schieder, 1975) and *Petunia* (Binding, 1975). According to Bajaj *et al.* (1978), there could be three reasons for the regeneration of diploid and polyploid plants from haploid protoplasts: 1) some of the protoplasts begin as diploids due to endopolyploidization in somatic cells of the haploid plant, 2) spontaneous fusion may occur during protoplast isolation, and 3) repeated endomitosis and nuclear fusion may occur in callus cultures. Although the phenomenon of genetic instability in plants regenerated from haploid protoplasts is undesirable for maintenance of the haploid condition, it can be usefully exploited to amplify genetic variability.

3. Factors influencing haploid production

3.1. *Genetic factors and developmental stages of microspores*

The most important parameter that determines success of anther, ovule or microspore culture is the genetic element, i.e., the intrinsic capacity of a genotype to undergo embryogenic (androgenic or gynogenic) induction. In *Nicotiana*, there is a wide range of variability from highly embryogenic species (*N. tabacum*, *N. sylvestris*) to poorly embryogenic species (*N. rustica*, *N. alata*, *N. glutinosa*) to nearly recalcitrant species (*N. acuminata*, *N. bonariensis*, *N. glauca*, *N. langsdorffii*, *N. longiflora*, *N. repanda*, *N. suaveolens*) (Nitsch, 1969). In anther culture even of highly embryogenic species, however, the percentage of microspores undergoing induction has always been low (2.0–2.5%). Of about 40,000 microspores within an anther, typically only 80–100 will be induced. Considering the genetic differences among microspores and that only a particular genotype could produce embryos, Devreux *et al.* (1975) cultivated anthers of a homozygous line of *N. tabacum* obtained by chromosome doubling of a haploid. Under such conditions, it could be hypothesized that genetically similar microspores would be uniformly embryogenic. On the contrary, results corroborated the same low production of embryos. The limited inductive response of microspores was attributed to the effect of inhibiting agents released during senescence and degeneration of anther somatic tissues.

Another important aspect of androgenesis is that the anther donor plants should be grown under controlled conditions, with proper fertilization and free of insects and pathogens. In this respect, anthers formed *in vivo* or *in vitro* show different expressions of androgenetic capacity (Bridgen & Veilleux, 1988). Trinh & Tran Thanh Van (1985), after obtaining *in vitro* flower formation on thin cell layers from floral branches of haploid and diploid plants of *N. tabacum* and *N. plumbaginifolia*, found that anthers from *in vitro* flowers of diploid plants gave a higher percentage of haploid plants than anthers from *in vivo* flowers. Anthers formed *in vitro* on haploid thin cell layers, however, gave rise to plants whereas anthers formed *in vivo* on haploid plants did not. Compared to *in vivo* anthers, *in vitro* anthers have fewer microspores that are more sensitive to embryogenic induction (Sorrentino *et al.*, 1987) (Plate 1(1–3)). Among plants regenerated from *in vitro* anthers of haploid plants, Trinh & Tran Thanh Van (1985) found hypohaploids and many morphological and biochemical abnormalities. In *N. tabacum*, some hypohaploids were albino or had variegated leaves with laciniate petals and petaloid or stigmatoid anthers. They also found that peroxidase isoenzyme contents and activity were lower in hypohaploids than in haploids or diploids. Similar observations were made in *N. plumbaginifolia* in which the leaves of hypohaploids had a single vein and the ovules were exposed due to reduction in the size of the carpels.

Most researchers agree that the most suitable microspore developmental stage for embryogenic induction is the uninucleate or early binucleate stage (i.e., just after the first haploid mitosis). In *N. tabacum* this stage coincides with a specific developmental stage of the flower bud in which corolla and sepals are the same length (stage 2, according to Nitsch, 1969).

3.2. *Role of the anther wall*

Embryogenic induction in microspores occurs more efficiently inside the anthers than in isolated pollen culture thus implicating a role of the anther wall in the embryogenic process. Nitsch & Norreel (1973) observed that the aqueous extract of cultured anthers had a positive effect on embryo induction in isolated *D. innoxia* microspore cultures, due to the presence of a number of amino acids in the extract. The addition of glutamine and serine to a synthetic culture medium (Nitsch, 1974a) induced a higher incidence of embryogenic microspores. In a study on amino-acid levels in embryogenic anthers of *N. tabacum*, Horner & Pratt (1979) reported a four- or five-fold increase in glutamine during anther culture compared with *in vivo* anthers. No significant variation in serine levels was observed. Aruga & Nakajima (1985) compared starch granules, soluble sugar and free amino acid content of *in vitro* anthers with *in vivo* cultured anthers; they found that starch granules disappeared rapidly from cultured anther tissues. Soluble sugar decreased soon after the onset of culture, although sucrose was supplemented by the medium. After eight days of anther culture, however, the amount of soluble sugar (sucrose) increased. In the initial stage of culture, conditions of sugar starvation inside the locules favoured induction of pollen embryogenesis (Ghandimathi, 1982; Imamura *et al.*, 1982; Aruga *et al.*, 1985). Free amino acid analysis showed that on the tenth day of culture, during embryo formation, high levels of glutamine and asparagine were present in cultured anthers. Since no amino acid was supplied by the medium, those necessary for embryo formation were provided to microspores by cultured anthers.

For *N. tabacum* and *D. innoxia*, Villanueva *et al.* (1985) found that spermidine was the major polyamine present during gametogenesis whereas marked differences between the species in proteins, RNA and polyamines were evident during meiosis and at the first haploid mitosis. At the beginning of the first haploid mitosis, an unknown amine (X60) and a rapid increase in RNA and protein concentrations were observed in *N. tabacum* whereas a large increase in amine and RNA levels, followed by an arginine peak, were observed in *D. innoxia*. During embryo formation in anther culture, putrescine and spermidine were the major polyamines in both species. During this phase, an unknown amine (X81) was also observed in *N. tabacum*, with a peak during the developmental stages of embryogenesis. In *D. innoxia*, androgenic induction was marked by an arginine peak followed by an increase in the putrescine and spermidine levels associated with maximal RNA content. The authors found that these biochemical events were correlated with structural changes during pollen development.

Heberle-Bors & Reinert (1977) demonstrated some negative effects as a result of the release of inhibiting agents during cell wall senescence and degradation. From anthers cultured in liquid medium, pollen fractions extracted at regular intervals and transferred to fresh medium produced more embryos compared to either isolated microspores or anthers cultured continuously in the same medium (Sunderland & Roberts, 1977; Tyagi *et al.*, 1979). But, what happens to the cells of the anther wall during senescence and degradation?

In *N. tabacum*, the anther tapetal cells generally undergo two or more mitoses without cell division to produce polynucleate cells. Then, each nucleus synthesizes more DNA, reaching high levels of endopolyploidy (Scarascia, 1952). As degeneration of the tapetum occurs at the same time as haploid mitosis, DNA from the tapetum comes into contact with the microspores. To verify the possible presence of embryogenic granules localized within the anthers, corresponding to high concentrations of DNA, Devreux *et al.* (1975) made histological sections that showed a completely random distribution of embryos inside the locules. It seems probable that, initially, the role of the anther wall is positive as a result of the release of promoters in the first stages of its degradation. However, in subsequent stages, the presence of increasing concentrations of one or more inhibitors may prevent and nullify the action of the promoters. Oligosaccharide promoters are found in some polymers that compose the pectic and hemicellulosic fractions of the cell walls (Tran Thanh Van *et al.*, 1985). These substances (oligosaccharines) are thought to be responsible for a range of biological activities, including the promotion of microspore embryogenesis.

Growth regulators may promote the formation of undifferentiated haploid callus tissue but inhibit the development of newly formed embryos. Plants regenerated from such callus generally display numerous chromosome alterations and various ploidy levels (Nitsch, 1977). By early isolation of microspores from the anther and their subsequent culture in medium free of growth regulators, they may retain their embryogenic capacity and produce haploid plants in larger quantities. It is therefore advisable to isolate microspores as soon as induction occurs, i.e., just after the first cell division (Nitsch, 1977).

3.3. *Cold pretreatment*

Of several factors that can improve the androgenic induction of microspores, cold pretreatment has been among the most effective. Nitsch & Norreel (1973) observed that cold pretreatment at 3°C for 48 h was essential for success in *D. innoxia*, and 5°C treatment for 72 h in *N. tabacum* pollen cultures (Nitsch, 1974a,b). With the latter species, however, Sunderland & Roberts (1977) obtained best results at 7–8°C for 12 days. With *N. rustica*, Sunderland & Wildon (1979) reported better response at 15°C for 15 days, whereas Chowdhury (1984) found that 9°C for 15 days gave the best results when microspore development stage was near haploid mitosis. Chen *et al.* (1985), working with *N. alata*, *N. otophora*, *N. sylvestris* and *N. plumbaginifolia*, found that *N. alata* anthers were unresponsive to cold pretreatment whereas anthers of the other species produced embryos or pollen-derived callus after cold pretreatment.

N. tabacum microspores receiving cold treatment in the culture medium retained their viability for longer than those receiving cold treatment before culture (Duncan & Heberle, 1976). In some cases, pretreated anthers, both before and after culture, were most responsive when the premitotic microspore stage was selected. In *H. niger*, the most responsive stage was the early binucleate in pretreated anthers and the uninucleate in non-pretreated anthers (Nitsch, 1977; Sunderland & Wildon, 1979; Tyagi *et al.*, 1979).

The biological process of cold pretreatment remains unclear. In Nitsch's view (1974b), cold pretreatment increases the number of microspores that divide into two identical nuclei and maintains their viability but this was not corroborated by Duncan & Heberle (1976). Another hypothesis maintains that cold pretreatment causes the destruction of the microtubules that are known to be involved in the formation of the mitotic spindle. It is conceivable that cold pretreatment, by disrupting the determination of polarity, may alter the organization of the spindle. Since centrifugation was known to bring about the disorganization of microtubules, Sangwan-Norreel (1977) substituted low temperature with centrifugation of the floral buds of *D. innoxia* and found an increase of 6% more embryos than the control; however, the response increased by 31% when centrifugation and cold pretreatment were combined.

The enhancement of embryogenic induction by microtubule disruption is not yet understood. Bajaj (1978) suggested that cold pretreatment increases embryogenesis indirectly by prolonging microspore viability and slowing down senescence of the anthers, thus preventing premature abortion of the spores. All these changes lead to an increase in the number of microspores competent for induction.

3.4. *Culture medium*

During the initial attempts at haploid production, various tissue culture media (White, 1943; Nitsch & Nitsch, 1956; Gautheret, 1959; Murashige & Skoog, 1962; Linsmaier & Skoog, 1965; Gamborg *et al.*, 1968) supplemented with many ingredients, often of undefined composition (coconut milk, plum juice, etc.) were used (Guha & Maheshwari, 1966; Bourgin & Nitsch, 1967; Nakata & Tanaka, 1968; Niizeki & Oono, 1968). Experimental results demonstrated that medium effects were practically uninfluential, especially during the inductive stage of microspores.

Nitsch & Nitsch (1970) demonstrated that the first phases of embryogenesis can take place in a simple 2% sucrose solution. Under such conditions, however, embryo development stopped at the globular stage. The complete development of embryos has been obtained only with the addition of macro- and microelements. No embryos developed without iron (10^{-4} M Fe-EDTA, ethylene diamine tetraacetic acid). In *N. tabacum* anthers cultured on media containing 0–30 μM FeEDTA, Vagera *et al.* (1979) obtained only globular embryos, whereas on media with 40–150 μM, complete plants were regenerated. The lower limit of 40 μM to induce the development of globular embryos into plants, was the same both in light and dark. Darkening of the anthers, a general prerequisite to androgenic response, was directly proportional to FeEDTA concentration in the medium. Anthers cultured in the dark on a medium with 100 μM FeEDTA show a high androgenic capacity (Forche & Neumann, 1977).

The addition of auxins is not necessary but 100 $\mu g\ l^{-1}$ IAA (indole-3-acetic acid) has enhanced the frequency of embryo production. However, 2,4-D (2,4-dichlorophenoxyacetic acid), unlike IAA, brought about the development of plantlets with various cotyledon anomalies (Nitsch, 1969). The addition of 0.5–3.0 $mg\ l^{-1}$ GA_3 (gibberellic acid) did not increase the production of embryos but caused cell extension, leading to the formation of thin, weak, elongated plants (Nitsch, 1969). Low concentrations (1–10 μM) of ABA (abscisic acid) did not hinder the formation of embryos but slowed their development, resulting in dormancy similar to that occurring in zygotic embryos. This effect could be removed by 10^{-3} M glutathione (Nitsch, 1969).

Now, the more common basal medium usually employed in anther or isolated pollen cultures of *N. tabacum* is commonly known as NN-H (Nitsch & Nitsch, 1969) which contains: a) *Macronutrients*: (in $mg\ l^{-1}$: KNO_3 950; NH_4NO_3 750; $MgSO_4 \cdot 7H_2O$ 185; $CaCl_2$ 166; and KH_2PO_4 68); b) *Micronutrients*: (in $mg\ l^{-1}$: $MnSO_4 \cdot 4H_2O$ 25; $ZnSO_4 \cdot 7H_2O$ 10; H_3BO_3 10; $Na_2Mo_4 \cdot 2H_2O$ 0.25; and $CuSO_4 \cdot 5H_2O$ 0.025; c) *Iron*: (5 $ml\ l^{-1}$ of a solution obtained by dissolving 5.57 g $FeSO_4 \cdot 7H_2O$ and 7.45 g Na_2EDTA in 1 liter of sterile distilled water); d) *Vitamins*: (in $mg\ l^{-1}$: myo-inositol 100; nicotinic acid 5; folic acid 0.5; glycine 2; pyridoxine-HCl 0.5; biotin 0.05; and thiamine-HCl 0.5); e) *Sugar*: sucrose 2%; and f) *Agar*: Difco-bacto agar 0.8–1.0%.

3.5. *Culture conditions*

Culture conditions are mainly related to temperature, light and photoperiod. Cultures of anthers, microspores and ovules of *N. tabacum* have been incubated at 23–28°C, with the higher temperatures used during the first 2–3 days of culture. Usually, 4,000–5,000 lux light intensity has been provided by fluorescent and incandescent lamps, either separately or combined. Microspores of *N. tabacum* perform better under red light compared with white or blue light or darkness (Nitsch, 1977). Other species (*D. innoxia*), however, have not shown this preference (Sopory & Maheshwari, 1976).

Microspore embryogenesis occurs more effectively under light/dark treatment compared with continuous light (Sunderland, 1971; Corduan, 1975; Sopory & Maheshwari, 1976). Therefore, a 16–18 h photoperiod has usually been provided with a night temperature of 23–24°C and a day temperature of 27–28°C. The incubation period normally lasts 3 to 5 weeks. After this period, embryos start growing into plantlets (Plate 1(5,6).

4. Methods of obtaining doubled haploids (DHs) and their field performance

After a few weeks on a growth regulator-free medium with reduced sugar concentration (1%), anther-derived plantlets may be transferred to pots containing peat or other potting mix and grown in the greenhouse. The regenerated haploid plants appear to be uniform but smaller (about 1/3–1/4) than parental lines. They usually flower abundantly without setting seed. Flowers are often less colourful and about 1/4 smaller than the parental line. Anthers contain only shrivelled pollen grains. Their photoperiod requirements are similar to those of the anther donors. Thus, for example, flowering of *N. sylvestris* (a long day species) haploid plants requires long days whereas *N. tabacum* haploid plants, such as "Maryland Mammoth" (a short-day type), require short days for flowering (Nitsch, 1969).

4.1. *Diploidization*

Haploid plants are sterile because of their meiotic abnormalities. To restore fertility, it is necessary to double their chromosome number, which has been done in two ways: 1) by using endomitotic substances (colchicine, acenaphthene, and others); or 2) by using tissue culture of stem medullar tissue to regenerate plants from callus.

The use of colchicine has perhaps been the most universally adopted technique for chromosome duplication, even though it may result in chimeric plants among the regenerants. Rooted haploid plantlets, about 2 cm high, are dipped in an aqueous colchicine solution (0.1–0.5%) for 24–48 h. The duration of treatment depends on the concentration used, on plant development and on the specific response of the cultivar or line. This is followed by their immersion into distilled water for another 24 h; then they are transplanted into a pot, transferred to a greenhouse, and subjected to continuous light of 5,000–6,000 lux for 15–20 days at 27–28°C. Although it is possible, by this procedure, to obtain 15–20% DHs, the percentage of plants with cytological anomalies has been high (Sorrentino *et al.*, 1978). Colchicine may also be applied to flower buds of haploid plants at the developmental stage preceding that when meiosis occurs in normal diploid plants. With this technique, it has been possible to obtain some seed production; however, the yield of DHs was low (Sorrentino *et al.*, 1978).

Alternatively, chromosome duplication can be achieved by *in vitro* spontaneous endomitosis of stem medullary tissue. Usually, 20–30% of regenerated plants have been DH (Sorrentino *et al.*, 1978). Chromosome duplication can occur using any somatic tissue (leaf disks, midribs, etc.) of a haploid tobacco plant after induction of callus. With these tissues, however, production of DHs has generally been reduced compared with stem medullary tissue (Sorrentino *et al.*, 1978).

The ploidy of the regenerated plantlets may be assessed by Feulgen staining of mitotic cells in root apices (Plate 3). The approximate level of ploidy in *N. tabacum* can also be assessed by counting the number of chloroplasts in leaf guard cells. This number has been correlated with genome size (Sorrentino *et al.*, 1987) (Plate 4).

4.2. *Field performance of DH lines*

Since microspore-derived plants exhibit the results of meiotic segregation, they represent a major source of genetic variability. This is particularly true for heterozygous cross-pollinated species or when microspores come from an F_1 hybrid. In these cases, DH populations obtained from anther or isolated microspore culture appear genetically variable.

Since tobacco is a self-pollinated species and most of its cultivars are pure lines with high levels of inbreeding, DH lines derived from tobacco cultivars were expected to be nearly identical to each other and to the parental cultivar. Although early experiments were discordant, it was soon confirmed that DH lines derived from pure line cultivars were significantly different from each other and from the parental cultivar whereas seed-propagated families of DHs were extremely uniform. This implied that unexpected genetic changes occurred before chromosome duplication.

According to Evans *et al.* (1984), gametoclonal variation was defined as the genetic variation that originates from anther or isolated microspore cultures. Reports on field performance of DH lines have been somewhat discordant. From an F_1 hybrid between LN38 and NC95 flue-cured tobacco cultivars, Burk *et al.* (1972) obtained DH lines that were superior to both parental lines and noticed that each family derived by selfing DHs appeared to be stable despite differences among the DHs. Collins *et al.* (1974a) obtained similar results from a cross between LA Burley 21 and Burley 21, without loss of vigor in DH lines compared to parental cultivars. Conversely, Burk & Matzinger (1976) obtained 46 DH lines from anther culture of flue-cured tobacco cultivar Coker 139 (C139), a pure line that had been self-pollinated on a single plant basis for 15 generations. The DHs displayed reduced vigor and growth but higher concentrations of total alkaloids and reducing sugars compared with C139. Bowman & Wernsman (1989) compared DH lines obtained paternally (PDH) by anther or isolated microspore culture with

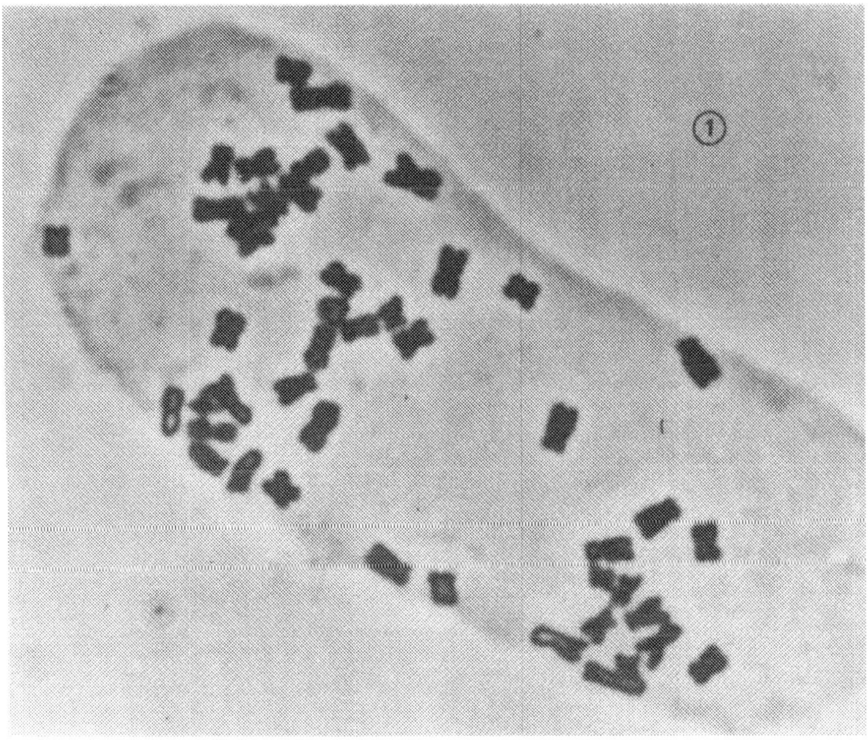

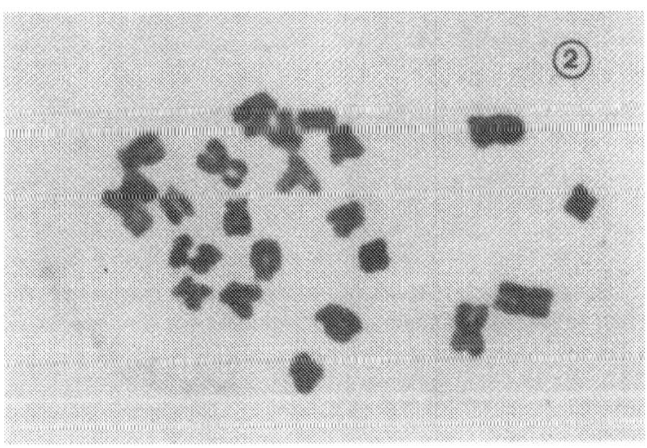

Plate 3. 1) Metaphase in a root apex cell of a diploid plant ($2n = 4x = 48$) of *N. tabacum* (975×)
2) Metaphase in a root apex cell of a haploid plant ($2n = 2x = 24$) of *N. tabacum* (975×)

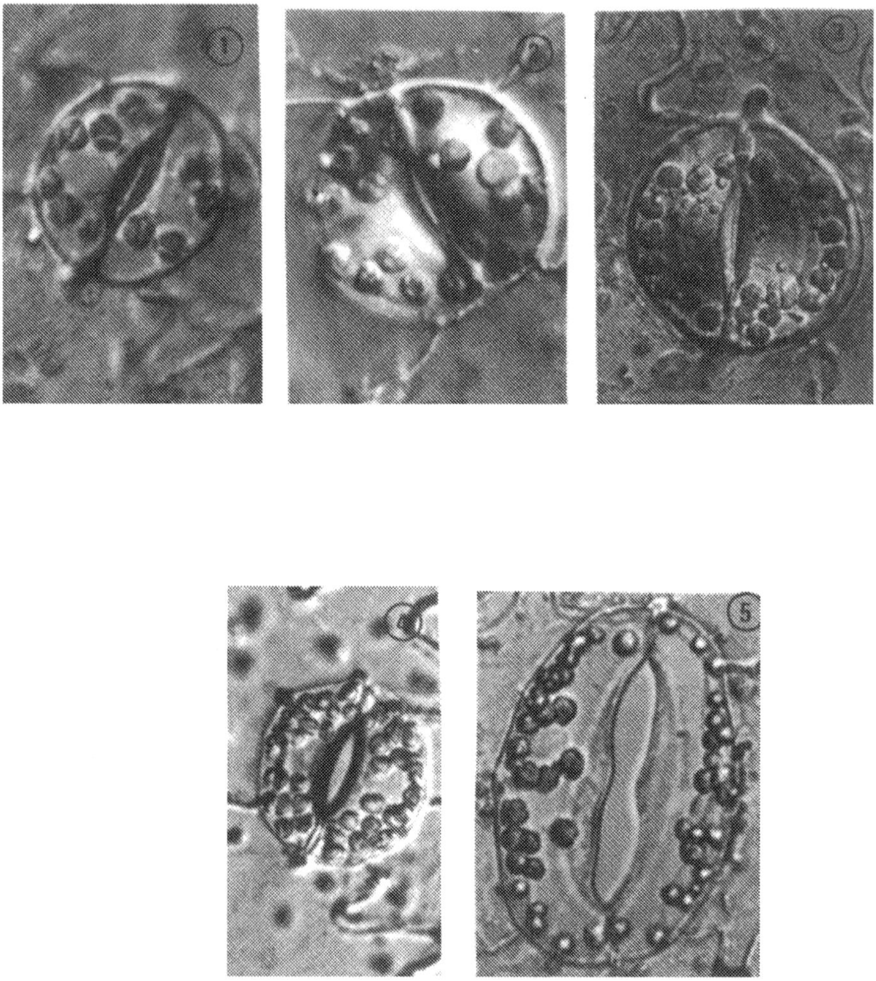

Plate 4. Leaf guard cells surrending stomata of *N. tabacum*: 1) Haploid plant ($2n = 2x = 24$) (315×). 2) Aneuploid plant ($2n - x$) (225×). 3) Diploid plant ($2n = 4x = 48$) (225×). 4) Aneuploid plant ($4n - x$) (135×). 5) Tetraploid plant ($4n = 96$) (233×).

those obtained maternally (MDH) after crossing several flue-cured tobacco cultivars with *N. africana* (Merxmüller & Buttler, 1975; Burk *et al.*, 1979). They found that MDH lines were as vigorous as conventional self-pollinated progeny. Of eleven traits examined, only two (number of ground suckers and reducing sugar concentration) differed significantly between the MDHs and their parents. In a comparison between a pure line cultivar, Burley 21, and two generations of anther-derived DH lines of the same cultivar, Collins *et al.* (1974b) found that DH lines differed from Burley 21 for several agronomic traits but the means of first and second generation DHs did not differ. Stability within DH lines and significant genetic differences between conventional and anther-derived DH lines have been found in several other reports (Devreux & Laneri, 1974; Oinuma & Yoshida, 1974; De Paepe *et al.*, 1977; Schnell *et al.*, 1980; Kasperbauer *et al.*, 1983; Wernsman *et al.*, 1984; Kumashiro & Oinuma, 1985; Deaton *et al.*, 1986a; and others).

Generally speaking, there is sufficient experimental evidence to assert that: a) DHs are highly stable within each family but genetically variable among them; b) PDH lines obtained from highly inbred cultivars are agronomically inferior compared with self-fertilized progeny of the parental cultivar; c) MDH lines do not show any significant loss of vigour compared with the parental cultivar.

In order to explain the unexpectedly low productivity of DHs, which can reach 15% (Arcia *et al.*, 1978), and gametoclonal variation among DH lines, several possibilities have been hypothesized. The more accredited are: the mutagenic effect of colchicine used for doubling the chromosome number; inbreeding depression of completely homozygous DH lines compared to parental cultivars that may retain some residual heterozygosity; chromosomal aberrations and altered relationship between nucleus and cytoplasm; DNA amplification.

About the mutagenic effect of colchicine, Burk & Matzinger (1976) compared spontaneously doubled DHs with colchicine-doubled DHs and found similar performance of both groups, although the range of variability among the spontaneously doubled was less than that among the colchicine-doubled DH group. Similar results were obtained by Kumashiro & Oinuma (1986). They also found no difference in variability between spontaneously and colchicine-doubled DHs although, in the latter group, there were many DH lines that differed significantly from the parental cultivar, also because of reduced productivity.

Two approaches have been taken to determine if residual heterozygosity in parental cultivars underlies their superior performance compared with completely homozygous DHs. The first involved derivation of a second cycle of DHs by anther culture and chromosome doubling to evaluate whether additional variation occurred. Significant genetic variation was found by Brown *et al.* (1983) among first and second-cycle androgenic DH lines and source cultivars of flue-cured tobacco. Both cycles of anther culture resulted in approximately equal reductions in vigour. Anther culture was also associated with the creation of significant genetic variability among lines arising from a single homozygous source, regardless of cycle number (Brown *et al.*, 1983). Similar results were also obtained by Kasperbauer *et al.* (1983) among second-cycle DH lines of burley tobacco cultivars, whereas no significant differences were observed by Deaton *et al.* (1986b) between first and second cycle DHs of burley tobacco cultivars. Wernsman *et al.* (1989) found that a second cycle of MDH line production resulted in no further yield reduction. The second approach was a comparison of field performance between MDHs and PDHs. Using ten anther-derived DH lines (PDHs), ten maternally derived DH lines (MDHs) and selfed progenies of the parental plant, Wernsman *et al.* (1984) found that MDH lines were agronomically superior to PDH lines. Both PDH and MDH lines were lower yielding, by 20% and 8% respectively, than selfed progenies of the parental cultivars. In a similar experiment, Wernsman *et al.* (1989) found that PDH and MDH lines were significantly different for all of the eight characters measured. The MDH lines were agronomically superior to PDH lines and more closely resembled the parental cultivars. The PDHs and MDHs yielded 79.9% and 91.5%, respectively, of the cultivars. A model of heterozygosity and inbreeding depression in parental cultivars could explain the lower yield of MDH lines; however, this would not account for differences between PDH and MDH lines.

The cause of gametoclonal variation has also been attributed to chromosomal aberrations, rearrangements, linkage effects, altered relationship between the nucleus and cytoplasm, nuclear gene mutation during anther culture (Kumashiro & Oinuma, 1985) or cytoplasmic mutation (Matzinger & Burk, 1984). Both mitotic and meiotic irregularities have been found in anther-derived plants (Bridgen & Veilleux, 1988). Reed *et al.* (1991) found aberrant cytological behaviour in hybrids between tobacco androgenic DHs and their parental cultivars (NC95 and C139). Observations on chromosome and spindle behaviour in mitotic anaphase, meiotic anaphase I, and tetrad cells revealed the presence of a low frequency of spindle errors in the hybrids. The genotypic effect was not significant for the frequency of spindle errors in the hybrids but highly significant for the frequency of abnormal anaphase I cells. In F_1 hybrids between PDHs and parental cultivars, Gerstel *et al.* (1974) found normal chromosome pairing and no multivalents at metaphase I, to demonstrate that chromosome structural changes were not responsible for variation observed in PDHs. Although the above phenomena are implicated in causing genetic variation, none can account for the magnitude of genetic variation that still occurs among DH lines.

Another possible cause, i.e., nuclear DNA amplification in androgenic DH lines, has been studied by many researchers (De Paepe *et al.*, 1982, 1990; Dhillon *et al.*, 1983; Reed & Wernsman, 1983, 1989; Reed & Burns, 1989; Reed *et al.*, 1992, 1994). This subject has been reviewed in the second volume of this series (Reed, 1996). Here, we only summarize the conclusions about amplified DNA in tobacco DH lines as follows: the source of amplified DNA is still unknown; vegetative nuclei of *Nicotiana* microspores have more than the C amount of DNA (D'Amato *et al.*, 1965; De Paepe *et al.*, 1990); since tobacco haploids arise from the vegetative nucleus, it could be the origin of amplified DNA in DHs (Dhillon *et al.*, 1983); the high metabolic activity of the vegetative nucleus adds credibility to this hypothesis. Because conclusive evidence is still lacking, however, we can only speculate as to the nature of the amplified DNA in tobacco.

5. Conclusions

The genus *Nicotiana* retains a privileged position in plant breeding compared to many other cultivated species, due to its capacity to generate haploid and DH plants easily. *N. tabacum* is a model plant to study the numerous factors involved in androgenic induction.

DH lines of tobacco are a formidable tool for geneticists and breeders to accelerate progress in genetic improvement programmes. Haploids and DHs provide opportunities for the instruments of classic genetics as well as the more modern instruments of molecular biology and biotechnology. Current problems such as: disease resistance (Bourgin, 1983; Garcia & Rey, 1985; Rufty *et al.*, 1985; Cseplo *et al.*, 1986); growth regulator physiology (Kerbauy *et al.*, 1986); floral morphogenesis (Altamura *et al.*, 1986); drought-resistance (Zhu & Xu, 1986); gamma-radiation effect (Hell, 1983; Nielsen *et al.*, 1985); genetic transformation (Bezdek *et al.*, 1984; Durand-Tardif *et al.*, 1985; Pollock *et al.*, 1985); alkaloid content (Ostrem *et al.*, 1986); enzymatic activity (Trihn & Tran Thanh Van, 1985); genetic variability (Wernsman *et al.*, 1984; Kumashiro & Oinuma, 1985); pollen embryogenic capacity (Bridgen & Veilleux, 1988; Misoo & Watanabe, 1985); somatic hybrids (Potrykus *et al.*, 1983; Morrison *et al.*, 1985); photosynthetic efficiency (Medrano & Primo-Millo, 1985); herbicide resistance (Singer & McDaniel, 1985); virus resistance (Witherspoon *et al.*, 1991; Yung *et al.*, 1991); as well as quality improvement, adaptation to marginal areas, etc., have benefitted from availability of haploid tobacco germplasm.

The loss of agronomic performance associated with DH lines appears to be due to amplification of DNA. Low performing DH lines generally had more DNA than the more productive ones, although the correlation between these two variables was not significant (Reed & Wernsman, 1989). The availability and study of monosomic lines (Cameron, 1959) may help us to understand the causes of such reduced productivity.

6. References

Akehurst, B.C., 1981. Tobacco. Humanities Press, New York.

Altamura M.M., G. Pasqua & B. Monacelli, 1986. *In vitro* floral morphogenesis in a doubled haploid tobacco. Plant Sci. 46: 69–75.

Arcia M.A., E.A. Wernsman & L.G. Burk, 1978. Performance of anther-derived dihaploids and their conventionally inbred parents as lines, in F_1 hybrids, and in F_2 generations. Crop Sci. 18: 413–418.

Aruga, K. & T. Nakajima, 1985. Role of anther on pollen embryogenesis in anther culture of *Nicotiana tabacum* L. Jpn. J. Breed. 35: 390–397.

Aruga, K., T. Nakajima & K. Yamamoto, 1985. Embryogenic induction in pollen grain of *Nicotiana tabacum* L. Jpn. J. Breed. 35: 50–58.
Bajaj, Y.P.S., 1978. Regeneration of haploid tobacco plants from isolated pollen grown in drop culture. Indian J. Exp. Biol. 16: 407–409.
Bajaj, Y.P.S., G. Gosch, M. Ottma, A. Weber & A. Gröbler, 1978. Production of polyploid and aneuploid plants from anthers and mesophyll protoplasts of *Atropa belladonna* and *Nicotiana tabacum*. Indian J. Exp. Biol. 16: 947–953.
Barfield, D.G., S.J. Robinson & R. Shields, 1985. Plant regeneration from protoplasts of long term haploid suspension cultures of *N. plumbaginifolia*. Plant Cell Rep. 4: 104–107.
Bezdek, M., J. Reich, V. Kuhrova, P. Karlovsky & B. Vyskot, 1984. Crown-gall transformation of *Nicotiana plumbaginifolia* haploid cells and characteristics of tumor lines. In: F.J. Novak, L. Havel & J. Dolezel (Eds.), Plant Tissue and Cell Culture Application to Crop Improvement, pp. 423–424. Czechoslovak Academy of Sciences, Prague.
Binding, H., 1975. Regeneration von Haploiden und Diploiden Pflanzen aus Protoplasaten von *Petunia hybrida* L. Z. Pflanzenphysiol. 74: 327–356.
Bourgin, J.P., 1983. Selection of tobacco protoplast-derived cells for resistence to amino acids and regeneration of resistant plants. In: P.F. Lurquin & A. Kleinhofs (Eds.), Genetic Engineering in Eukaryotes, pp. 195–214. Plenum Press, New York.
Bourgin J.P. & J.P. Nitsch, 1967. Obtention de *Nicotiana* haploïdes à partir d'étamines cultivées *in vitro*. Ann. Physiol. Vég̀. Paris 9: 377–382.
Bowman D.T. & E.A. Wernsman, 1989. Stability of maternal doubled haploid lines of flue-cured tobacco. Tobacco Sci. 33: 74–76.
Bridgen, M.P. & R.E. Veilleux, 1988. A comparison of *in vitro* flowers to *in vivo* flowers of haploid and diploid *Nicotiana tabacum* L. Plant Cell Tiss. Organ Cult. 13: 3–13.
Brown J.S., E.A. Wernsman & R.J. Schnell, 1983. Effect of a second cycle of anther culture on flue-cured lines of tobacco. Crop Sci. 23: 729–733.
Burbidge, N.T., 1960. The Australian species of *Nicotiana* L. (Solanaceae). Aust. J. Bot. 8: 342–380.
Burk, L.G., 1967. Preservation of *Nicotiana* species collection. Tobacco Sci. 11: 120–123.
Burk, L.G. & D.F. Matzinger, 1976. Variation among anther-derived doubled haploids from an inbred line of tobacco. J. Hered. 67: 381–384.
Burk, L.G., G.R. Gwynn & J.F. Chaplin, 1972. Diploidized haploids from aseptically cultured anthers of *Nicotiana tabacum*. J. Hered. 63: 355–360.
Burk, L.G., D.U. Gerstel & E.A. Wernsman, 1979. Maternal haploids of *Nicotiana tabacum* L. from seed. Science 206: 585.
Cameron, D.R., 1959. The monosomics of *Nicotiana tabacum*. Tobacco Sci. 3: 164–166.
Cersosimo, A., 1996. Haploidy in *Vitis*. In: S.M. Jain, S.K. Sopory & R.E. Veilleux (Eds.), *In Vitro* Haploid Production in Higher Plants, Vol. 3: Important Selected Plants, pp. 209–227. Kluwer Academic Publishers, Dordrecht.
Chaplin, F.J. & L.G. Burk, 1979. Plant propagation. In: R.D. Durbin (Ed.), *Nicotiana*: Procedure for Experimental Use, pp. 28–32. United States Department of Agriculture. Technical Bulletin no. 1586.
Chen, C-C., C-R. Huang & K-Y. To, 1985. Anther cultures of four diploid *Nicotiana* species and chromosome numbers of regenerated plants. Bot. Bull. Acad. Sin. 26: 147–153.
Chowdhury, M.K.U., 1984. An improved method for dihaploid production in *Nicotiana rustica* through anther culture. Theor. Appl. Genet. 69: 199–204.
Clausen, R.E. & D.H. Cameron, 1944. Inheritance in *Nicotiana tabacum*. XVIII. Monosomic analysis. Genetics 29: 447–477.
Clausen, R.E. & W.E. Lammerts, 1929. Interspecific hybridization in *Nicotiana*. X. Haploid and diploid merogony. Am. Nat. 63: 279–282.
Collins G.B., P.D. Legg & M.J. Kasperbauer, 1974a. Use of anther-derived haploids in *Nicotiana*. I. Isolation of breeding lines differing in total alkaloid content. Crop Sci. 14: 77–80.
Collins G.B., P.D. Legg & C.C. Litton, 1974b. The use of anther derived haploid lines obtained by conventional breeding methods. Tobacco Sci. 18: 40–42.

Corduan G., 1975. Regeneration of anther-derived plants from *Hyoscyamus niger* L. Planta 127: 27–36.

Cseplo, A., P. Medgyesy & L. Marton. 1986. *In vitro* induction, isolation and transfer of chloroplast mutations in *Nicotiana*. Nuclear techniques and *in vitro* culture for plant improvement. In: Proc. Symp. Int. Atomic Energy Agency and the Food and Agriculture Organization of the United Nations, Vienna, 19–23 August 1985, pp. 137–146.

D'Amato, F., 1985. Cytogenetics of plant cell and tissue cultures and their regenerates. Crit. Rev. Plant Sci. 3: 73–112.

D'Amato, F., M. Devreux & G.T. Scarascia-Mugnozza, 1965. The DNA content of the nuclei of the pollen grain in tobacco and barley. Caryologia 18: 377–382.

Deaton, W.R., G.B. Collins & M.T. Nielsen, 1986a. Vigor and variation expressed by anther-derived doubled haploids of burley tobacco (*Nicotiana tabacum* L.). I. Comparison of sexual and doubled-haploid populations. Euphytica 35: 33–40.

Deaton, W.R., G.B. Collins & M.T. Nielsen, 1986b. Vigor and variation expressed by anther-derived doubled haploids of burley tobacco (*Nicotiana tabacum* L.). II. Evaluation of first- and second-cycle doubled haploids. Euphytica 35: 41–48.

De Paepe C., C. Nitsch, M. Godard & J. Pernes, 1977. Potential from haploid and possible use in agriculture. In: N. Barz, E. Reinhart & M.H. Zenk (Eds.), Plant Tissue Culture and its Biotechnological Application, pp. 341–352. Springer-Verlag, New York.

De Paepe, C., D. Prat & T. Huguet, 1982. Heritable nuclear DNA changes in doubled haploid plants obtained by pollen culture of *Nicotiana sylvestris*. Plant Sci. Lett. 28: 11–28.

De Paepe, C., A. Koulon, J.L. Phamand & S.C. Brown, 1990. Nuclear DNA content and separation of *Nicotiana sylvestris* vegetative and generative nuclei at various stages of male gametogenesis. Plant Sci. 70: 255–265.

Devreux, M. & U. Laneri, 1974. Anther culture, haploid plants, isogenic lines and breeding research in *Nicotiana tabacum* L. In: Polyploidy and Induced Mutations in Plant Breeding, pp. 101–107. International Atomic Energy Agency, Vienna.

Devreux, M., U. Laneri & P.D. Martinis, 1975. Reflexions sur nos cultures d'anthères de plantes cultivées. G. Bot. Ital. 109: 335–349.

Dhillon, S.S., E.A. Wernsman & J.P. Miksche, 1983. Evaluation of nuclear DNA content and heterochromatin changes in anther-derived dihaploids of tobacco (*Nicotiana tabacum*) cv. Coker 139. Can. J. Genet. Cytol. 25: 169–173.

Duncan, E.J. & E. Heberle, 1976. Effect of temperature shock on nuclear phenomena in microspores of *Nicotiana tabacum* and consequently of plantlet production. Protoplasma 90: 173–177.

Durand-Tardif, M., R. Broglie, J. Slightom & D. Tepfer, 1985. Structure and expression of Ri T-DNA from *Agrobacterium rhizogenes* in *Nicotiana tabacum*: Organ and phenotypic specificity. J. Mol. Biol. 186: 557–564.

Durr, A. & J. Fleck, 1980. Production of haploid plants of *Nicotiana langsdorffii*. Plant Sci. Lett. 18: 75–79.

Evans, D.A., W.R. Sharp & H.P. Medina-Filho, 1984. Somaclonal and gametoclonal variation. Am. J. Bot. 71: 759–774.

Forche, E. & K.H. Neumann, 1977. Der Einfluß verschiedener Kulturfaktoren auf die Gewinnung haploider Pflanzen aus Antheren von *Datura innoxia* und *Nicotiana tabacum* ssp. Z. Pflanzenzüchtg. 79: 250–255.

Gamborg, O.L., R.A. Miller & K. Ojima, 1968. Nutrient requirements of suspension cultures of soybean root cells. Exp. Cell Rep. 50: 151–158.

Garcia, A. & X. Rey, 1985. Use of haploid-diploid method for obtaining isogenic lines resistant to blue mould. Ciencia -y- Tecnica en la Agricoltura, Tobacco 8: 63–66.

Gautheret, R.J., 1959. La Culture des Tissus Végétaux, Techniques et Réalisations. Mason Cie., Paris.

Gerstel, D.U., 1961. Essay on the origin of tobacco. Tobacco Sci. 5: 15–17.

Gerstel, D.U., J.A. Burns & L.G. Burk, 1974. Meiosis in F_1 hybrids between haploid-derived diploids and their parents. Tobacco Sci. 18: 122–124.

Ghandimathi, H., 1982. Direct pollen culture of *Nicotiana tabacum*. In: A. Fujiwara (Ed.), Plant Tissue Culture 1982, pp. 527–528. Maruzen, Tokyo.
Goodspeed, T.H., 1954. The genus *Nicotiana*. Chronica Bot. 16: 1–536.
Gray, J.C., S.D. Kung, S.G. Wildman & S.J. Sheen, 1974. Origin of *Nicotiana tabacum* L. detected by polypeptide composition of fraction I protein. Nature (London) 252: 226–227.
Guha S. & S.C. Maheshwari, 1964. *In vitro* production of embryos from anthers of *Datura*. Nature (London) 204: 497.
Guha, S. & S.C. Maheshwari, 1966. Cell division and differentiation of embryos in the pollen grains of *Datura in vitro*. Nature (London) 212: 97–98.
Heberle-Bors, E. & J. Reinert, 1977. Factors of haploid production by isolated pollen cultures. Naturwissenschaften 64: 100–101.
Hell, K.G., 1983. Survival of *Nicotiana tabacum* L. cv. Wisconsin 38 plants regenerated from gamma-irradiated tissue cultures. Environ. Exp. Bot. 23: 139–142.
Horner, M. & M.L. Pratt, 1979. Amino acid analysis of *in vivo* and androgenic anthers of *Nicotiana tabacum*. Protoplasma 98: 279–282.
Hsu, C.L. & J.McD. Stewart, 1976. Callus induction by (2–chloroethyl)-phosphonic acid on cultured cotton ovules. Physiol. Plant. 36: 150–153.
Imamura, J., E. Okabe, M. Kyo & H. Harada, 1982. Embryogenesis and plantlet formation through direct culture of isolated pollen of *Nicotiana tabacum* cv. Samsun and *Nicotiana rustica* cv. Rustica. Plant Cell Physiol. 23: 713–716.
Kameya, T. & K. Hinata, 1970. Induction of haploid plants from pollen grains of *Brassica*. Jpn. J. Breed. 20: 82–87.
Kasperbauer, M.J., P.D. Legg & T.G. Sutton, 1983. Growth, development, and alkaloid content of doubled haploids vs. inbreds of burley tobacco. Crop Sci. 23: 965–969.
Kehr, A.E., 1951. Genetic tumors in *Nicotiana*. Am. Naturalist 85: 51–64.
Kerbauy, G.B., J.A. Peters & K.G. Hell, 1986. Cytokinin autotrophy and differentiation in tissue cultures of haploid *Nicotiana tabacum* L. Plant Sci. 45: 125–132.
Kostoff, D., 1929. An androgenic *Nicotiana* haploid. Z. Zellforsch. Mikrosk. Anat. 9: 640–642.
Kostoff, D., 1943. Cytogenetics of *Nicotiana*. States Printing House, Sofia.
Kumashiro, T. & T. Oinuma, 1985. Comparison of genetic variability among anther-derived and ovule-derived doubled haploid lines of tobacco. Jpn. J. Breed. 35: 301–310.
Kumashiro, T. & T. Oinuma, 1986. Genetic variation of colchicine induced and spontaneous doubled haploids obtained by anther culture of an inbred tobacco. Jpn. J. Breed. 36: 355–362.
Kuo, C.S., 1982. The preliminary studies on culture of unfertilized ovary of rice *in vitro*. Acta Bot. Sin. 24: 33–38.
Kyo, M. & H. Harada, 1985. Studies on conditions for cell division and embryogenesis in isolated pollen culture of *Nicotiana rustica*. Plant Physiol. 79: 90–94.
Linsmaier, E.M. & F. Skoog, 1965. Organic growth factor requirements of tobacco tissue cultures. Physiol. Plant. 18: 100–127.
Maheshwari, N. 1961. *In vitro* culture of excised ovules of *Papaver somniferum* L. Phytomorphology 11: 307–314.
Maheshwari, S.C., A.K. Tyagi & K. Malhotra, 1980. Induction of haploidy from pollen grains in Angiosperms – the current status. Theor. Appl. Genet. 58: 193–206.
Maheshwari, S.C., A. Rashid & A.K. Tyagi, 1982. Haploids from pollen grains – retrospect and prospect. Am. J. Bot. 69: 865–879.
Matzinger, D.F. & L.G. Burk, 1984. Cytoplasmic modification by anther culture in *Nicotiana tabacum* L. J. Hered. 75: 167–170.
McComb, J.A. & A.J. McComb, 1977. The cytology of plantlets derived from cultured anthers of *Nicotiana sylvestris*. New Phytol. 79: 679.
Medrano, H. & E. Primo-Millo, 1985. Selection of *Nicotiana tabacum* haploids of high photosynthetic efficiency. Plant Physiol. 79: 505–508.
Melchers, G., 1974. Polyploids and Induced Mutations in Plant Breeding. Int. Atomic Energy Agency, Vienna.

Merxmüller, H. & K.P. Buttler, 1975. *Nicotiana* in der Afrikanischen Namibein pflanzengeographisches und phylogenetisches Rätsel. Mitteilungen der Botanischen Staatssammlung Muenchen 12: 91–104.

Misoo, S. & N. Watanabe, 1985. Studies on the mechanisms of pollen embryogenesis. VI. Anther culture in *Nicotiana rustica* and its ancestral species, *N. paniculata* and *N. undulata*. Sci. Rep. Fac. Agric. Kobe Univ. 16: 385–390.

Morrison, R.A., D.A. Evans & R.E. Koning, 1985. Anther culture of *Nicotiana* somatic hybrid. Am. J. Bot. 72: 877.

Mothes, K., 1955. Physiology of alkaloids. Ann. Rev. Plant Physiol. 6: 393–432.

Mullins, M.G. & C. Srinivasan, 1976. Somatic embryos and plantlets from an ancient clone of grapevine (cv. Cabernet Sauvignon) by apomixy *in vitro*. J. Exp. Bot. 27: 1922–1030.

Murashige, T. & F. Skoog, 1962. A revised medium for rapid growth and bioassays with tobacco tissue cultures. Physiol. Plant. 15: 473–497.

Nakata, K. & M. Tanaka, 1968. Differentiation of embryoids from developing germ cells in anther culture of tobacco. Jpn. J. Genet. 43: 65–71.

Nielsen, E., E. Selva, C. Schirinzetti & M. Devreux, 1985. The mutagenic effect of gamma rays on leaf protoplasts of haploid and dihaploid *Nicotiana plumbaginifolia*, estimated by valine resistance mutation frequencies. Theor. Appl. Genet. 70: 259–264.

Niizeki, H. & K. Oono, 1968. Induction of haploid rice plants from anther culture. Proc. Jpn. Acad. 44: 554–557.

Nitsch, C., 1974a. La culture de pollen isolé sur milieu synthetique. C.R. Acad. Sci. Paris 270: 1031–1034.

Nitsch, C., 1974b. Pollen culture – a new technique for mass production of haploids and homozygous plants. In: K.J. Kasha (Ed.), Haploids in Higher Plants: Advances and Potential, pp. 123–135. The University of Guelph, Guelph, Ontario.

Nitsch, C., 1977. Culture of isolated microspores. In: J. Reinert & Y.P.S. Bajaj (Eds.), Applied and Fundamental Aspects of Plant Cell Tissue and Organ Culture, pp. 268–278. Springer-Verlag, Berlin.

Nitsch, J.P. & C. Nitsch, 1956. Am. J. Bot. 43: 839–851.

Nitsch, J.P., 1969. Experimental androgenesis in *Nicotiana*. Phytomorphology 19: 389–404.

Nitsch, J.P. & C. Nitsch, 1969. Haploid plants from pollen grains. Science 163: 85–87.

Nitsch, J.P. & C. Nitsch, 1970. Obtention de plantes haplöides à partir de pollen. Bull. Soc. Bot. France 117: 339–360.

Nitsch, J.P. & B. Norreel, 1973. Effet d'un choc thermique sur le pouvoir embryogène du pollen de *Datura innoxia* cultivé dans l'anthère ou isolé de l'anthère. C.R. Acad. Sci. Paris 276: 303–306.

Ohaski, Y., 1976. *Nicotiana kawakamii*: a new species of the genus *Nicotiana*. In: Sixth International Tobacco Science Congress Proceedings, Tokyo, pp. 146–147.

Oinuma, T. & T. Yoshida, 1974. Genetic variation among doubled haploid lines of burley tobacco varieties. Jpn. J. Breed. 24: 211–216.

Ostrem, J.A., C.C. Litton & G.B. Collins, 1986. Isolation of dark tobacco breeding lines differing in alkaloid concentration via anther-derived haploids. Z. Pflanzenzüchtg. 96: 224–231.

Pelletier, G., 1973. Les conditions et les premièrs stades de l'androgénèse *in vitro* chez *Nicotiana tabacum*. Soc. Bot. France Memoires 1973, Colloq. Morphologie, pp. 261–268.

Pollock, K., D.G. Barfiel, S.J. Robinson & R. Shields, 1985. Transformation of protoplast-derived cell colonies and suspension cultures by *Agrobacterium tumefaciens*. Plant Cell Rep. 4: 202–205.

Potrykus, I., R.D. Shillito, J. Jia & G.B. Lazar, 1983. Auxotroph complementation via protoplast fusion in *Hyoscyamus muticus* and *Nicotiana tabacum*. In: P.F. Lurquin & and A. Kleinhofs (Eds.), Genetic Engineering in Eukaryotes, pp. 253–263. Plenum Press, New York.

Raghavan, V., 1976. The role of the generative cell in androgenesis in henbane. Science 191: 388–389.

Raina, S.K. & R.D. Iyer, 1973. Differentiation of diploid plants from pollen callus in anther cultures of *Solanum melongena*. L. Z. Pflanzenzüchtg. 70: 275–280.
Ran, B.D., 1980. Induction of haploid plants from unfertilized tobacco ovules. Zhongguo Yancao 3: 25–26.
Reed, S.M., 1996. DNA amplification in doubled haploids of *Nicotiana*. In: S.M. Jain, S.K. Sopory & R.E. Veilleux (Eds.), *In Vitro* Haploid Production in Higher Plants, Vol. 2, pp. 111–125. Kluwer Academic Publishers, Dordrecht.
Reed, S.M. & J.A. Burns, 1989. Cytological consequences of DNA amplification in an anther-derived doubled haploid line of *Nicotiana tabacum* L. Am. J. Bot. 76: 958–965.
Reed, S.M. & E.A. Wernsman, 1983. The doubled haploid experience in *Nicotiana*. Agron. Abstr. 77. American Society of Agonomy, Madison, Wisconsin.
Reed, S.M. & E.W. Wernsman, 1989. DNA amplification among anther-derived doubled haploid lines of tobacco and its relationship to agronomic performance. Crop Sci. 29: 1072–1076.
Reed, S.M., E.A. Wernsman & J.A. Burns, 1991. Aberrant cytological behavior in tobacco androgenic doubled haploid × parental cultivar hybrids. Crop Sci. 31: 97–101.
Reed, S.M., J.A. Burns & E.A. Wernsman, 1992. Cytological comparison of amplified chromosome segments in four tobacco doubled haploids. Crop Sci. 32: 362–366.
Reed, S.M., E.A. Wernsman & J.A. Burns, 1994. Cytological evaluation of tobacco doubled haploids for presence of amplified homologous chromosomal regions. Crop Sci. 34: 236–239.
Rufty, R.C., E.A. Wernsman & G.V. Gooding, 1985. Evaluation of tobacco haploids for resistance to TMV, PVY and *Meloidogyne incognita* using detached leaves. Phytopathology 75: 11, 1335.
Sangwan-Norreel, B.S. 1977. Androgenic stimulating factors in the anther and isolated pollen grain culture of *Datura innoxia*. Mill. J. Exp. Bot. 28: 843–852.
Scarascia, G.T., 1952. Sviluppo del tappeto dell'antera in *Nicotiana tabacum* var. Bright. Tobacco 636: 1–23.
Schieder, O., 1975. Regeneration von Haploiden und Diploiden *Datura innoxia* Mill. Mesophyll-Protoplasten zu pflanzen. Z. Pflanzenphysiol. 76: 462–466.
Schnell II, R.J., E.A. Wernsman & L.G. Burk, 1980. Efficiency of single seed-descent vs. anther-derived dihaploid breeding methods in tobacco. Crop Sci. 20: 619–622.
Sharp, W.R., R.S. Raskin & H.E. Sommer, 1972. The use of nurse culture in the development of haploid clones in tomato. Planta 104: 357–361.
Sheen, S.J., 1972. Isozymic evidence bearing on the origin of *Nicotiana tabacum* L. Evolution 26: 143–154.
Singer, S.R. & C.N. McDaniel, 1985. Selection of glyphosate-tolerant tobacco calli and the expression of this tolerance in regenerated plants. Plant Physiol. 78: 411–416.
Smith, H.H., 1943. Effects of genome balance, polyploidy, and single extra chromosomes on size in *Nicotiana*. Genetics 28: 227–236.
Smith, H.H., 1979. The genus as a genetic resource. In: R.D. Durbin (Ed.), *Nicotiana*: Procedures for Experimental Use, pp. 1–16. United States Department of Agriculture. Technical Bulletin no. 1586.
Sopory, S.K. & S.C. Maheshwari, 1976. Development of pollen embryoids in anther cultures of *Datura innoxia*. 1. General observations and effects of physical factors. J. Exp. Bot. 27: 49–57.
Sopory, S.K. & B.H. Tan, 1979. Regeneration and cytological studies of anther pollen calli of dihaploid *Solanum tuberosum*. Z. Pflanzenzüchtg. 82: 31–35.
Sorrentino, C., A. Di Muro & G. Lombardo, 1978. Diploidizzazione di piante aploidi ottenute da antere, a partire da ibridi F_1. Annali Istituto Sperimentale per il Tabacco – Vol. V, 1978: 71–77.
Sorrentino, C., A. Tonini, A. Di Muro, R. Nicoletti & M. Contiero, 1987. Coltura *in vitro* di ovuli ed ovari non fecondati di *Nicotiana tabacum*. Annali Istituto Sperimentale per il Tabacco – Vol. XIII 1987: 147–156.

Steinberg, R.A., 1957. Some growth requirements in the genus *Nicotiana*. Tobacco Sci. 1: 99–102.

Steinberg, R.A., 1959. Factors influencing reproduction of some *Nicotiana* species in the greenhouse and field. Tobacco Sci. 3: 131–135.

Sunderland, N., 1971. Anther culture: a progress report. Sci. Progr. 59: 527–549.

Sunderland, N. & J.M. Dunwell, 1977. Anther and pollen culture. In: H.E. Street (Ed.), Plant Tissue and Cell Culture, pp. 223–265. Blackwell Scientific, Oxford.

Sunderland, N. & M. Roberts, 1977. New approach to pollen culture. Nature 270: 236–238.

Sunderland, N. & D.C. Wildon, 1979. A note on the pretreatment of excised flower buds in float culture of *Hyoscyamus* anthers. Plant Sci. Lett. 15: 169–175.

Tran Thanh Van, K., P. Toubart, A. Cousson, A.G. Darvill, D.J. Gollin, P. Chaleff & P. Albersheim, 1985. Manipulation of the morphogenetic pathways of tobacco explants by oligosaccharins. Nature (London) 314: 615–617.

Trinh, T.H. & K. Tran Thanh Van, 1985. Differing expression of androgenetic capacity of anther formed *in vivo* or *in vitro* in *Nicotiana*. Bull. Soc. Bot. France, Actual. Bot. 132(3/4): 167.

Tso, T.C., 1972. Physiology and Biochemistry of Tobacco Plants. Dowden, Hutchinson, and Ross, Stroudsburg, Pa.

Tyagi, A.K., A. Rashid & S.C. Maheshwari, 1979. High frequency production of embryos in *Datura innoxia* from isolated pollen grains by combined cold treatment and serial culture of anthers in liquid medium. Protoplasma 99: 11–17.

Vagera, J., P. Havrànek & Z. Opatrny, 1979. Regulation of *in vitro* androgenesis of tobacco: Relationship between concentration of iron ions and kinetin. Biochem. Physiol. Pflanzen 174: 752–760.

Villanueva, V.R., V. Mathivet & R.S. Sangwan, 1985. RNA, proteins and polyamines during gametophytic and androgenetic development of pollen in *Nicotiana tabacum* and *Datura innoxia*. Plant Growth Reg. 3: 293–308.

Wells, P.V., 1960. Variation in section Trigonophyllae of *Nicotiana*. Madrono 15: 148–151.

Wernsman, E.A., D.F. Matzinger & R.C. Rufty, 1984. Characterization of androgenetic and gynogenic doubled haploids from inbred cultivars. In: 8th International Tobacco Scientific Congress, 7–12 October 1984, Vienna.

Wernsman, E.A., D.F. Matzinger & R.C. Rufty, 1989. Androgenetic vs. gynogenetic doubled haploids of tobacco. Crop Sci. 29: 1151–1155.

White, P.R., 1943. A Handbook of Tissue Culture. Ronald Press, New York.

Witherspoon, Jr., W.D., E.A. Wernsman, G.V. Gooding, Jr. & R.C. Rufty, 1991. Characterization of a gametoclonal variant controlling virus resistance in tobacco. Theor. Appl. Genet. 81: 1–5.

Wu, B.J. & K.C. Cheng, 1982. Cytological and embryological studies on haploid plant production from cultured unpollinated ovaries of *Nicotiana tabacum* L. Acta Bot. Sin. 24: 125–129.

Wu, H.S. & Z.C. Zhu, 1988. Observation on the ploidy level and characters of the plants regenerated from tobacco unpollinated ovaries *in vitro*. Genet. Manip. Crops Newsl. 4: 69–75.

Yung, C.H., E.A. Wernsman & G.V. Gooding, Jr., 1991. Characterization of potato virus Y resistance from gametoclonal variation in flue-cured tobacco. Phytopathology 81: 887–891.

Zagorska, N.A., M.D. Abadjieva & H.A. Georgiev, 1982. Inducing regeneration in anther cultures of tomatoes (*Lycopersicum esculentum* Mill.). C.R. Acad. Bulg. Sci. 35: 97.

Zhu, Q.L. & Y. Xu, 1986. Selection of hydroxy-L-proline-resistant variants of tobacco from cultured haploid callus and study on their physiological and biochemical characteristics. J. Agric. Sci. China. 2 (Suppl.): 42–54

Zhu, Z.C., H.S. Wu, K. Anq & Z.Y. Liu, 1980. *In vitro* induction of haploid plantlets by tissue culture from unpollinated ovaries of male steriles in *Nicotiana*. Hereditas (Beijing) 2(5): 36.

Zhu, Z.C., H.S. Wu, Z.Y. Liu & Q.K. An, 1983. Production of Haploid Plantlets from Unpollinated Ovaries. Cell and Tissue Culture Techniques for Cereal Crop Improvement. Science Press, Beijing.

9. *In vitro* induction of haploidy in cotton

Y.P.S. BAJAJ and MANJEET S. GILL

Contents

1. Introduction

Cotton is the most important natural source of fibre used in the textile industry, and is also a valuable source of oil. It belongs to the genus *Gossypium* (family Malvaceae) which comprises 50 species (Fryxell, 1992). The common species are the two old World diploids ($2n = 2x = 26$), *G. arboreum* L. and *G. herbaceum* L., and two American tetraploids ($2n = 4x = 52$), *G. hirsutum* L. and *G. barbadense* L. (Wendel *et al.*, 1992); the tetraploids now dominate cotton production. Genetic improvement of cotton is necessary, both for increased yield and for upgrading the quality of lint. Cotton is propagated through seeds, and sometimes multiplied by cuttings. Extensive efforts have been made to develop plant types that can cope with various environmental hazards, insects, and diseases, and can give stable yield from year to year. However, conventional methods alone do not seem sufficient, and thus biotechnological approaches have been introduced (Bajaj & Gill, 1986, 1992).

Production of haploids in cotton is valuable for obtaining homozygous lines in one generation, and these homozygous lines can be evaluated in the field. Because doubled haploids possess additive variance only, early generation testing of segregating populations is possible. Haploids can also be used to increase the efficiency of existing breeding methods through improved reliability of selection. The conventional methods of haploid production in cotton are polyembryony and semigamy. Polyembryony (Blank & Allison, 1963; Lee, 1970), in which one or both members of a pair of twin seedlings may be haploid, is a rare phenomenon. Semigamy (Turcotte & Feaster, 1974; Chaudhari, 1978) is restricted to the two tetraploid species, *G. barbadense* and *G. hirsutum*, and does not work for the diploid species. Agronomic evaluations of semigametically derived doubled haploids of *G. hirsutum* in *G. barbadense* cytoplasm revealed superiority for some traits but inferiority for others, compared to the cultivars from which they were derived

S.M. Jain, S.K. Sopory & R.E. Veilleux (eds.), In Vitro *Haploid Production in Higher Plants, Vol. 5,* 165–174.

(Mahill *et al.*, 1984). Parthenogenesis has also been induced in *G. hirsutum* (Zhou *et al.*, 1991).

The *in vitro* culture of anthers can be employed to produce haploids under defined conducive conditions (Bajaj, 1983, 1990). In early attempts to induce androgenesis in cotton through anther culture, callus was produced which mostly contained diploid and a few haploid cells (Barrow *et al.*, 1978; Baluch, 1979; Hsi & Wu, 1981). Later, Shamina *et al.* (1986), working with anther culture of *G. hirsutum*, and Bajaj & Gill (1989), working with the cultivated diploid, *G. arboreum*, reported the induction of both pollen-embryos and callus with haploid cells as well as a wide range of variation in chromosome number (Bajaj & Gill, 1985). Further improvements to the medium for anther culture of cotton have been made to optimize the viability of microspores (Barrow, 1986; Turaev & Shamina, 1986, 1993; Turaev *et al.*, 1990); however, successful regeneration of haploid plants from anther culture of cotton has not been reported. The present chapter is primarily based on our published work (Bajaj, 1982; Bajaj & Gill, 1985, 1989).

2. Anther culture

Flower buds from field-grown plants of a diploid Asiatic cotton (*Gossypium arboreum* cv. G27) were collected, and the ones containing anthers at the uninucleate pollen stage were used for culture. The buds were sterilized with freshly prepared chlorine water for 8–10 min, and washed twice with sterilized distilled water. Anthers were excised aseptically and 20–25 of them were cultured per test-tube on MS medium (Murashige & Skoog, 1962) containing various concentrations and combinations of indoleacetic acid (IAA), kinetin (kin), benzyladenine (BA), naphthalene acetic acid (NAA), 2,4-dichlorophenoxy acetic acid (2,4-D), casein hydrolysate (CH) and coconut water (CW). Twenty replicates were made for each treatment, ten of which were kept at 4–5°C in a refrigerator for 48 h in the dark prior to transfer to the culture room. For androgenic studies, the anthers were periodically fixed in acetic alcohol (1:3), and stained with acetocarmine (1%). For chromosome counts, fast-growing callus was subcultured on a medium containing 3% glucose instead of sucrose, and was fixed in acetic alcohol for 1 h. Staining was done by subjecting the callus to 60–70°C in 2% acetic orcein for 2–3 min. After this the callus was kept in the stain for 3–10 days. Squash preparations were made in 45% acetic acid. The slide was heated to 60–70°C with two changes in acetic acid, and studied under the microscope.

2.1. *Establishment of callus*

The cultured anthers increased in size within seven days (Fig. 1A). Callus formation was observed during the second week (Fig. 1B) and a mass of

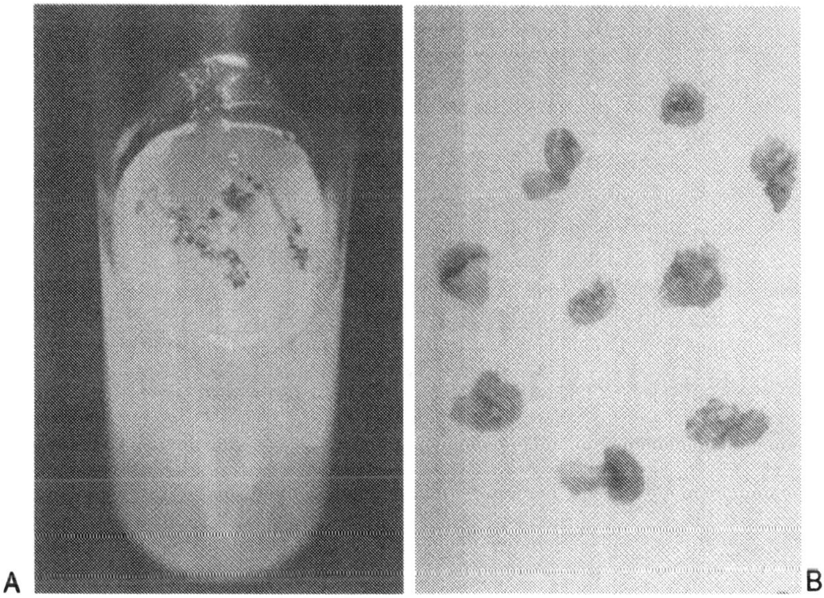

Figure 1. (A,B) Callus formation from excised anthers of *Gossypium arboreum* cultured at the uninucleate pollen stage. (A) Proliferating anthers one week after culture on MS + NAA (2 mg l^{-1}) + BA (1 mg l^{-1}) + sucrose 3% (Bajaj, 1982). (B) A two-week-old culture of proliferating anthers (Bajaj, 1982).

callus was formed in four weeks. The percentage of cultured anthers that formed callus on the various media is shown in Table 1. Callus induction was greatest on MS + 2 mg l^{-1} NAA + 1 mg l^{-1} BA and MS + 2 mg l^{-1} 2,4-D + 0.2 mg l^{-1} kin + 7% CW, with 45 and 37% of cultured anthers forming callus, respectively. Two other media, MS + 4 mg l^{-1} IAA + 2 mg l^{-1} kin and MS + 1 mg l^{-1} IAA + 0.2 mg l^{-1} kin + 500 mg l^{-1} CH, also produced good callus growth. Callus that formed on medium containing NAA and BA was compact and pinkish; however, callus on 2,4-D medium was friable and off-white. Callus subcultured on media containing glucose (3%) grew rapidly.

2.2. *Pollen embryogenesis*

Cold treatment (4°C for 48 h in the dark) influenced the response of cotton anther cultures in two ways. The cold treated anthers produced less callus than the untreated ones (Table 1) but, on the other hand, formed multinucleate and multicellular microspores (Table 2).

On MS + 2 mg l^{-1} 2,4-D + 0.2 mg l^{-1} kin + 7% CW, the microspores showed various modes of development (Fig. 2A-E). Pollen-embryos and

Table 1. Effect of various modifications to Murashige & Skoog (1962) (MS) medium and cold treatments (4°C for 48 h) in the dark on callus induction in cultured anthers of *G. arboreum* (Bajaj & Gill, 1989)

Supplements[1] to MS medium (mg l^{-1})	Without cold treatment			Cold treated		
	Number of anthers cultured	Number of anthers callusing	% Callusing	Number of anthers cultured	Number of anthers callusing	% Callusing
2 IAA + 0.5 Kin	369	29	8	340	19	6
2 IAA + 0.05 Kin	290	14	5	250	11	4
4 IAA + 2 Kin	315	79	25	275	48	17
1 IAA + 0.2 Kin + 500 CH	336	91	27	270	52	19
2 2,4-D + 0.2 Kin + 70 ml l^{-1} CW	285	105	37	270	59	22
2 NAA + 1 BA	415	185	45	294	81	28

[1] IAA = indole acetic acid; Kin = kinetin; CH = casein hydrolysate; 2,4-D = 2,4-dichlorophenoxyacetic acid; CW = coconut water; BA = benzyladenine.

Table 2. Effect of cold treatment (4°C for 48 h) on pollen embryogenesis in excised anthers of *G. arboreum* cultured on MS + 2.0 mg l^{-1} 2,4-D + 0.2 mg l^{-1} kin + 7% CW + 3% sucrose (Bajaj & Gill, 1989)

	No. of pollen studied	Uninucleate	Binucleate	Tetranucleate	Multinucleate	Pollen embryos
Control	880	839 (95%)	41 (5%)	–	–	–
Cold-treated	895	766 (85%)	96 (11%)	11 (1.2%)	14 (1.5%)	8 (0.9%)

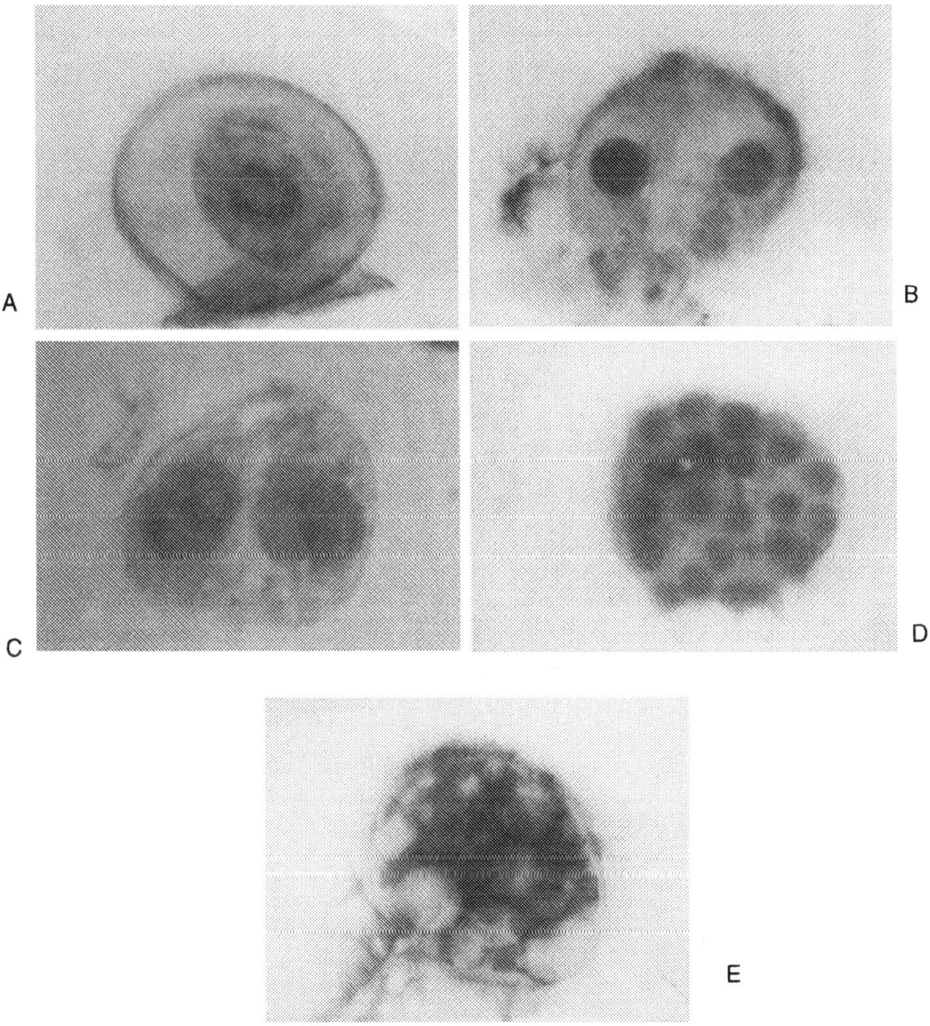

Figure 2. (A-E) Various stages of pollen embryogenesis in excised anthers of *Gossypium arboreum* cultured on MS + 2 mg l^{-1} 2,4-D + 0.2 mg l^{-1} kin + 7% CW + 3% sucrose. (A-C) Uni-, bi-, and tetra-nucleate pollen from anthers cultured for one, two, and three weeks, respectively. (D) A multinucleate pollen from 5-week-old culture (Bajaj & Gill, 1989). (E) A pollen-embryo from a 5–week-old cultured anther (Bajaj & Gill, 1989).

multinucleate microspores were detected only in the cold-treated anthers (Table 2). The percentage of embryo formation was low (0.9%). Most of the microspores were still uninucleate; however, binucleate (10.7%), tetranucleate (1.2%), and multinucleate microspores (1.5%) were observed.

2.3. *Chromosomal variation*

Cytological examination of the callus revealed a wide range of chromosome numbers varying from haploid ($2n = x = 13$) to hexaploid ($2n = 6x = 78$) (Figs. 3 and 4). The haploid cells occurred at a low frequency (3.5%). Most of the cells were diploid (38%) or tetraploid (35%); however, triploid (15%), pentaploid (7%) and hexaploid (1.5%) cells were also present (Fig. 4). Aneuploid cells that were observed were grouped with the nearest euploid category. Some of the cells at anaphase showed laggards.

The occurrence of only a low frequency of haploid cells may indicate that most of the anther-derived callus was derived from somatic tissue, although instability of haploid callus may also be responsible for this phenomenon. Endomitosis and other factors have resulted in the gain or loss of chromosomes in anther-derived callus of wheat (Hu, 1986).

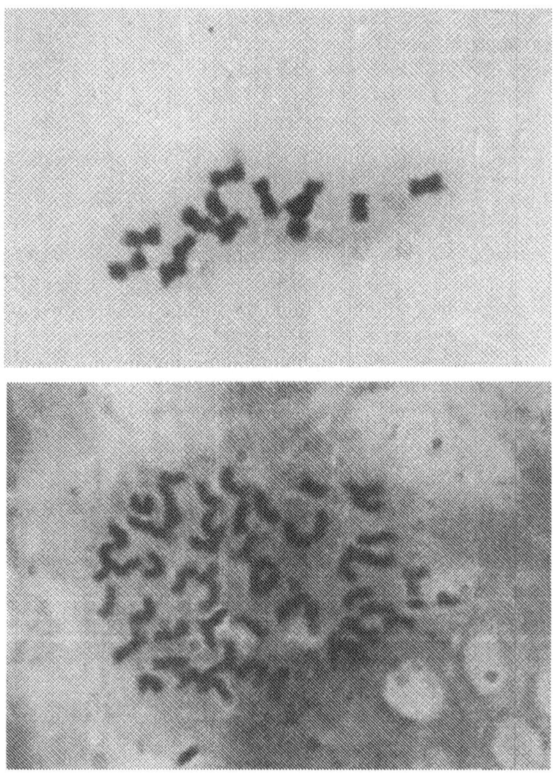

A

B

Figure 3. (A,B) Range of chromosome numbers in anther-derived callus of *Gossypium arboreum*. (A) Haploid number of chromosomes ($2n = 1x = 13$). (B) tetraploid cell ($2n = 4x = 52$).

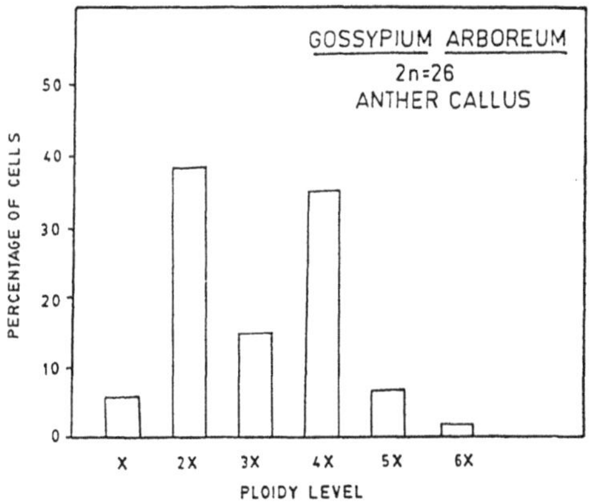

Figure 4. Histogram showing the extent of genetic variability in anther-derived callus; the chromosome number varied from haploid to hexaploid (Bajaj & Gill, 1985).

3. Cryopreservation of anther-derived callus

The conservation of the germplasm by cryogenic methods is important especially for relatively unstable haploid cultures (Bajaj, 1979, 1995). The significance of haploid cell cultures of cotton in mutation and biochemical genetics is evident (Barrow *et al.*, 1978). The results on the freeze preservation of anther-derived cotton callus in liquid nitrogen (−196°C), based on earlier work, are summarized (Bajaj, 1982).

Flower buds obtained from field-grown plants of two species (*G. hirsutum* L. cv F414, and *G. arboreum* cv G 27) were cut open under sterile conditions. Twenty anthers were cultured per test tube containing MS (Murashige & Skoog, 1962) and SH (Schenk & Hildebrandt, 1972) media supplemented with 5 mg l^{-1} 2,4-D. After 10–15 days, the proliferating anthers (Fig. 1B) were again transferred to SH medium with 5 mg l^{-1} 2,4-D + 3% DMSO and incubated further for 4 days. A total of 100 proliferating anthers was pooled together in a Petri dish on a sterile filter paper, moistened with cryoprotectants (mixture of 5% each of sucrose, glycerol and DMSO), and maintained for 2–3 h. During this period, petri dishes were placed in ice. The specimens were wrapped in sterile aluminium foil and subjected to freezing treatment for 5 min in liquid nitrogen. After freezing treatment, the aluminium wrapped material was thawed in warm water (35–38°C), and retrieved anthers were recultured on SH medium containing 5 mg l^{-1} 2,4-D at 23–25°C in diffused light. The resumption of growth in retrieved anthers was considered as a selection survival parameter of anthers.

The proliferating anthers of both species resumed growth. Of all the media tested, the best callus growth was obtained on SH medium supplemented with 5 mg l^{-1} 2,4-D. In control, anthers initiated proliferation within a week

Table 3. Effect of various cryoprotectants on survival of callus cultures of *Gossypium arboreum* frozen in liquid nitrogen. Data based on the resumption of growth (Bajaj, 1982)

Cryprotectant	Survival% of control cultures	
	Anther callus	Ovule callus
Sucrose 8%	7	9
Glycerol 10%	15	19
DMSO 10%	29	27
Sucrose + glycerol + DMSO (5% each)	34	42

and callus formed within 2 weeks of culture. Callus formation initiated by bursting the anthers, which seemed to be of pollen origin, and its texture was compact, whereas callus originating from other parts of anthers was more friable.

Callus growth was dependent on genotype. The callus of *G. arboreum* grew faster than that of *G. hirsutum*. In the latter, callus became spongy and died. Cell suspension obtained from the retrieved callus resumed active growth, and contained highly cytoplasmic aggregates of cells.

The effect of various cryoprotectants on the survival of cells is shown in Table 3. A mixture containing 5% each of sucrose, glycerol and DMSO increased viability of cells compared to chemicals used individually. Anther-derived callus was more sensitive than that obtained from the ovules (Bajaj, 1982).

These cryopreservation studies confirm the earlier work on anther-derived haploid callus of tobacco (Bajaj, 1976), and androgenic anthers of *Atropa* (Bajaj, 1978), rice (Bajaj, 1980), and *Primula* (Bajaj, 1981).

4. Conclusion

In cotton, haploids are conventionally obtained by polyembryony, semigamy, and parthenogenesis. *In vitro* culture of anthers adds yet another dimension to haploid production. Although there has been considerable speculation about the potential of anther-derived haploids in cotton improvement programs (Bajaj & Gill, 1985, 1986), there is a lack of knowledge regarding the factors influencing androgenesis. Among the various pretreatments given to flower buds, low temperature shock has been observed to be the most effective. Pollen embryogenesis and multinucleate pollen were induced through cold treatment. However, no anther-derived haploid plants have yet been reported in cotton.

The low frequency of haploid cells may be the result of haploid callus originated from somatic tissues, and *in vitro* instability of haploid callus. Endomitosis process can form polyploids and aneuploids, and other factors may result in gain or loss of chromosomes in anther callus (Hu, 1986). Since the success of a crop improvement programme depends on the extent of genetic diversity in the base population, some of the *in vitro*-derived genetic variability (Bajaj & Gill, 1985) or gametoclonal variation could be utilized in cotton breeding. Furthermore, methods of micropropagation of cotton have been developed (Bajaj & Gill, 1992) and could be utilized in large-scale multiplication of cotton haploid plants.

The cryopreservation of haploid cultures is of considerable interest for maintaining the genetic stability of cotton haploid cells (Bajaj, 1982).

5. References

Bajaj, Y.P.S., 1976. Regeneration of plants from cell suspensions frozen at −20, −70 and −196°C. Physiol. Plant. 37: 263–268.

Bajaj, Y.P.S., 1978. Effect of super-low temperature on excised anthers and pollen embryos of *Atropa*, *Nicotiana* and *Petunia*. Phytomorphology 28: 171–176.

Bajaj, Y.P.S., 1979. Technology and prospects of crypreservation of germplasm. Euphytica 28: 267–285.

Bajaj, Y.P.S., 1980. Induction of androgenesis in rice anthers frozen at −196°C. Cereal Res. Commun. 8: 365–369.

Bajaj, Y.P.S., 1981. Regeneration of plants from ultra-low frozen anthers of *Primula obconica*. Sci. Hortic. 14: 93–95.

Bajaj, Y.P.S., 1982. Survival of anther-, and ovule-derived cotton callus frozen in liquid nitrogen. Curr. Sci. 51: 139–140.

Bajaj, Y.P.S., 1983. *In vitro* production of haploids. In: D.A. Evans, W.R. Sharp, P.V. Ammirato & Y. Yamada (Eds.), Handbook of Plant Cell Culture, Vol. 1, pp. 228–287. MacMillan Press, New York.

Bajaj, Y.P.S., 1990. Biotechnology in Agriculture and Forestry, Vol. 12. Haploids in Crop Improvement I. Springer-Verlag, Berlin.

Bajaj, Y.P.S., 1995. Biotechnology in Agriculture and Forestry, Vol. 32. Cryopreservation of Plant Germplasm I. Springer-Verlag, Berlin.

Bajaj, Y.P.S. & M.S. Gill, 1985. *In vitro* induction of genetic variability in cotton (*Gossypium* spp.). Theor. Appl. Genet. 70: 363–368.

Bajaj, Y.P.S. & M.S. Gill, 1986. Biotechnology of cotton improvement. In: O.J. Crocomo, W.R. Sharp, D.A. Evans, J.E. Bravo, F.C.A. Tavares, & E.F. Paddock (Eds.), Biotechnology of Plants and Microorganisms, pp. 118–151. Ohio State University Press, Columbus, Ohio.

Bajaj, Y.P.S. & M.S. Gill, 1989. Pollen-embryogenesis and chromosomal variation in anther culture of a diploid cotton (*Gossypium arboreum* L.). SABRAO J. 21: 57–63.

Bajaj, Y.P.S. & M.S. Gill, 1992. Micropropagation of cotton (*Gossypium* species). In: Y.P.S. Bajaj (Ed.), Biotechnology in Agriculture and Forestry, Vol. 19. High-Tech and Micropropagation III, pp. 483–504. Springer-Verlag, Berlin.

Baluch, Z.A.M., 1979. *In-vitro* anther culture of *Gossypium* species. Pakistani Cottons 2: 161–163.

Barrow, J.R., 1986. The conditions required to isolate and maintain viable cotton (*Gossypium hirsutum* L.) microspores. Plant Cell Rep. 5: 405–408.

Barrow, J., F. Katterman & D. Williams, 1978. Haploid and diploid callus from cotton anthers. Crop Sci. 18: 619–622.

Blank, M.L. & D.C. Allison, 1963. Frequency of polyembryony in certain strains of *Gossypium hirsutum* L. Crop Sci. 3: 97–98.

Chaudhari, H.K., 1978. The use of semigamy in the production of cotton haploids. Bull. Torrey Bot. Club 105: 98–103.

Fryxell, P.A., 1992. A revised taxonomic interpretation of *Gossypium* L. (Malvaceae). Rheedea 2: 108–165.

Hsi, Y.L. & K.I. Wu, 1981. Callus formation induced from the anther culture of Upland cotton (*Gossypium hirsutum* L.). In: Abstr. Proc. Symp. Plant Tissue Culture, pp 239–240. Pitman Publ. Ltd, London.

Hu, H, 1986. Wheat: Improvement through anther culture. In: Y.P.S. Bajaj (Ed.), Biotechnology in Agriculture and Forestry, Vol. 2. Crops I, pp. 55–72. Springer-Verlag, Berlin.

Lee, J.A., 1970. On the origin of haploid/diploid twinning in cotton. Crop Sci. 10: 453–454.

Mahill, J.F., J.N. Jenkins, J.C. McCarty Jr. & W.L. Parrott, 1984. Performance and stability of doubled haploid lines of Upland cotton derived via semigamy. Crop Sci. 24: 271–277.

Murashige, T. & F. Skoog, 1962. A revised medium for rapid growth and bioassays with tobacco tissue cultures. Physiol. Plant. 15: 473–497.

Schenk, R.U. & A.C. Hildebrandt, 1972. Medium and techniques for induction and growth of monocotyledonous and dicotyledonous plant cell cultures. Can. J. Bot. 50: 199–204.

Shamina, Z.B., A.M. Turaev & D.A. Musaev, 1986. Capacity for induction of callus formation in a cotton anther culture. Cytol. Genet. 20: 40–44.

Turaev, A.M. & Z.B. Shamina, 1986. Optimization of the medium for cotton anther culture. Sov. Plant Physiol. 33: 439–444.

Turaev, A.M. & Z.B. Shamina, 1993. Culture of cotton microspores in agarose. Russ. Plant Physiol. 40: 294–297.

Turaev, A.M., A.N. Mun, Z.B. Shamina & R.G. Butenko, 1990. Isolation, fractionation, and cultivation of cotton microspores. Dokl. Bot. Sci. Akad. Nauk SSSR 311: 24–27.

Turcotte, E.L. & C.V. Feaster, 1974. Methods of producing haploids: semigametic production of cotton haploids. In: K.J. Kasha (Ed.), Haploids in Higher Plants – Advances and Potential, pp. 53–64. Univ. Guelph Press, Guelph, Ontario.

Wendel, J.F., C.L. Brubaker & A.E. Percival, 1992. Genetic diversity in *Gossypium hirsutum* and the origin of Upland cotton. Am. J. Bot. 79: 1291–1310.

Zhou, S.Q., D.Q. Qian & X.Y. Cao, 1991. Induction of parthenogenesis, and chromosome behaviour in plants of parthenogenetic origin in cotton (*Gossypium hirsutum*). Genome 34: 255–260.

10. Gynogenic haploids *in vitro*

G. LAKSHMI SITA

Contents

1. Introduction

The importance of *in vitro* induction of haploids in crop improvement has been extensively reviewed (Bajaj, 1990; Hu & Yang, 1986). During the last decade or so, tremendous progress has been made in this area. Induction of haploids by androgenesis has been successfully and extensively exploited for crop breeding. Haploid plants thus obtained have the gametophytic number of chromosomes and have been useful in studies on the induction of mutations and also for the production of homozygous plants. Recent advances in plant tissue culture have resulted in the successful induction of haploid plants from ovary and ovule culture. This would indicate that megaspores or female gametophytes of angiosperms can be triggered *in vitro* to sporophytic development, thus opening new vistas for studies on crop improvement and haploid breeding. To distinguish them from androgenic plants derived from microspores, megaspore-derived plants have been described as gynogenic. *In vitro* culture of unpollinated ovaries and ovules represents an alternative for the production of haploid plants in species for which anther culture has either given unsatisfactory results (e.g., yielded too many albinos) or has proven insufficient, as for *Gerbera* (Cappadocia & Vieth, 1990).

Several reviews (Maheshwari & Rangaswamy, 1965; Rangan, 1982, and references therein) have described initial efforts on ovary and ovule culture. At first, ovary culture was used to gain a better understanding of several aspects of fruit physiology, such as morphogenesis and physiological and biochemical changes. With ovary culture, these could be studied under controlled environmental and nutritional conditions. Ovaries from several species have been grown *in vitro* with variable success. Difficulty in growing very young or minute embryos led to attempts to culture ovules. As early as 1932, White cultured ovules of *Antirrhinum*. Subsequently, ovules of several species have been cultured, mostly with a view to understand the

S.M. Jain, S.K. Sopory & R.E. Veilleux (eds.), In Vitro *Haploid Production in Higher Plants, Vol. 5,* 175–193.

factors leading to development of the zygote to a mature embryo (Maheshwari & Rangaswamy, 1965; Rangan, 1982) or to reduce the time between pollination and maturation of seeds. Hayashi *et al.* (1986) and Straathof *et al.* (1987) exploited possibilities of ovule and ovary culture as alternatives to embryo rescue. Van Tuyl *et al.* (1991) refined these methods to include *in vitro* pollination, fertilization and embryo rescue and thereby reduced the time between pollination and *in vitro* culture from 40 days after pollination (DAP) with embryo rescue, to seven DAP with ovary slice culture in *Lilium*.

Until the late 1970s, ovary and ovule culture were underexploited (Rangan, 1984a,b). Despite at least 100 successful reports of embryos from the haploid components of the embryo sac, their potential application was long neglected. Subsequently, several reports suggested that unfertilized ovule culture may be a promising approach to obtain gynogenic haploids (Yang & Zhou, 1982). Likewise, a significant offshoot of research on ovule culture has been the technique of test tube pollination and fertilization, which has been effectively employed to raise interspecific and intergeneric hybrids (Maheshwari & Rangaswamy, 1965; Zenkteler, 1970). The present paper summarizes results obtained with ovary and ovule culture to obtain gynogenic haploids and examines the potential of gynogenic haploids in crop improvement.

2. History of ovary and ovule culture

Early work on culture of unfertilized ovules has been previously reviewed (Rangan, 1982; Yang & Zhou, 1982). Summarizing the work of that period, Maheshwari & Rangaswamy (1965) concluded that "ovule culture may also prove useful in the artificial induction of parthenogenesis. A direct handling of the egg of angiosperms is by no means easy because of the problems involved in removing them without injury. However, there is no such bar to the use of unfertilized ovules. While ovules excised at the zygote or 2–4 celled proembryo stage can be reared *in vitro*, unpollinated and unfertilized ovules have not been proved amenable". While ovule and ovary culture suffered setbacks, Guha & Maheshwari (1964) made a breakthrough in anther culture diverting attention to androgenic haploids. As a result, culture of female tissues was neglected. However there were scattered reports during the 1970s of studies on gynogenic haploids wherein developmental changes were observed. For example, Uchimiya *et al.* (1971) cultured unpollinated ovaries of *Zea mays* and ovules of *Solanum melongena* and observed the division of haploid cells in callus tissues. They believed that *in vitro* induction of haploid plants from angiosperm megagametophytes was possible. Jensen *et al.* (1977) found that the unfertilized polar nuclei could undergo free nuclear division and subsequent precocious cell formation; degeneration of one of the synergids occurred after pollination. The first successful report on the induction of gynogenic haploids was in barley by San Noeum (1976, 1979). Subsequently, haploid plants were obtained from ovary or ovule cultures of wheat and tobacco (Chu & Wu, 1979), rice (Asselin de Beauville, 1980; Zhou & Yang, 1981a,b,c, 1991) barley (Wang & Kuang, 1981; Huang *et al.* 1982) maize (Ao *et al.* 1982; Truong Andre & Demarly, 1984) sunflower (Cai & Zhou, 1984) rubber, mulberry, etc. (Table 1).

3. Gynogenesis in crop plants

In barley, a greater frequency of normal plants has been obtained through ovule and ovary cultures compared to anther culture, where 99% albinos have been regenerated. This more than compensates for the comparatively low frequency of gynogenic haploids because all regenerated plants are haploid and green. Anther culture and interspecific crossing with *Hordeum bulbosum* have also been used to obtain haploids of barley; however, both methods have limitations. Anther culture has resulted in polysomaty in both pollen calli and regenerated plants. In addition, more than 20% of the regenerated plants have been reported to be sterile and grass-like with narrow leaves, multiple tillers and short stature (Mix *et al.*, 1978). Anther culture has been more efficient than the *H. bulbosum* method for winter barley, whereas the *H. bulbosum* method has been more efficient for spring barley. The low morphogenetic capacity of pollen calli and high frequency of albinism still restrict the application of anther culture for barley improvement.

San Noeum (1976) used 9.0 μm 2,4-dichlorophenoxyacetic acid (2,4–D) to induce callus from barley ovaries. Regeneration was not possible if the calli remained more than 2–3 months on regeneration medium, perhaps due to chromosomal aberrations induced in culture. Although a balance between auxin and cytokinin seems necessary, their total concentration should be as low as possible since an excess of growth regulators may induce chromosomal variation in culture. Haploid cell lines have a greater tendency to increase their ploidy than diploid ones. Castillo & Cistué (1993) improved the frequency of haploid production from cultured ovaries of barley by the use of a combination of plant growth regulators and selection of ovaries at a specific stage of development. According to them, 4-chloro-2-methyl phenoxyacetic acid (MCPA) at 0.6 μm, indole acetic acid (IAA) at 2.8 μm and benzyladenine (BA) at 4.4 μm in N_6 medium gave the optimal response. Further increase in the rate of callus formation and the number of green plants was possible with the addition of 90 g l^{-1} sucrose and 100 g l^{-1} coconut water. Of the two cytokinins tried, BA was better than kinetin (kin) for producing green haploid plantlets. The rates of callus induction and regenerated plants were also higher for MCPA than for 2,4-D, whether used alone or in combination with another growth regulator. All plants obtained by direct embryogenesis or from callus during the first 45 days of culture were haploid and green. However, calli that were kept on regeneration medium longer than two months were mixoploid. Haploids from ovary culture of barley have also been reported by Huang *et al.* (1982) and Gu & Zheng (1984).

Table 1. Some important examples of ovary/ovule culture

Rice (*Oryza sativa*)	Ovary	Zhou & Yang, 1981a
		Zhou & Yang, 1981c
		Zhou *et al.*, 1983
		Zhou *et al.*, 1986
		Asselin de Beauville, 1980
Barley (*Hordeum vulgare*)	Ovary/ovule	San Noeum, 1976, 1979
		Wang & Kuang, 1981
		Huang *et al.*, 1982
		Gu & Zheng, 1984
		Gu & Cheng, 1983
Wheat (*Triticum vulgare*)	Ovary	Chu & Wu, 1979
Maize (*Zea mays*)	Ovary	Truong-Andre & Demarly, 1984
		Genovesi, 1990
Onion (*Allium cepa*)	Flower buds	Cohart, 1994
	Ovule	Campion & Alloni, 1990
	Ovules/ovaries and flower buds	Keller, 1990
	Ovary	Muren, 1989
	Flower buds	Smith *et al.*, 1991
Sugar beet (*Beta vulgaris*)	Ovules	Ferrant & Bouharmont, 1994
	Ovules	Bossoutrot & Hosemans, 1985
		D'Halluin & Keimer, 1985
	Ovules and ovaries	Doctrinal *et al.*, 1989
		Hosemans & Bossoutrot, 1983
		Lux *et al.*, 1990
		Van Geyt *et al.*, 1987
		Sangwan Noreel & Duhoux, 1982
Sunflower (*Helianthus annuus*)	Ovary	Sitbon, 1981
		Cai & Zhou, 1984
		Mix, 1985
		Ahmin & Vieth, 1986
Gerbera (*Gerbera jamesonii*)	Ovule	Ahmin & Vieth, 1986
		Sitbon, 1981
		Cappadocia *et al.*, 1988
		Meynet & Sibi, 1984
		Honkanen *et al.*, 1992
Poplar (*Populus* spp.)	Unpollinated ovaries	Wu & Xu, 1984
		Wu & Nagarajan, 1990
Rubber (*Hevea brasiliensis*)	Unpollinated ovaries	Chen *et al.*, 1987
Mulberry (*Morus alba*)	Unpollinated ovaries	Lakshmi Sita & Ravindran, 1991

For obtaining gynogenic haploids of rice, several systematic studies have addressed: a) various factors which affect ovary culture, b) embryological studies on the development of gynogenic processes in cultured ovaries, and c) cytological studies of regenerated plants (Zhou & Yang, 1981a,b; Tian & Yang, 1983; Zhou *et al.*, 1983; Liu & Zhou, 1984). Zhou *et al.*, 1986; Zhou & Yang, 1991) have reviewed *in vitro* culture of unpollinated ovaries in rice. Regeneration of gynogenic plants was dependent on the genotype of the donor cultivar with Japonica types more responsive. The developmental stage of the embryo sac was also important, with optimal response occurring when ovaries were cultured at stages ranging from uninucleate to four-nucleate embryo sacs. As the embryo sac matured, the induction frequency decreased. Unlike androgenesis, cold treatment to embryo sacs before culture was ineffective. Retention of floral parts was beneficial and culture of unhusked flowers with intact pistil, stamens, and sterile glumes gave positive results. Zhou & Yang (1981a) used float culture on liquid media for rice and sunflower. Among the growth regulators used, MCPA was better than other auxins, with low concentrations (0.1–0.5 mg l^{-1}) capable of inducing gynogenesis. Increase in MCPA to 2 mg l^{-1} or more resulted in callusing. Other growth regulators, including 4-amino-3,5,6-trichloro picolinic acid (picloram) and naphthalene acetic acid (NAA), and increasing sucrose from 1–2% to 3–6% were also beneficial (unpublished). Plantlets were mostly obtained by organogenesis from callus cultures. Origin of gynogenic embryos in cultured rice ovaries seemed to be through apogamy of one of the synergids. Antipodals usually formed multicellular glandular structures in cultured ovaries. By comparison, in barley, the egg or antipodal cells developed into embryos whereas synergids only resulted in callus. Among the 111 gynogenic rice plants obtained by ovary culture, 77.5% were haploid and 22.5% diploid.

While anther culture has been reasonably successful in maize and wheat, ovary/ovule culture has had only limited success (Chu & Wu, 1979; Ao *et al.* 1982; Truong-Andre & Demarly, 1984). Genovesi (1990) briefly discussed the salient features of ovule culture in his review on haploids in maize. According to him, induction is strongly genotype-dependent. The degree of success is also dependent on the season of the year and thus the physiology of the donor plants. Both N_6 and MS (Murashige & Skoog, 1962) basal media were capable of initiating responses. Sucrose concentration in the range of 5–12% could be used. Physiologically mature ovules were necessary to induce a response. Of 175 cultured ovules, six responded but embryos germinated directly from only two (Ao *et al.*, 1982). Truong-Andre & Demarly (1984) obtained six plants, one of which yielded selfed seed, from 317 cultured ovules.

4. Gynogenesis in vegetables and ornamentals

Among vegetable crops and ornamentals, most success with gynogenesis has been achieved with sunflower (*Helianthus annuus* L.), sugar beet (*Beta vulgaris*) Gerbera daisy (*Gerbera jamesonii* H. Bolus ex Hook), and onion (*Allium cepa*); hence, these will be discussed. Many new lines of sunflower are now available which are superior in combining ability, agronomic type, oil percentage and disease and insect resistances. To combine various characters in a breeding program requires at least 4–6 years compared to a single year or so by anther/ovary/ovule culture. Yang *et al.* (1986) reported results of ovule culture on eight cultivars of sunflower. Ovules were cultured on N_6 medium supplemented with 0.5–2 mg l^{-1} MCPA and 6% sucrose. After 45–55 days of culture, small globular or rod-shaped embryos could be dissected from the ovules. A total of 7576 ovules was cultured in liquid medium. Among these, 360 contained embryos. After transfer to solid medium, 96 embryos germinated precociously, i.e., only root development was observed and shoot development was incomplete. Although they observed haploid cells cytologically, Yang *et al.* (1986) failed to regenerate well-developed plantlets. Thus the potential of obtaining haploids in sunflower by ovule culture was demonstrated but this potential was not realized. Mix (1985) cultured unpollinated ovaries of different cultivars of sunflower. Callus was induced from all the cultivars on MS medium supplemented with 0.25 mg l^{-1} 2,4-D, 1 mg l^{-1} kin, 1 mg l^{-1} gibberellic acid (GA_3) and 0.5 mg l^{-1} thiamine-HCl. Regeneration was obtained for two cultivars on MS medium supplemented with 0.1–0.4 mg l^{-1} thiamine-HCl and 0.5 mg l^{-1} BA or MS with 1 mg l^{-1} each of IAA, NAA, BA, 0.5 mg l^{-1} kin and 10% coconut milk. Although 19 plants were regenerated, all showed diploid metaphases. Yang *et al.* (1986) obtained haploid embryos from cultured sunflower ovules; however, development of these embryos was arrested by 40 days after inoculation of the unfertilized ovules.

In vitro culture of unfertilized ovules has been the most efficient and reliable technique for obtaining haploid and doubled haploid plants of sugar beet (Doctrinal *et al.*, 1990). The first gynogenic haploid beets were obtained using a male sterile genotype (Hosemans & Bossoutrot, 1983; Van Geyt *et al.*, 1987; Lux *et al.*, 1990). Gynogenesis was more efficient than pollination of male sterile *Beta vulgaris* with *B. vulgaris* ssp. *esculenta*, i.e., red beet, for obtaining haploids. Bossoutrot & Hosemans (1985) reported gynogenesis using male fertile plants. Two plants were obtained from 1000 cultured ovules. Histological studies revealed that embryos came from antipodal cells. The gynogenic plants displayed endopolyploidization at the root meristem whereas the shoot meristem remained haploid. Doctrinal *et al.* (1989, 1990) later published a detailed analysis of various parameters for the production of gynogenic beet. They increased the percentage of ovule-derived haploids to approximately 10% of cultured ovules, compared with the conventional technique where only approximately 0.01–0.02% success was common. In a recent report, Ferrant & Bouharmont (1994) indicated that the maturity of the gametophyte was reached a few days before anthesis and therefore ovules from unopened flowers were suitable for culture. The majority of plants regenerated were haploid. Sugar beet breeding should be improved by the availability of homozygous inbred lines that may hasten the development of new cultivars in this allogamous crop.

Gerbera jamesonii, an allogamous species, is characterized by a chromosome number of $2n = 50$. Genetic studies performed thus far have not determined whether *Gerbera* is a true diploid or a tetraploid. Because of its allogamous mode of reproduction, it may be advantageous to produce homozygous diploids for use in breeding. Sitbon (1981) was the first to report haploid production from capitulum explants of *Gerbera*. An improved method of haploid production from unfertilized ovules in *Gerbera jamesonii* was reported by Meynet & Sibi (1984). About 200 plants were regenerated from 8000 ovules of 19 cultivars. Regeneration was indirect via organogenesis from callus, each step needing different conditions. The mineral composition was critical for production of callus, with MS medium better than other media used (Sitbon, 1981). Among the growth regulators tried, two cytokinins (BA at 2 mg l^{-1} and kin at 2 mg l^{-1}) used simultaneously and auxin at 0.5 mg l^{-1} gave the most satisfactory result, which the authors attributed to a synergistic effect of the two cytokinins (Sitbon, 1981). Meynet & Sibi (1984) found that a combination of IAA (0.05 mg l^{-1}) and BA (0.3 mg l^{-1}) was appropriate for callus induction whereas a medium containing BA (3 mg l^{-1}), IAA (0.5 mg l^{-1}) and GA_3 (0.1 mg l^{-1}) was able to sustain callus growth. Regeneration was obtained on medium containing IAA (0.25 mg l^{-1}) and BA (1 mg l^{-1}). Ahmin & Vieth (1986) also produced callus from unfertilized ovules by transferring them to a medium with IAA (0.1 mg l^{-1}) and BA (2 mg l^{-1}). Cappadocia *et al.* (1988) also examined several combinations of growth regulators.

Onion is another allogamous crop for which the main limitation of conventional breeding has been the time required to develop inbred lines to be combined in hybrids. Initial experiments to produce gynogenic plants were unsuccessful according to Campion *et al.* (1992). Campion & Alloni (1990) first demonstrated that onion haploids could be obtained *in vitro*. Gynogenic plants have also been reported by Cohart (1994). However the percentage of ovules producing embryos has been small (0.28%). The use of adenine sulphate to reduce flower degeneration has given rise to a remarkable increase in the number of culturable ovules per flower. Campion *et al.* (1992) further increased the yield of gynogenic plants and tried to identify the parameters. According to them, gynogenic embryos from ovule culture required approximately 25 days to develop and become visible, including 10–14 days of flower preculture. By comparison, with ovary and flower bud culture, gynogenic embryos were visible after 50–55 and 65–70 days, respectively. The yield of gynogenic embryos from ovule culture was significantly increased when: 1) 10% sucrose was supplied in both preculture and culture media, and 2) 2,4-D and BA were both used at 2 mg l^{-1} in the preculture medium. Embryo yield from ovary culture was significantly improved in four cultivars by using a modified BDS medium containing 100 g l^{-1} sucrose, 1 mg l^{-1} NAA and 2 mg l^{-1} BA; a high percentage of embryos was induced from flower bud culture by using a modified BDS medium containing 100 g l^{-1} sucrose, 2 mg l^{-1} 2,4-D and 2 mg l^{-1} BA. Statistical analysis of the data revealed a close relationship between the kind of cultured organs (ovary or flower) and the type of auxin in the medium. Ovary culture gave the best yield in the presence of NAA whereas both ovary and flower bud culture responded well to 2,4-D. Whole plantlets were regenerated from 36% of the embryos and 88.3% of the regenerants were haploid. The capacity of each type of organ to produce embryos was also influenced by the genotype and environmental conditions of the donor plant. Campion *et al.* (1992) also suggested the use of both ovary and flower bud culture.

5. Gynogenesis in tree crops

Tree improvement has been hampered due to many problems common among tree species, including long life cycles, heterozygosity, polyploidy, and dioecy, resulting in extensive outcrossing. Stettler *et al.* (1969) recognized the significance of haploids for tree crops and reported the induction of haploids by parthenogenesis in cottonwood (Stettler & Bawa, 1971). *In vitro* production of haploids to produce homozygous doubled haploids is extremely important to take advantage of heterosis in tree improvement. Thorpe & Biondi (1984) stated that "trees of poor or heterogeneous quality and slow or unreliable propagation method are no longer acceptable. There is an urgent need to make broad and concerted effort to domesticating and breeding trees with superior wood quality, optimal stem form and uniformity, rapid growth rates, short rotations and high production index, resistance to diseases and pests and ability to adapt to climates and extreme environmental variables, ability to respond well to sylvicultural practices, etc.".

It is now possible to obtain haploids and homozygous diploid plants with superior genotypes by anther culture and unpollinated ovule culture (Chen, 1990). Significant contributions have been made in obtaining pollen plantlets of eight families, ten genera, and more than 20 species of economic forest and fruit trees (Chen, 1986). A complete procedure for haploid breeding biotechnology has been established by study of the culture medium, culture conditions, optimal time of inoculation of anthers, transfer of callusing anthers and pollen embryos, and transplantation techniques of pollen plantlets. However, success with ovule culture has been more limited, although Chen *et al.* (1987) reported haploid induction from unpollinated ovules of rubber. Four plants were obtained. Unpollinated ovules were inoculated onto a medium which induced the haploid cells in the ovule to produce calli or embryos. The MB (Chen *et al.*, 1987) medium supplemented with 1–1.5 mg

l^{-1} kin or 0.5–1.0 mg l^{-1} BA, 1–2 mg l^{-1} NAA and 1 mg l^{-1} 2,4–D was effective for callus induction. For differentiation, callus was subcultured to 0.5–1.0 mg l^{-1} NAA, 0.5–0.8 mg l^{-1} GA, 8% sucrose, and 0.5% agar. Embryos were obtained and retained on the same medium for about 90 days. Cytological studies confirmed the plantlets obtained by unpollinated ovule culture were haploid and originated from the embryo sac.

In 1974, haploid production was first reported for *Populus trichocarpa* by androgenesis whereas Stettler & Bawa (1971) reported parthenogenesis in *Populus trichocarpa* somewhat earlier. In the parthenogenetic method of reproduction, the embryo develops directly from an unfertilized egg. If the chromosome number of the gamete has been reduced in the normal manner at meiosis and chromosome doubling of the unfertilized gamete does not occur, the apomictically produced embryo and the plant developing from it will be haploid. Wu & Xu (1984) produced haploid poplars by *in vitro* ovary culture. Wu & Nagarajan (1990) have reviewed the work on haploids in poplars. Plantlets were obtained both directly and via callus phase. Direct regeneration simplified the culture procedures by reducing the time and eliminating the complications of chromosomal aberrations due to a prolonged culture phase.

Gynogenic plants have recently been reported by Lakshmi Sita & Ravindran (1991) in mulberry (*Morus indica*). Mulberry (*Morus* spp.) is important in the silk industry to nourish the silkworm (*Bombyx mori*). The conventional method of propagating mulberry is by stem cuttings. Its perennial nature coupled with a prolonged juvenile period limit the speed of improvement in this crop. Dioecy is also a barrier to genetic improvement by conventional techniques. Moreover, haploid induction by anther culture was found to be difficult.

Flowers of mulberry are unisexual and occur on separate plants. The female flower consists of four perianth lobes and the pistil consists of ovary, style and forked stigma. In the ovary, there is a single ovule. Multiple or composite fruits are formed from several free ovaries of the whole inflorescence. The rachis, bracts, and the fleshy perianth fuse to form the edible parts. Each polygonal area, called a sorosis, represents a flower. Individual ovaries or soroses were cultured on MS medium supplemented with growth regulators. A total of 65 plantlets was obtained by direct regeneration without a callus phase. In some media, e.g., that with 1 mg l^{-1} BA and 1 mg l^{-1} kin, four plantlets were obtained from a single ovary. Usually plantlets emerged within three weeks. Figure 1 shows different stages of the emerging plantlets with cotyledons and hypocotyls. Plantlets were complete with a well-developed tap root, suggesting direct embryogenesis. Since histology was not done, the exact origin was not determined. Plantlet growth was slow. Initially, when embryos were transferred to media containing GA_3 (1 mg l^{-1}) and kin (0.3 mg l^{-1}), growth was normal. After 15 days, plantlets were subcultured to half-strength MS salts supplemented with 0.5–1.0 mg l^{-1} indolebutyric acid (IBA). Profuse rooting was observed after this transfer. When plantlets

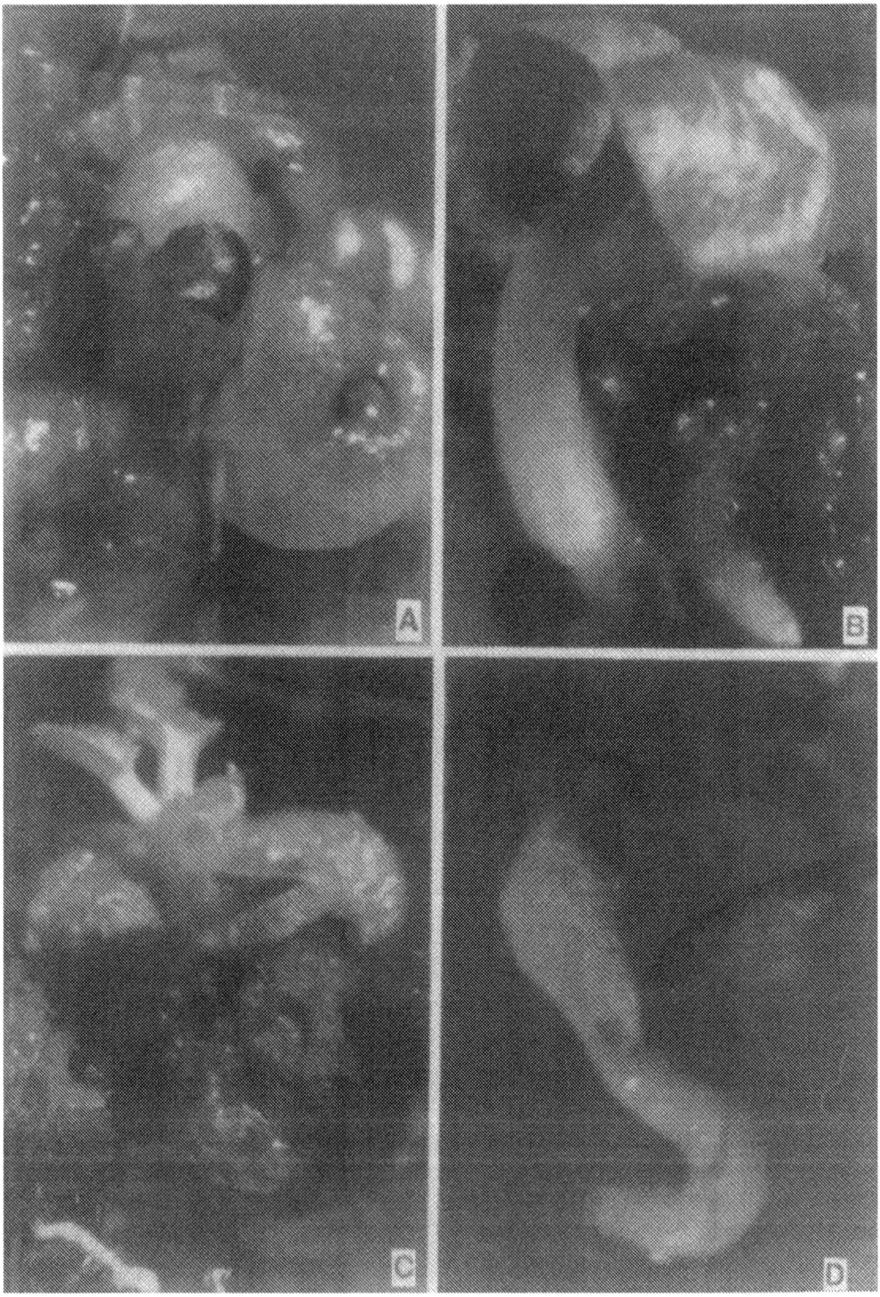

Figure 1. Induction of gynogenic plants in mulberry. (A-D) showing different stages of emerging gynogenic plants.

were 5–7 cm with 3–5 nodes, nodal cuttings were taken and cultured on MS media containing 0.1 mg l^{-1} BA and 0.2 mg l^{-1} kin on which well-developed plants were formed. Occasionally 3–4 multiple shoots were obtained on media containing cytokinins. These in turn could either be rooted or used as source material for nodal explants. Cytological preparations were made from the root tips. Some exhibited the haploid number of chromosomes (n = 14) whereas others were diploid. On media supplemented with 3 mg l^{-1} BA, 1 mg l^{-1} kin, and 0.5 mg l^{-1} NAA, plantlets regenerated consistently. On media supplemented with 1 mg l^{-1} BA and mg l^{-1} kin alone, four embryos were obtained. The authors concluded that the plants obtained from the cultured ovaries were of haploid origin since the ovaries were not pollinated and 3–4 embryos could emerge from a single ovary, suggesting the participation of embryo sac cells. For a normal fertilized ovary, only a single plantlet would have been expected. Also there were no male trees in the vicinity and no seeds were observed in the fruits. The regenerated diploids ($2n$ = 28) could have been due to chromosome doubling during embryo development, as in lettuce and sunflower (San Noeum & Gelebart, 1986). During gynogenesis, the rate of spontaneous diploids has been higher in many plants compared to androgenesis (San Noeum & Gelebart, 1986). The haploid chromosome number has often been observed only in the first root tip of the plantlets; after transfer, all root tips analyzed were diploid. We observed a similar phenomenon in mulberry.

Gynogenesis in gymnosperms has been reported in many species (Rohr, 1987). Although the results have shown that regeneration by gynogenesis is potentially possible, apogamous plants have remained small and did not survive after transfer from the *in vitro* environment to soil due to lack of proper vascular connections between shoot and root. In comparison, mulberry plantlets obtained by ovary culture were complete with well-developed shoot and root systems which formed normal lateral roots when established in pots. So far, *in vitro* induction of haploid trees has been more successful by androgenesis rather than gynogenesis.

6. Gynogenesis in gymnosperms

The importance of haploids to produce homozygous diploids in gymnosperms has been well-recognized by geneticists. Breeders have long sought haploid and doubled haploid plants as a tool for tree improvement. In gymnosperms, there are two major sources of haploid cells, pollen and the female gametophyte. Each microspore in the microsporangium is genetically unique; therefore a callus derived from more than one microspore will reflect a mixture of genotypes. On the other hand, the megagametophyte is formed from a single cell, i.e., the megaspore; therefore, its tissues are isogenic and eminently suited to the production of haploid or doubled haploid plants. The more frequent route to haploidy in gymnosperms, however, has been androgenesis. There are a number of reviews on haploid induction in tree species (Chen, 1987; Rohr, 1987; Sangwan-Norreel & Duhoux, 1982; Bonga *et al.*, 1988). Little information is available on gynogenesis or megagametophyte culture of gymnosperms. Haploid callus has been derived from megagametophyte tissue culture of *Ginkgo biloba* (Tulecke, 1964). Haploid cultures from other gymnosperms include *Gnetum ula* (Vasil, 1993) from which buds and proembryos formed after several passages. In *Zamia integrifolia*, Norstog (1965) reported embryos and shoots. Bonga (1974) cultured megagametophytes of *Pinus nigra* var. *austriaca* and *P. mugo* var. *mughus*. Callus from *P. mugo* remained haploid for a year after which it started to double slowly; one complete plantlet was obtained. There were also reports of haploid callus in *Taxus* (Zenkteler & Guzowska, 1970) and *Sequoia* (Ball, 1987). This work was limited to the induction of callus with only occasional regeneration. Many calli were capable of organogenesis, particularly when the initial culture medium, often rich in auxin and cytokinin was replaced on subculture by media reduced in growth regulators. The simplest media have often supported the most successful apogamy. Norstog (1982) obtained 59% organogenesis in megagametophyte cultures of *Zamia* on a more complex medium, i.e., modified White's medium with amino acids (alanine, glutamine, asparagine), an auxin, and a cytokinin. Organogenesis was obtained with 2,4–D but was better with IAA. In *Picea abies*, organogenesis was stimulated by a combination of the polyamines, putrescine, spermidine and spermine (Huhtinen *et al.*, 1981).

Rapidity of response in terms of organogenesis has also varied among species. For example, roots developed within two weeks in *Ephedra foliata* (Singh *et al.*, 1981) whereas roots did not appear until after two years in cultures of *Cycas revoluta* (La Rue, 1950). Thus early megagametophyte cultures of gymnosperms were limited to the induction of callus and regeneration by organogenesis. These results showed the potential of gynogenesis in gymnosperms but apogamous plantlets remained small and did not survive transfer to soil (Norstog, 1982). One major problem has been the lack of proper vascular connections between shoots and roots (Huhtinen *et al.*, 1981).

Nagmani & Bonga (1985) first reported the induction of plantlets by embryogenesis from culture of megagametophytes of *Larix decidua* on medium without growth regulators. Small plantlets were transferred to soil. A few formed needles, but none survived after a few weeks in soil. Regeneration of *Larix decidua* was achieved from both pollinated and unpollinated megagametophytes (Von Aderkas *et al.*, 1990). Von Aderkas *et al.* (1990) believed that the composition of the medium, with high concentrations of organic supplements such as casein hydrolysate and glutamine, was critical. They also found that maturation of embryos was possible only with abscisic acid (ABA). The number of genotypes capable of forming green embryos was small. Only two seedlings survived after one year, both derived from the same genotype of *Larix decidua* originally cultured by Von Aderkas *et al.*

(1990). In *Picea abies* organogenesis was reported from callus cultures by Simola & Honkanen (1983).

According to Rohr (1987) and Von Aderkas *et al.* (1990), the embryogenic competence of megagametophyte tissue was dependent on the genotype, environmental factors affecting tree physiology and megagametophyte development, as well as seasonal conditions. Also, the inherent difficulty of dissecting the megagametophytic tissue limits the widespread usage of megagametophyte culture. Gynogenesis can be obtained either by organogenesis or embryogenesis, as long as the ploidy of resulting plantlets is haploid and they are derived from the female gametophyte. It appears that growth regulators are not always required for the initiation of andro- or gynogenesis.

7. Use of irradiated pollen for haploid induction

Induction of *in situ* gynogenesis has been obtained by combining the use of irradiated pollen with *in vitro* culture of pollinated ovaries or immature seeds. Winton & Einspahr (1968) irradiated *Populus alba* pollen with 100 rads of gamma irradiation and pollinated untreated female flowers of *P. tremuloides*. One slow-growing plant from this cross revealed a haploid chromosome number, but later gave rise to a fast-growing diploid branch. Raquin (1985) first used this method of haploidization successfully in *Petunia* and this was followed by successful reports in melon (Sauton & Dumas de Vaulx, 1987; Cuny *et al.*, 1993), carrot (Rode *et al.*, 1987), and rose (Meynet *et al.*, 1994).

To overcome the problem of low yields of plantlets obtained by *in situ* gynogenesis and difficulties associated with *in vitro* androgenesis Zhang *et al.* (1988) sought other methods of haploid induction in apple. Two cultivars of apple, Lodi and Erovan, were used as female parents for pollination with irradiated pollen. The pollen was collected from a clone that was homozygous (*RR*) for the marker gene *R* that codes for anthocyanin production and hence gives a red phenotype to true hybrids. Pollen was irradiated by gamma rays from cobalt 60 at 45 Gy to give total doses of 500, 1000 and 1500 Gy. Pollination was done inside an isolation cage. Ovules from young fruits (1–4 weeks after pollination) were cultured on MS medium supplemented with NAA, BA and GA. The haploid plants obtained were all green (*rr*) compared to the normal red diploid plants. Chromosome counts from apices of adventitious shoots of *in vitro* cultures showed the haploid number of 17.

In most *Brassica* spp., anther culture has been successfully used for haploid induction; however, cabbage has been more recalcitrant. Doré (1989) successfully used the technique of irradiated pollen. The experiments included the use of three early nuclear genetic markers. Pollination with pollen irradiated with 200 to 1200 Gy followed by *in vitro* ovary culture resulted in two haploid plants (approximately 2%). The few ovules per ovary is a limiting factor for this procedure. Pollination with irradiated pollen can be considered as an alternate technique for the induction of gynogenic plants when other methods have been unsuccessful.

8. Conclusions

In vitro androgenic haploids have been successfully employed for the improvement and release of cultivars in agricultural crops, such as wheat, rice, maize, barley, and *Brassica*. Although the potential of gynogenic haploids has been well-documented, they have not been used nearly as much as androgenic haploids for crop improvement. Ovary culture may result in a higher frequency or green regenerants, as for barley. Positive gynogenic responses may be obtained by using a range of developmental and physiological stages of embryo sac and various culture conditions. The enzyme maceration technique developed recently is an exciting and promising approach for isolating the female gametophyte of angiosperms (Zhou & Yang, 1986). Also embryo sac culture itself may lead to a better understanding of the origin of gynogenic plants. Some work along these lines has already been initiated (Zhou & Yang, 1986). Embryo sac culture may represent a refinement of unpollinated ovary and ovule culture just as microspore culture has been developed after success with anther culture (Zhou & Yang, 1986). *In vitro* culture of unfertilized organs still remains complex. In sugar beet, unfertilized ovule culture has proved to be an efficient and reliable technique for obtaining haploid and doubled haploid plants (Doctrinal *et al.* 1990). From a low haploid frequency of 0.01%–0.02% haploids by conventional breeding, the percentage was increased to 10% by ovule culture. In rice, the percentage of success varied from 1.5–12%. In poplar, the frequency of haploid production by ovary culture was much less than 10%. The induction frequency was generally lower than for anther culture, but the actual rate of induced embryo sacs was greater than that of induced pollen grains, considering that there is only one embryo sac per ovary compared with thousands of pollen grains per anther. In mulberry, haploid induction frequency was also 3–6% by ovary culture. To increase the percentage of gynogenic haploids, attention should be paid to various parameters, like genotype, embryo sac stage, effect of cold treatment, effect of floral parts, such as pedicel, calyx or glumes, exogenous growth regulator concentrations, effect of solid and liquid media, sucrose concentration, etc. Another novel approach to gynogenic haploidy has been the use of irradiated pollen for haploid induction. Tree improvement probably may benefit by the use of ovary, ovule and female gametophyte cultures. A positive beginning has been made in gymnosperms with successful induction of gynogenic plants in larch. The future is promising for tree improvement by the various methods described here, which may provide germplasm that would otherwise not be possible.

9. References

Ahmin, M. & J. Vieth, 1986. Production de plantes haploïdes de *Gerbera jamesonii* par culture *in vitro* d'ovules. Can. J. Bot. 64: 2355–2357.

Ao, G.M., S.X. Zhao & G.H. Li, 1982. *In vitro* induction of haploid plantlets from unpollinated ovaries of corn (*Zea mays* L.) Acta Genet. Sin. 9: 281–283.

Asselin de Beauville, M., 1980. Obtention d'haploïdes *in vitro* à partir d'ovaires non fecondés de riz, *Oryza sativa* L. C.R. Acad. Sci. 290D: 489–492.

Bajaj, Y.P.S, 1990. *In vitro* production of haploids and their use in cell genetics and plant breeding. In: Y.P.S. Bajaj (Ed.), Biotechnology in Agriculture and Forestry: Haploids in Crop Improvement I, Vol. 12, pp. 3–44. Springer-Verlag, Berlin.

Ball, E.A, 1987. Tissue culture multiplication of sequoia. In: J.M. Bonga & D.J. Durzan (Eds.), Cell and Tissue Culture in Forestry, Vol. 3, pp. 146–158. Martinus Nijhoff Publishers, Dordrecht.

Bonga, J., 1974. *In vitro* culture of microsporophylls and megagametophyte tissue of *Pinus*. In Vitro 9: 270–277.

Bonga, J.M., P. Von Aderkas & D. James, 1988. Potential application of haploid cultures of tree species. In: J.W. Hanover & D.E Keathley (Eds.), Genetic Manipulation of Woody Plants, pp. 57–77. Plenum Press, New York.

Bossoutrot, D. & D. Hosemans, 1985. Gynogenesis in *Beta vulgaris* L.: From *in vitro* culture of unpollinated ovules to the production of doubled haploid plants in soil. Plant Cell Rep. 4: 300–303.

Cai, D.T. & C. Zhou, 1984. *In vitro* production of haploid embryoids and plantlets from unpollinated young florets and ovules of *Helianthus annuus* L. Kexue Tongbao 29: 680–682.

Campion, B. & C. Alloni, 1990. Induction of haploid plants in onion (*Allium cepa* L.) by *in vitro* culture of unpollinated ovules. Plant Cell Tiss. Organ Cult. 20: 1–6.

Campion, B., M.T. Azzimonti, E. Vicini, M. Sachari & A. Falavigna, 1992. Advances in haploid plant induction in onion (*Allium cepa* L.) through *in vitro* gynogenesis. Plant Sci. 86: 97–104.

Cappadocia, M., L. Chrètien & G. Laublin, 1988. Production of haploids in *Gerbera jamesonii* via ovule culture: Influence of fall versus spring sampling on callus formation and shoot regeneration. Can. J. Bot. 66: 1107–1110.

Cappadocia, M. & J. Vieth 1990. *Gerbera jamesonii* H. Bolus ex Hook: *In vitro* production of haploids. In: Y.P.S. Bajaj (Ed.), Biotechnology in Agriculture and Forestry: Haploids in Crop Improvement I, Vol. 12, pp. 417–427. Springer-Verlag, Berlin.

Castillo, A.M. & L. Cistué, 1993. Production of gynogenic haploids of *Hordeum vulgare* L. Plant Cell Rep. 12: 139–143.

Chen, Z., 1986. Induction of androgenesis in woody plants. In: H. Hu & H. Yang (Eds.), Haploids of Higher Plants *in Vitro*, pp. 42–66. Springer-Verlag, Berlin.

Chen, Z., 1987. Induction of androgenesis in hardwood trees. In: M.R. Ahuja (Ed.), Somatic Cell Genetics of Woody Plants, pp. 247–268. Kluwer Academic Publishers, Dordrecht.

Chen, Z. 1990. Rubber (*Hevea brasiliensis* Muell. Arg.): *In vitro* production of haploids. In: Y.P.S. Bajaj (Ed.), Biotechnology in Agriculture and Forestry: Haploids in Crop Improvement I, Vol. 12, pp. 215–236. Springer-Verlag, Berlin.

Chen, Z., L. Wenbin, Z. Lihua, X. Xuen & Z. Shijie, 1987. Production of haploid plantlets in cultures of unpollinated ovule of *Hevea brasiliensis* Muell-Arg. In: M.R. Ahuja (Ed.), Somatic Cell Genetics of Woody Plants, pp. 39–42. Kluwer Academic Publishers, Dordrecht.

Chu, C.C. & H.S Wu, 1979. *In vitro* production of haploid plantlets from the unpollinated ovaries of *Triticum aestivum* and *Nicotiana tabacum*, spring wheat and tobacco, cultivars. Acta Genet. Sin. 6: 181–183.

Cohart, J., 1994. Production of gynogenetic plants by *in vitro* culture of flower buds in shallot (*Allium cepa* var. *aggregatum*). Agronomie 14: 299–304.

Cuny, F., M. Grotte, R. Dumas de Vaulx & A. Rieu, 1993. Effects of gamma irradiation of

pollen on parthenogenetic haploid production in muskmelon (*Cucumis melo* L.). Environ. Expt. Bot. 33: 301–312.
D'Halluin, K. & B. Keimer, 1985. Production of haploid sugarbeets (*Beta vulgaris* L.) by ovule culture. In: W. Horn, C.J. Jensen, W. Odenbach & O. Schieder (Eds.), Genetic Manipulation in Plant Breeding: Proc. Intern. Symp. EUCARPIA, Sept. 8–13, 1985, Berlin, pp. 307–309. W. de Gruyter, Berlin.
Doctrinal, M., R.S. Sangwan & B.S. Sangwan-Norreel, 1989. *In vitro* gynogenesis in *Beta vulgaris* L.: Effect of plant growth regulators, temperature, genotypes and season. Plant Cell Tiss. Organ Cult. 17: 1–12.
Doctrinal, M., R.S. Sangwan & B.S. Sangwan-Norreel, 1990. Sugarbeet (*Beta vulgaris* L.): *In vitro* induction of haploids. In: Y.P.S. Bajaj (Ed.), Biotechnology in Agriculture and Forestry: Haploids in Higher Plants I, Vol. 12. pp. 346–357. Springer-Verlag, Berlin.
Doré, C., 1989. Obtention de plantes haploïdes de chou cabus (*Brassica oleracea* L. var. *capitata*) après culture *in vitro* d'ovules pollini ses par du pollen irradié. C.R. Acad. Sci. Paris 309: 729–734.
Ferrant, V. & J. Bouharmont, 1994. Origin of gynogenetic embryos of *Beta vulgaris* L. Sex. Plant Reprod. 7: 12–16.
Genovesi, A.D, 1990. Maize (*Zea mays* L.): *In vitro* production of haploids. In: Y.P.S. Bajaj (Ed.), Biotechnology in Agriculture and Forestry: Haploids in Crop Improvement I, Vol. 12, pp. 176–203. Springer-Verlag, Berlin.
Gu, Z.P. & K.C. Cheng, 1983. *In vitro* induction of haploid plantlets from unpollinated young ovaries of lily and its embryological observations. Acta Bot. Sin. 25: 24–28.
Gu, Z.P. & Zheng, 1984. *In vitro* induction of haploid plantlets from unpollinated ovaries of highland barley. Acta Bot. Sin. 26: 549–551.
Guha, S. & S.C. Maheshwari, 1964. *In vitro* production of embryos from anthers of *Datura*. Nature (London) 204: 497.
Hayashi, M., K. Kanoh, Y. Serizawa & E. Yoon, 1986. Ovary slice culture of *Lilium formosanum* Wallace. Jpn. J. Breed. 36: 304–308.
Honkanen, J., A. Aapola, P. Seppanen, T. Tormala, J.C. Wit, L.J.M. de Stravers & H.F. Esendam, 1992. Production of doubled haploid *Gerbera* clones. Acta Hortic. 300: 341–346.
Hosemans, D. & D. Bossoutrot, 1983. Induction of haploid plants from *in vitro* culture of unpollinated beet ovules (*Beta vulgaris* L.). Z. Pflanzenzüchtg. 91: 74–77.
Hu, H. & H.Y. Yang, 1986. Haploids of Higher Plants *in Vitro*. China Academic Publishers, Beijing/Springer-Verlag, Berlin.
Huang, Q.F., H.Y Yang & C. Zhou, 1982. Embryological observations on ovary culture of unpollinated young flowers in *Hordeum vulgare* L. Acta Bot. Sin. 25: 24–28.
Huhtinen, O., L.K. Honkanen & Simola, 1981. Effects of genotype and nutrient media on callus production and differentiation of Norway spruce endosperms cultured *in vitro*. In: Proc. IUFRO Sect. S2O15. Int. Workshop on *in Vitro* Cultivation for Tree Species, Fontainebleau, pp. 307–311.
Jensen, W.A., P. Schulz & M.E. Ashton, 1977. An ultrastructural study of early endosperm development and synergid changes in unfertilized cotton ovules. Planta 133: 179–189.
Keller, J., 1990. Culture of unpollinated ovules, ovaries, and flower buds in some species of the genus *Allium* and haploid induction via gynogenesis in onion (*Allium cepa* L.). Euphytica 47: 241–247.
Lakshmi Sita, G. & S. Ravindran, 1991. Gynogenic plants from ovary cultures of mulberry (*Morus indica*). In: J. Prakash & R.L.M. Pierik (Eds.), Horticulture: New Technologies and Applications, pp. 225–229. Kluwer Academic Publishers, Dordrecht.
La Rue, C.D., 1950. Regeneration in *Cycas*. Am. J. Bot. 37: 664.
Liu, Z.L. & C. Zhou, 1984. Investigation on the chromosome and other characters of the gynogenic and androgenic plants in *Oryza sativa* L. Acta Genet. Sin. 11: 113–119.
Lux, H., L. Herrmann & C. Wetzel, 1990. Production of haploid sugar beet (*Beta vulgaris* L.) by culturing unpollinated ovules. Plant Breed. 104: 177–183.
Maheshwari, P. & N.S. Rangaswamy, 1965. Embryology in relation to physiology and genetics.

In: R.D. Preston (Ed.), Advances in Botanical Research, Vol. 2, pp. 319-321. Academic Press, London.
Meynet, J., R. Barrade, A. Duclos & R. Siadous, 1994. Dihaploid plants of roses (*Rosa* × *hybrida* cv. Sonia) obtained by parthenogenesis induced using irradiated pollen and *in vitro* culture of immature seeds. Agronomie 14: 169–75.
Meynet, J. & M. Sibi, 1984. Haploid plants from *in vitro* culture of unfertilized ovules in *Gerbera jamesonii*. Z. Pflanzenzüchtg. 93: 78-85.
Mix, G., H.M. Wilson & B. Foroughi-Wehr, 1978. The cytological status of barley plants regenerated from microspore callus. Z. Pflanzenzüchtg. 80: 89–99.
Mix, G., 1985. Antherren und Ovar Kultur von Sonnenblumen (*Helianthus* spp.): Landbauforsch. Volkenrode 35: 153–156.
Murashige, T. & F. Skoog, 1962. A revised medium for rapid growth and bioassays with tobacco tissue cultures. Physiol. Plant. 15: 473–497.
Muren, R.C., 1989. Haploid plant from unpollinated ovaries in onion. HortScience 24: 833–834.
Nagmani, R. & J.M. Bonga, 1985. Embryogenesis in subcultured callus of *Larix decidua*. Can. J. For. Res. 15: 1088–1091.
Norstog, K., 1965. Induction of apogamy in the gametophyte of *Zamia integrifolia*. Am. J. Bot 52: 993–999.
Norstog, K., 1982. Experimental embryology of gymnosperms. In: B.M. Johri (Ed.), Experimental Embryology of Vascular Plants, pp. 25–51. Springer-Verlag, Berlin.
Rangan, T.S, 1982. Ovary, ovule and nucellus culture In: B.M. Johri (Ed.), Experimental Embryology of Vascular Plants, pp. 105–129. Springer-Verlag, Berlin.
Rangan, T.S, 1984a. Culture of ovaries. In: I.K. Vasil (Ed.), Cell Culture and Somatic Cell Genetics of Plants, Vol. 1, pp. 221–226. Academic Press, New York.
Rangan, T.S., 1984b. Culture of ovules. In: I.K. Vasil (Ed.), Cell Culture and Somatic Cell Genetics of Plants, Vol. 1, pp. 227–231. Academic Press, New York.
Raquin, C., 1985. Induction of haploid plants by *in vitro* culture of *Petunia* ovaries pollinated with irradiated pollen. Z. Pflanzenzüchtg. 94: 166–169.
Rode, J.C., R. Dumas & D. DeVaulx, 1987. Obtention de plantes haploïdes de carotte (*Daucus carota* L.) issues de parthenogenèse induite *in situ* par du pollen irradiet culture *in vitro* des graines immatures. C.R. Acad. Sci. Paris 305: 225–229.
Rohr, R., 1987. Haploids (Gymnosperms). In: J.M. Bonga & D.J. Durzan (Eds.), Cell and Tissue Culture in Forestry, Vol. 2, pp. 230–246. Martinus Nijhoff Publishers, Dordrecht.
Sangwan-Noreel, B.S. & E. Duhoux, 1982. Les conditions de la gynogenèse et de l'androgenèse *in vitro* chez les arbres. Rev. Cytol. Biol. Veget. Bot. 5: 171–187.
San Noeum, L.H., 1976. Haploïdes d'*Hordeum vulgare* L. par culture *in vitro* d'ovaires non fecondés. Ann. Amélior. Plantes 26: 751–754.
San Noeum, L.H., 1979. *In vitro* induction of gynogenesis in higher plants. In: A.M. van Harten & A.C. Zeven (Eds.), Proc. Conf. Broadening the Genetic Base of Crops, pp. 327–329. Pudoc, Wageningen.
San Noeum, L.H. & P. Gelebart, 1986. Production of gynogenic haploids. In: I.K. Vasil (Ed.), Cell Culture and Somatic Cell Genetics of Plants, Vol. 3, pp. 305–322. Academic Press, New York.
Sauton, A. & R. Dumas de Vaulx, 1987. Obtention de plantes haploïdes chez le melon (*Cucumis melo* L.) par gynogenèse induite par du pollen irradié. Agronomie 7: 141–147.
Simola, L.K. & J. Honkanen, 1983. Organogenesis and fine structure in megagametophyte callus lines of *Picea abies*. Physiol. Plant. 59: 551–561
Singh, M.N., R.N. Konar & S.P. Bhatnagar, 1981. Haploid plantlet formation from female gametophytes of *Ephedra foliata* Boiss. *in vitro*. Ann. Bot. 48: 215–220.
Sitbon, M., 1981. Production of haploid *Gerbera jamesonii* plants by *in vitro* culture of unfertilized ovules. Agronomie 1: 807–812.
Smith, B.M., R.M. Godwin, E. Harvey & C.P. Werner, 1991. Gynogenesis from whole flower

buds in bulb onions (*Allium cepa*) and leeks (*Allium porreum* L.). J. Genet. Breed. 45: 353–358.

Stettler, R.F. & K.S. Bawa, 1971. Experimental induction of haploid parthenogenesis in black cottonwood. Silvae Genet. 20: 15–25.

Stettler, R.F., K.S. Bawa & G.K. Livingston, 1969. Experimental induction of haploid parthenogenesis in forest trees. In: Symp. Nature Induction Util. Mutations Plants Proc., pp. 611–619.

Straathof, T.P., J.M. Van Tuyl, C.J. Keijzer, M.J. Wilms, A.A.M. Kwakkenbos & M.P. van Dien, 1987. Overcoming postfertilization barriers in Lillium by ovary and ovule culture. Plant Cell in Comp. Newsletter 11: 69–74.

Tian, H.Q. & H.Y. Yang, 1983. Synergid apogamy and egg cell anomalous division in cultured ovaries of *Oryza sativa* L. Acta Bot. Sin. 25: 403–408.

Thorpe, T.A & S. Biondi, 1984. Conifers. In: W.R. Sharp, D.A. Evans, P.V. Ammirato & Y. Yamada (Eds.), Handbook of Plant Cell Culture, Vol. 2, Crop Species, pp. 435–470. Macmillan Publishing Co., New York.

Truong-Andre, I. & Y. Demarly, 1984. Obtaining plants by *in vitro* culture of unfertilized maize ovaries (*Zea mays* L.) and preliminary studies on the progeny of a gynogenetic plant. Z. Pflanzenzüchtg. 92: 309–320.

Tulecke, W.R., 1964. A haploid tissue culture from the female gametophyte of *Ginkgo biloba* L. Nature (London) 203: 94–95.

Uchimiya, H., T. Kameya & N. Takahashi, 1971. *In vitro* culture of unfertilized ovules in *Solanum melongena* and ovaries in *Zea mays*. Jpn. J. Breed. 21: 247–250.

Van Geyt, J., G.J. Speckmann Jr., K. D'Halluin & M. Jacobs, 1987. *In vitro* induction of haploid plants from unpollinated ovules and ovaries of sugarbeet (*Beta vulgaris* L.). Theor. Appl. Genet. 73: 920–925.

Van Tuyl, J.M., M.P Van Diën, M.G.M. Van Creij, T.C.M. Van Kleinwee, J. Franken & R.J. Bino, 1991. Application of *in vitro* pollination, ovary culture, ovule culture and embryo rescue for overcoming incongruity barriers in interspecific *Lilium* crosses. Plant Sci. 74: 115–126.

Vasil, V., 1963. *In vitro* culture of embryos of *Gnetum ula* Brongn. In: P. Maheswari & N.S. Rangaswamy (Eds.), Plant Tissue and Organ Cutlure - A Symposium, pp. 278–280. Int. Soc. Plant Morphologists, Delhi.

Von Aderkas, P., K. Klimaszewska & J.M. Bonga, 1990. Diploid and haploid embryogenesis in *Larix leptolepis*, *L. decidua* and their reciprocal hybrids. Can. J. Res. 20: 9–14.

Wang, C.C. & B.J. Kuang, 1981. Induction of haploid plants from the female gametophyte of *Hordeum vulgare*. Acta Bot. Sin. 28: 329–330.

Winton, L. & D. Einspahr, 1968. The use of heat treated pollen for sperm haploid induction. For. Sci. 14: 406–407.

Wu, K. & P. Nagarajan, 1990. Poplars (*Populus* spp.): *In vitro* production of haploids. In: Y.P.S. Bajaj (Ed.), Biotechnology in Agriculture and Forestry, Haploids in Crop Improvement I, Vol. 12, pp. 237–249. Springer-Verlag, Berlin.

Wu, K. & M. Xu, 1984. Induction of maternal haploid plants from unpollinated ovaries of poplar *in vitro*. Acta Genet. Sin. 11: 47–51.

Yang, H., C. Zhou, D. Cai, H. Yan, Y. Wu & X. Chen, 1986. *In vitro* culture of unfertilized ovules in *Helianthus annuus* L. In: H. Hu & H.Y. Yang (Eds.), Haploids of Higher Plants *in Vitro*, pp. 182–191. China Academic Publishers, Beijing/Springer-Verlag, Berlin.

Yang, H.Y. & C. Zhou, 1982. *In vitro* induction of haploid plants from unpollinated ovaries and ovules. Theor. Appl. Genet. 63: 97–104.

Zenkteler, M., 1970. Test-tube fertilization of ovules in *Melandrium album* Mill. with pollen grains of *Datura stramonium* L. Experientia 26: 661–662.

Zenkteler, M.A. & I. Guzowska, 1970. Cytological studies on the regenerating mature female gametophyte of *Taxus baccata* L. and mature endosperm of *Tilia platyphyllos*, Scop. in *In vitro* culture. Acad. Soc. Bot. Polon. 39: 161–173.

Zhang, Y.X., Lespinasse & E. Chevreau, 1988. Obtention de plantes haploïdes pommier (*Malus*

X *domestica* Borkh.) issues de parthenogenèse induite *in situ* par du pollen irradie culture *in vitro* des pépins immatures. C.R. Acad. Sci. Paris 307: 451–457.

Zhang, Y.X., G. Salesses & Y. Lespinasse, 1988. Étude cytogenétique de quelque clones diploïdes et polyploïdes de pommier (*Malus* × *domestica* Borkh.). Considérations quant à l'origine du pommier. Cytologia 53: 739–748.

Zhou, C. & H.Y. Yang, 1981a. Induction of haploid rice plantlets by ovary culture. Plant. Sci. Lett. 20: 231–237.

Zhou, C. & H.Y. Yang, 1981b. Studies on the *in vitro* induction of callus from embryos sacs. Heriditas (Beijing) 3: 10–12.

Zhou, C. & H.Y. Yang, 1981c. *In vitro* embryogenesis in unfertilized embryo sacs of *Oryza sativa* L. Acta Bot. Sin. 23: 176–180.

Zhou, C. & H.Y. Yang, 1986. Isolation of embryo sacs by enzymatic maceration and its potential in haploid study. In: H. Hu & H.Y. Yang (Eds.), Haploids of Higher Plants *in Vitro*, pp. 192–203. China Academic Publishers, Beijing/Springer-Verlag, Berlin.

Zhou, C., H.Y. Yang, H. Tian, Z.L. Liu & H. Yan, 1986. *In vitro* culture of unpollinated ovaries in *Oryza sativa* L. In: H. Hu & H.Y. Yang (Eds.), Haploids of Higher Plants *in Vitro*, pp. 165–181. China Academic Publishers, Beijing/Springer-Verlag, Berlin.

Zhou, C., H.Y. Yang, H. Yan & S. Cai, 1983. Factors affecting callus formation in unpollinated ovary culture of rice: In: (Ed.), Cell and Tissue Culture Techniques for Cereal Crop Improvement, pp. 81–94. Science Press, Beijing/Gordon and Breach, New York.

Zhou, C. & H.Y. Yang, 1991. *In vitro* production of haploids in rice through ovary culture. In: Y.P.S. Bajaj (Ed.), Biotechnology in Agriculture and Forestry, Vol. 14, Rice, pp. 180–192. Springer-Verlag, Berlin.

11. Haploidy in *Citrus*

MARIA ANTONIETTA GERMANÀ

Contents

1. Introduction

Citrus fruits are cultivated in more than 80 countries and rank first among fruit crops for worldwide production with about 73 millions tons harvested annually (FAO, 1994). *Citrus* fruits are an important source of human nutrition. Due to lower production costs, the cultivation of *Citrus* in developing countries has increased from 50 to 65% of total global production during the last few years (Chang, 1992).

The most important cultivated *Citrus* species are orange (*C. sinensis* L. Osbeck) and tangerine (*C. unshiu* Marc., *C. nobilis* Lour., *C. deliciosa* Ten., *C. reticulata* Blanco and their hybrids) which account for more than 80% of land under *Citrus* cultivation; lemon (*C. limon* L. Burm. f.), lime (*C. aurantifolia* Christm. Swing.) and grapefruit (*C. paradisi* Macf.) account for the remainder in approximately equal proportions.

The processing industries utilize about 24 million tons (33% of total production) of *Citrus* for juice and frozen concentrates. The most commonly processed *Citrus* are: orange (more than 80% of the total production), grapefruit (about 8%), lemon (5%) and tangerine (5%).

S.M. Jain, S.K. Sopory & R.E. Veilleux (eds.), In Vitro *Haploid Production in Higher Plants, Vol. 5,* 195–217.

1.1. *Botany of* Citrus

The genus *Citrus* and its wild relatives belong to the family Rutaceae, subfamily Aurantioideae. The commonly cultivated *Citrus* fruits (genera: *Citrus*, *Fortunella*, *Poncirus*) belong to the tribe Citreae, subtribe Citrinae. The most commercially important *Citrus* species are indigenous to Southeast Asia. Recent taxonomic studies on chemical and biochemical characteristics, chloroplast DNA and restriction fragment length polymorphism (RFLPs) suggest that there are only three or four botanical species among the edible *Citrus* genus: *C. grandis*, *C. reticulata*, *C. medica* and *C. halimii* (Vardi, 1988a). Oranges and grapefruits are probably hybrids between pummelo (*C. grandis* L. Osbeck) and *C. reticulata*, whereas *C. limon* may be a trihybrid between *C. grandis*, citron (*C. medica* L.) and another unidentified gene source. *Citrus* trees are generally propagated asexually by grafting or nucellar embryos.

All cultivated *Citrus* and related genera (*Poncirus*, *Fortunella*) are diploid, $2n = 2x = 18$ (Frost, 1925) although triploid and tetraploid *Citrus* also exist. Two seedless triploid *C. grandis* × *C. paradisi* hybrids were produced from $2x - 4x$ hybridization (Soost & Cameron, 1980, 1985); and triploid hybrid *Citrus* plants were recovered by somatic embryogenesis *in vitro* from endosperm-derived calli (Gmitter *et al.*, 1990).

1.2. *Breeding*

The main objectives of *Citrus* breeders are to shorten the juvenile nonfruiting period, increase yield, extend the ripening season, develop seedless cultivars, improve external and internal quality of the fruit, and select new scions and rootstocks that are resistant or tolerant to biotic and abiotic stresses. Most *Citrus* spp. are characterized by polyembryony and produce one to forty adventive nucellar embryos (Fusurato, 1957) that generally result in two or more embryos in a single seed. Polyembryony is genetically controlled but also influenced by nutritional and environmental factors (Cameron & Frost, 1968). Nucellar embryos are theoretically identical to the mother tree, but often show variability, probably due to somatic meiosis similar to the ones observed in other plants (Nuti Ronchi *et al.*, 1992). The presence of uniform, disease-free and vigorous nucellar embryos is advantageous for *Citrus* propagation; however, it is a serious problem for breeders because zygotic embryos usually degenerate prior to seed maturation. Zygotic embryos from immature seed can be cultured *in vitro* for the production of hybrids (Rangan *et al.*, 1969).

Citrus breeding is based on either conventional methods – hybridization, selection, mutation – or biotechnology including the *in vitro* methods of regeneration from protoplasts, somatic hybridization, mutant selection, genetic transformation and haploid extraction by anther culture. An integrated approach is essential to improve *Citrus* by applying innovative and conventional technologies. Plants have been regenerated in *Citrus* and related genera from somatic embryos and protoplasts, including both somatic hybrids and cybrids (Grosser *et al.*, 1988; Vardi, 1988a,b, 1992; Grosser & Gmitter, 1990). Moreover, the production of transgenic *Citrus* by direct integration of genes into protoplasts or using *Agrobacterium*-mediated transformation of various explants has already been realized (Vardi *et al.*, 1990; Moore *et al.*, 1992; Hidaka & Omura, 1993; Kaneyoshi *et al.*, 1994).

2. Importance of haploids

Haploid plants have the gametophytic number of chromosomes in the sporophyte. The importance of haploids in plant breeding and genetic research has been recognized since the discovery of the first natural haploids in *Datura stramonium* and *Nicotiana* (Blakeslee *et al.*, 1922; Blakeslee & Belling, 1924; Kostoff, 1929). Haploids have potential applications in mutation breeding, *in vitro* selection, genetic analyses, production of inbred lines to exploit hybrid vigor (heterosis), and genetic transformation.

By doubling the chromosome number of haploids, homozygous diploid plants have been obtained for many crops, including tree crops, generally characterized by a long reproductive cycle, a high degree of heterozygosity, large size, and sometimes self-incompatibility. Actually, the absence of pure lines in many woody plants makes genetic studies difficult or even impossible. Segregation among microspores enables production of doubled haploids (DH) more rapidly compared with conventional inbreeding, and has resulted in the release of improved cultivars in some species, primarily in China and France (Veilleux, 1994). Furthermore, through fusion of haploid protoplasts, new forms of interspecific and intergeneric hybrids can be obtained.

3. The current status

Before 1960, pollination with irradiated pollen, interspecific hybridization, and selection of twin seedlings were the primary methods of haploid derivation. Spontaneous haploids were reported in over 100 angiosperms (Kasha, 1974). Most haploid fruit trees (especially pome and stone fruits) have been of spontaneous origin (Zhang *et al.*, 1990).

The discovery by Guha & Maheshwari (1964, 1966) that *in vitro* culture of immature anthers of *Datura innoxia* induced a change from normal gametophytic to sporophytic development revolutionized haploid research. *In vitro* androgenesis has been reported for over 200 plant species, especially among the Solanaceae, Cruciferae and Gramineae (Dunwell, 1986; Hu & Hongyuan, 1986; Sangwan-Norreel *et al.*, 1986; Bajaj, 1990; Raghavan, 1990; Wenzel *et al.*, 1995). Despite considerable effort, androgenesis or gynogenesis in woody plants has been less successful. Zhang *et al.* (1990) cultured anthers of several fruit trees, especially pome and stone fruits, and occasionally produced embryos that rarely germinated, whereas anther-derived haploid plants of poplar (*Populus nigra*), rubber tree (*Hevea brasiliensis* Muell. Arg.), grape (*Vitis vinifera* L.) and apple (*Malus pumila* Mill.) have been successfully transplanted into the field (Chen, 1986).

4. Research on haploid induction in *Citrus*

4.1. In vivo *methods*

A low frequency of spontaneous haploids occurs among seedlings of most plants. Among pome and stone fruit seedlings from open-pollinated flowers, haploid plants have been identified (Bouvier *et al.*, 1993), although none survived (Zhang *et al.*, 1990). In *Citrus natsudaidai* haploid seedlings were first obtained by application of gamma rays (Karasawa, 1971). In *Citrus*, one haploid embryo was obtained from an immature seed of a diploid *C. clementina* × diploid Pearl tangelo cross (Esen & Soost, 1972). Three *Citrus* haploids were obtained from two monoembryonic diploid × triploid crosses (Oiyama & Kobayashi, 1993). Restriction endonuclease analysis of nuclear ribosomal DNA and of chloroplast DNA determined the maternal origin of these haploids. Haploid and diploid embryos did not differ in size; however, haploid seedlings grew slowly in soil.

4.2. *Anther culture*

Haploid, diploid and mixoploid plantlets have been recovered by anther culture of cultivated *Citrus* spp. and their relatives (Table 1) including *Poncirus trifoliata* (Hidaka *et al.*, 1979; Vardi, 1988b) and *C. madurensis* (Chen *et al.*, 1980). Germanà *et al.* (1994) obtained haploid plantlets and embryogenic callus of *C. clementina*. The recovery of heterozygous anther-derived plantlets was reported for *C. aurantium* (Hidaka *et al.*, 1981; Germanà *et al.*, 1990), *C. sinensis* (Hidaka, 1984b), *C. aurantifolia* (Chaturvedi & Sharma, 1985), *C. madurensis* (Ling *et al.*, 1988), *C. reticulata* (Germanà *et al.*, 1994), *Poncirus trifoliata* and hybrid No. 14 of *C. ichangensis* × *C. reticulata* (Deng *et al.*, 1992). Similarly, Germanà *et al.* (1991) obtained haploid and diploid calli, embryos and leafy structures but no green plants by anther culture of *C. limon*. Anther culture produced embryonic structures in *C. sinensis* and *C. paradisi* (Starrantino, 1986) and mixoploid calli were obtained in *C. reticulata*, *C. deliciosa*, *C. sinensis*, *C. limon*, *C. paradisi* and Mapo tangelo (Geraci & Starrantino, 1990).

4.3. Citrus *anther culture technique*

Floral buds, with microspores at the early uninucleate stage of development, are surface-sterilized by immersion, first in 70% ethanol for 1–2 min, then in sodium hypochlorite solution (approx. 0.5% active chlorine in water) for

Table 1. Research on *in vitro* androgenesis in *Citrus* and its relatives

Genotype	Response	Basal medium + growth regulators (mg l⁻¹)[z]	Reference
Poncirus trifoliata (L.) Raf.	(x) plantlets	MS+IAA(0.2)+Kin(0.2)	Hidaka et al. (1979)
	(2x) callus	MS; NTH; Gamborg	Khuroshvili et al.(1982)
	(2x) plantlets	1/2MT+NAA(0.1)	Deng et al.(1992)
C. aurantium (L.)	(2x) plantlets	MS+IAA(2.0)(0.02)+Kin(0.02)	Hidaka et al. (1981)
	(2x) plantlets	MS+NAA(0.02)+ZEA(0.5)+Kin(0.5)	Germanà et al.(1990)
C. madurensis Lour.	(x) (2x) plantlets	N6+2,4D(0.1)+BA(1.0)	Chen et al.(1980) Chen(1985)
	(2x) plantlets	MS+IAA(2.0)+Kin(2.0)	Ling et al.(1988)
C. clementina Hort. ex Tan.	(x) plantlets	MS+NAA(0.02)+ZEA(0.5)+Kin(0.5)	Germanà et al.(1994)
	(x) plantlets	N6;NTH;NAA(0.02)+2,4D(0.02)+ ZEA(0.5)+Kin(1.0)+GA3(0.5)+BA(0.5)	Germanà et al.(unpublished)
	(x) plantlets	N6+NAA(0.02)+2,4D(0.02)+ ZEA(0.5)+Kin(1.0)+BA(1.0)	Germanà et al.(unpublished)
C. aurantifolia (Christm)Swing.	(2x) plantlets	MS+IAA(1.0)+BA(0.5)	Chaturvedi & Sharma(1985)
C. limon L. Burm. f.	callus	DB+NAA(1.0)+2,4D(1.0)+BA(0.1)	Drida & Benbadis(1975)
	callus	N6;MS+2,4D+NAA+IBA+BA	Chen(1985)
	(2x) callus	Sj-1+2,4D(1.0)+BA(0.5)	Starrantino(1986)
	mixoploid callus	1/2MT+NAA(1.0)+BA(1.0)	Geraci&Starrantino(1990)
	(x) callus	NTH+NAA(0.1)+ZEA(1.0)+Kin(2.0)	Germanà et al.(1991)
C. sinensis Osbeck	(2x) plantlets	MS+IAA(0.02)(0.002)+Kin(0.02)(0.2)	Hidaka(1984b)
	(2x) embryos	N6+2,4-D+NAA+IBA+BA	Chen(1985)
	embryonic structures	Sj-1+2,4-D(1.0)+Kin(10.0)	Starrantino(1986)
	mixoploid callus	1/2MT+NAA(1.0)+BA(1.0)	Geraci&Starrantino(1990)
	(2x) embryos	MS+NAA(0.02)+ZEA(0.5)+Kin(0.5)	Germanà et al.(1990)
C. reticulata Blanco	callus	N6+2,4-D+NAA+IBA+BA	Chen(1985)
	mixoploid callus	MT+2,4-D(0.5)+BA(0.1)	Geraci&Starrantino(1990)
	(2x) plantlets	MS+NAA(0.02)+ZEA(0.5)+Kin(0.1)	Germanà et al.(1994)
C. paradisi Macf.	embryos	Sj-1+2,4-D(10.0)+Kin(0.5)	Starrantino(1986)
	mixoploid callus	MT+2,4-D(0.5)+BA(1.0)	Geraci&Starrantino(1990)
	embryos	NTH+NAA(0.02)+2,4-D(0.02)+ZEA(0.5)+Kin(1.0)+ GA3(0.5)+BA(0.5)	Germanà et al.(unpublished)
C. grandis Osb.	callus	N6+2,4D(0.05)(0.1)+BA	Chen(1985)
C. medica L.	callus	DB+NAA(1.0)+2,4D(1.0)+BA(0.1)	Drida & Benbadis(1975)
	callus	Sj-1+2,4-D(1.0)+BA(0.5)	Starrantino(1986)
C. limetta Risso	callus	Sj-1+2,4-D(1.0)+BA(0.5)	Starrantino(1986)
C. reticulata x *C. grandis*	embryos	N6+2,4(0.1)+BA(1.0)	Chen(1985)
C. deliciosa x *C. paradisi*	callus	Sj-1+2,4-D(1.0)+BA(0.5)	Starrantino(1986)
	callus	MT+2,4-D(0.5)+BA(1.0)	Geraci&Starrantino(1990)
C. sinensis x *C. reticulata* (Szipusian)	callus	N6+2,4-D(0.05)(0.1)+BA(2.0)	Chen (1985)
C. ichangensis x *C. reticulata* No.14	(2x) plantlets	1/2MT+NAA(0.1)	Deng et al.(1992)

[z] For abbreviations and other details, see 'Factors affecting *in vitro* androgenesis' (section 5).

7–15 min. Before removing the petals with small forceps, the floral buds are rinsed two or three times in sterile distilled water. The anthers can then be dissected under aseptic conditions and placed on the medium. The stage of microspore development can be determined by staining one or more anthers per bud in acetocarmine or Schiff's reagent.

For *Citrus* anther culture, 20–25 × 100–150 mm culture tubes (Hidaka *et al.*, 1979, 1981; Hidaka, 1984a; Hidaka & Omura 1989b; Chaturvedi & Sharma, 1985; Starrantino, 1986; Geraci & Starrantino, 1990; Deng *et al.*, 1992), 55 mm diameter Petri dishes (Germanà *et al.*, 1990, 1991, 1994; Starrantino, 1986), and 50 ml flasks (Chen, 1985; Ling *et al.*, 1988) have been employed.

5. Factors affecting *in vitro* androgenesis

Although research on androgenesis has made significant progress in recent years, several aspects of this phenomenon remain unclear, especially the induction process and factors that control it. The identification of inhibitory and stimulatory factors is a fundamental aspect of further research in recalcitrant species like *Citrus*.

The rate of callus or embryo induction (calluses or embryos per hundred anthers) has been low in *Citrus* compared to other genera. Because estimates of the number of pollen grains per anther have been low in *Citrus*, e.g., 369–4,367 for lemon, 1,557–1,782 for grapefruit, 5,052–14,671 for orange (Frau, 1973), and about 2,600 for clementine (unpublished data) compared to 40,000 for *Nicotiana* (Heberle-Bors, 1985), it may be better to refer to calluses per hundred pollen grains to estimate androgenic response.

In many androgenic species (*Nicotiana*, *Hordeum*, *Paoenia*, etc.) pollen dimorphism has been observed (Sunderland, 1971, 1974, 1983; Heberle-Bors, 1980; Sangwan & Sangwan-Norreel, 1990), resulting in distinct classes of pollen: a) non-androgenic – large and densely stained by acetocarmine, and b) androgenic – small with a thinner exine and only lightly staining (P-grains). P-grains found *in vivo* are interpreted as a particular form of male sterility (Heberle-Bors, 1982). Their frequency depends on the growth conditions of the donor plants (Heberle-Bors, 1985). P-grain formation *in vivo* or *in vitro* seems to be related to nitrogen starvation (Heberle-Bors, 1983). Sangwan & Sangwan-Norreel (1990) hypothesized that the difference between embryogenic and non-embryogenic grains was at the molecular or cellular level in *Datura*. The presence of P-grains has not been documented in *Citrus*.

5.1. *Genotype*

Vasil (1980) has stated that, in most species, only a few genotypes can be induced to undergo androgenic development. *Citrus* is no exception. Almost all research carried out on androgenesis in *Citrus* has recognized the influence of genotype (Hidaka, 1984a; Chen, 1985; Germanà *et al.*, 1991, 1994; Khuroshvili *et al.*, 1982). Our research (Germanà *et al.*, 1991, 1994, and unpublished data) conducted simultaneously on several *Citrus* cultivars (four clementine, two mandarin, four sweet orange, four sour orange, five lemon, four grapefruit) resulted in plantlets from haploid embryogenic callus only in *C. clementina* cv. Nules, haploid callus in one cultivar of *C. limon*, and diploid and embryogenic callus from two cultivars of *C. reticulata* (Avana and Tardivo di Ciaculli) and two cultivars of *C. aurantium* (A.A. C.N.R. 10 and A.A. C.N.R. 23). All experiments were carried out under similar growing conditions, pretreatments (4°C for four days), using 11 different media.

5.2. *Stage of pollen development*

The pollen developmental stage has generally been found to affect the success of anther culture. Selection of the most responsive stage varies with the crop or genotype used. Generally, pollen grains between the uninucleate and early binucleate stage have been most responsive in anther culture. After microspores have begun to accumulate storage reserves, they lose their embryogenic capacity and follow the gametophytic developmental pathway (Heberle-Bors, 1989; Raghavan, 1990).

Hidaka *et al.* (1979) studied the effect of different developmental stages of *P. trifoliata* pollen grains on embryo, pseudobulbil, and callus formation and demonstrated that the early uninucleate stage was most suitable for embryo formation in anther culture. Anthers with other developmental stages from pollen mother cell (PMC) to binucleate stage produced only calli. In *C. aurantium*, Hidaka *et al.* (1981) obtained embryos from anthers only at the late uninucleate stage and callus production from anthers at all other developmental stages except for PMC in meiosis and the tetrad stage.

Various researchers have used different developmental stages of microspores for *Citrus* anther culture including uninucleate stage (Deng *et al.*, 1992; Hidaka, 1984a,b; Hidaka & Omura, 1989b; Germanà *et al.*, 1990, 1991, 1994; Starrantino, 1986; Geraci & Starrantino, 1990); tetrads at telophase (Chaturvedi & Sharma, 1985); and tetrads at anaphase (Chen, 1985). However, different developmental stages have also been reported among developing microspores within a single anther and between different anthers of the same flower bud in *Citrus*, *Poncirus* and many other genera (Hidaka *et al.*, 1979, 1981; Chen, 1985; Vasil, 1967; Shull & Menzel, 1977).

5.3. *Physiological condition of the donor plant*

Although the physiological conditions of the donor plant have been reported to affect the androgenic process of many, especially herbaceous, species (Sunderland & Dunwell, 1977; Vasil, 1980) this parameter has not been considered in the literature on *Citrus* anther culture.

5.4. *Pretreatment*

Pretreatments (such as chilling, high temperature, high humidity, water stress, anaerobic treatment, centrifugation) to excised floral buds or to anthers before culture has been reported to promote *in vitro* androgenesis of many species (Sunderland, 1974; Sangwan-Norreel *et al.*, 1986). Cold pretreatment is commonly used in *Citrus* to delay and modify pollen mitosis (two equal-sized vegetative-type nuclei are formed instead of one vegetative and one generative), block starch production, disintegrate microtubules (Nitsch, 1977), or maintain viability of the cultured P-grains (Heberle-Bors, 1985).

In *Citrus* anther culture, different cold pretreatments have been used, such as 5°C for 2 h (Starrantino, 1986), overnight treatment (Geraci & Starrantino, 1990), 3–5 days (Ling *et al.*, 1988), 4°C for four (Germanà *et al.*, 1991), six (Germanà *et al.*, 1994), or 2–6 days (Deng *et al.*, 1992). After providing a 3°C cold pretreatment to anther cultures of *C. madurensis* for 3–25 days, Chen (1985) concluded that a duration of 5–10 days was most favourable to induce callus and embryos.

In our laboratory (unpublished results), research on factors affecting androgenesis in *C. clementina* showed a negative effect on callus production of centrifugation (4000 rpm for five min) and high temperature (40° C for 24 h) compared with chilling (4°C for ten days).

5.5. *Culture medium*

Different species show preferences for basal media for induction of microspore-derived plants. The basal media that have been used in *Citrus* anther culture are: B5 (Gamborg *et al.*, 1968; Khuroshvili *et al.*, 1982); DB (Drira & Benbadis, 1975), MS (Murashige & Skoog, 1962; Khuroshvili *et al.*, 1982; Hidaka, 1984a,b; Hidaka & Omura, 1989b; Hidaka *et al.*, 1979, 1981; Starrantino, 1986; Ling *et al.*, 1988; Germanà *et al.*, 1994); modified MS (Chen, 1985; Chaturvedi & Sharma, 1985; Germanà *et al.*, 1994); Murashige & Tucker (1969) (Geraci & Starrantino, 1990; Deng *et al.*, 1992); N_6 (Chu, 1978; Chen, 1985; Starrantino, 1986; Germanà *et al.*, unpublished); SH (Schenk & Hildebrandt, 1972; Chaturvedi & Sharma, 1985); Sj-1 (Starrantino, 1986); Chaturvedi & Mitra (1974) medium (Germanà *et al.*, 1994); Nitsch & Nitsch (1969) medium (NTH) (Khuroshvili *et al.*, 1982; Germanà *et al.*, unpublished).

5.6. *Carbon source*

Carbon source and concentration are essential for embryo induction. The effect of concentration is probably related to osmotic pressure regulation during the induction phase (Sunderland & Dunwell, 1977; Sangwan & Sangwan-Norreel, 1990). Furthermore, high concentration of carbohydrates seems to be deleterious (Keller *et al.*, 1975). Sucrose has been the most common carbon source used in *Citrus* anther culture, commonly at 5% concentration (Hidaka *et al.*, 1979, 1981; Hidaka, 1984b; Hidaka & Omura, 1989b; Geraci & Starrantino, 1990), although other concentrations, e.g., 8% (Starrantino, 1986), 2% (Ling *et al.*, 1988) and 2.5–5% (Drira & Benbadis, 1975; Deng *et al.*, 1992) have also been reported. Hidaka (1984a) studied the effects of sucrose concentration (1–3–5–7–9%) on embryo and callus formation and found that 3% sucrose was ideal for embryo formation in *P. trifoliata*, 7% in *C. aurantium*, and 1% in *C. sinensis* (Trovita orange). Chen (1985) reported that 10% sucrose was optimal for *C. madurensis* whereas Germanà *et al.* (1991) used 3% and 6% for *C. limon* androgenesis. Germanà *et al.* (1994) tested the influence of two carbon sources (sucrose and glucose) on anther culture of two *C. clementina* and two *C. reticulata* cultivars and found that 5% sucrose was better, although the response varied with the species or cultivar.

Glycerol in combination with sucrose stimulated callus production in *C. clementina* whereas galactose was effective for the production of embryos from *Citrus* calli (Hidaka & Omura, 1989a) and had a positive influence on haploid production (Germanà *et al.*, unpublished).

5.7. *Plant growth regulators*

The effects of plant growth regulators have been extensively investigated in anther culture of *Citrus*. In *C. limon* and *C. medica* anther culture, Drira & Benbadis (1975) obtained callus on a medium containing 1 mg l^{-1} 2,4-dichlorophenoxyacetic acid (2,4–D), 1 mg l^{-1} α-naphthaleneacetic acid (NAA) and 0.1 mg l^{-1} 6-benzyladenine (BA). Hidaka *et al.* (1979) found that a medium having 0.2 mg l^{-1} each of indole-3–acetic acid (IAA) and kinetin (kin) was most efficient in embryo formation, while the addition of 2,4–D increased callus formation in *P. trifoliata* anther culture. Furthermore, Hidaka *et al.* (1981) produced embryos in *C. aurantium* by anther culture on a medium supplemented with 0.02 mg l^{-1} kin and 0.02 or 2.0 mg l^{-1} IAA. Hidaka (1984b) found that both *C. sinensis* and sour orange preferred lower concentrations (0.002 or 0.02 mg l^{-1}) of IAA and kin for pollen embryogenesis. Chaturvedi & Sharma (1985) induced androgenesis in *C. aurantifolia* on a medium supplemented with 0.5 mg l^{-1} BA and 1.0 mg l^{-1} IAA. Chen (1985) found that 1.0 mg l^{-1} BA and 0.1 mg l^{-1} 2,4–D was the best combination of growth regulators for embryo production, and the concentration of 2,4–D in the medium was crucial for embryo production. As the concentration of 2,4–D increased in the medium, callus formation increased and embryo development was inhibited in *C. madurensis*.

Starrantino (1986) tested numerous media containing an auxin (0.1–0.5–1.0–5.0–10.0 mg l^{-1} IAA or 2,4–D) in combination with a cytokinin (0.1–0.5–1.0–5.0–10.0 mg l^{-1} BA or kin) and obtained two embryonic structures from two *C. sinensis* anthers in medium containing 1.0 mg l^{-1} 2,4–D and

10.0 mg l^{-1} kin, and one embryonic structure from an anther of *C. paradisi* in medium supplemented with 10 mg l^{-1} 2,4–D and 0.5 mg l^{-1} kin. Ling *et al.* (1988) obtained the highest frequency (0.92%) of embryo formation for *C. madurensis* in media with 2.0 mg l^{-1} of both IAA and kin. Geraci & Starrantino (1990) obtained the highest percentage of callus proliferation (25%) in the presence of 1 mg l^{-1} BA and 0.5 mg l^{-1} 2,4–D in *C. reticulata*, *C. deliciosa*, *C. paradisi* and Mapo tangelo (*C. deliciosa* × *C. paradisi*) while in *C. sinensis* and *C. limon*, the highest callus proliferation rate (44.2%) was obtained with 1.0 mg l^{-1} each of NAA and BA. Deng *et al.* (1992) reported the best response in *P. trifoliata* and an ichang papeda hybrid (*C. reticulata* Blanco × *C. ichangensis* Swingle) on medium with 0.1 mg l^{-1} NAA.

Anther culture of *C. limon* was successful on medium containing 2.0 mg l^{-1} kin + 1.0 mg l^{-1} zeatin (zea) + 0.1 mg l^{-1} NAA (Germanà *et al.*, 1991). Germanà *et al.* (1994) found the best growth regulator combination to be 0.02 mg l^{-1} NAA + 0.5 mg l^{-1} zea + 0.5 mg l^{-1} kin for callus production (6–28% depending on the genotype) from anther culture of *C. clementina* and *C. reticulata*.

5.8. *Activated charcoal*

The addition of activated charcoal (0.5–2 g/l) in the medium has increased the efficiency of androgenesis in several genera. It seems to act by removing inhibitory substances from the medium, and presumably from the anther wall and in regulating the level of endogenous and exogenous growth regulators (Heberle-Bors, 1985; Reinert & Bajaj, 1977; Vasil, 1980). Deng *et al.* (1992) found that medium with activated charcoal was effective in *P. trifoliata* embryo induction. However, no positive response to activated charcoal (0.5 g/l) was observed in anther culture of several other *Citrus* species (Germanà *et al.*, 1994, and unpublished).

5.9. *Other substances*

The addition of glutamine was necessary to produce callus in *C. limon* anther culture (Drira & Benbadis, 1975). In our laboratory, we usually add casein (400 mg l^{-1}), biotin (0.5 mg l^{-1}) and sometimes coconut water (5–10% v:v) in addition to glutamine (200 mg l^{-1}) in the medium. Various natural undefined extracts, e.g., coconut water, have been used and they seem to improve androgenic response, although their mode of action is unclear.

5.10. *pH*

Stuart *et al.* (1987) found that pH influenced embryogenic processes in alfalfa. In *Citrus* anther culture, pH of the medium has usually been adjusted to 5.7–5.8 before autoclaving. The effect of the pH level (4–5–6–7–8) of the media on anther culture of *Poncirus trifoliata*, *C. aurantium* and *C. sinensis* was investigated by Hidaka (1984a) who found that both pH 5 and 6 were effective in embryo formation of all three species.

5.11. *Solidifying agents*

Usually, *Citrus* anther culture media have been solidified with agar. Chaturvedi & Sharma (1985) obtained diploid plantlet regeneration by floating *C. aurantifolia* anthers on a liquid medium, followed by embedding them in a semisolid medium. In preculturing anthers for pollen isolation, we obtained better results using a solid medium compared to liquid. In the liquid medium, anthers initially swelled, but later turned brown and sometimes shrivelled. Our research on *Fortunella margarita* and *C. paradisi* anther culture showed the beneficial effect of potato starch as a gelling agent on callus production (unpublished).

5.12. *Conditions of incubation*

The conditions of incubation, especially light quality and photoperiod, have not received much attention in *Citrus* anther culture. Hidaka (1984a) induced embryos at 24°C and 28°C in trifoliate orange, sour orange and "Trovita" orange. Chen (1985) observed that temperature seemed to be more important than light in *Citrus* androgenesis and obtained embryos at 20–25°C especially in the dark (2.21% induction rate). In *Citrus*, the most commonly used temperatures have been 25 ± 1°C (Ling *et al.*, 1988) and 27–28 ± 1°C (Hidaka *et al.*, 1979, 1981; Hidaka, 1984b; Hidaka & Omura, 1989b; Chaturvedi & Sharma, 1985; Starrantino, 1986; Geraci & Starrantino, 1990; Germanà *et al.*, 1990, 1991, 1994; Deng *et al.*, 1992). Several light intensities have also been used: 500 lux with 16 h photoperiod (Hidaka 1984a,b; Hidaka *et al.*, 1979, 1981); 500–800 lux with 12 h photoperiod (Chen, 1985); 3000 lux with 15–16 h photoperiod (Chaturvedi & Sharma, 1985; Ling *et al.*, 1988); 1000 lux with 16 h photoperiod (Starrantino, 1986; Geraci & Starrantino, 1990); and 3500 lux with 16 h photoperiod (Germanà *et al.*, 1991, 1994; Deng *et al.*, 1992). A short inductive period in darkness may be beneficial.

6. Embryo development from microspores and origin of haploids

Plantlet formation from cultured anthers may occur either directly through embryogenesis of microspores or indirectly through organogenesis of microspore-derived callus. Calli can differentiate embryos (Figs. 1–3) or buds. It has been difficult to induce buds from calli which differentiated roots first. The embryos develop normally through the globular, heart, torpedo and cotyledonary stages and often produce secondary embryos. Sometimes green and spherical pseudobulbils, with or without callus, have been produced.

Figure 1. A cluster of embryos in *Citrus reticulata* (Tardivo di Ciaculli NL 19).

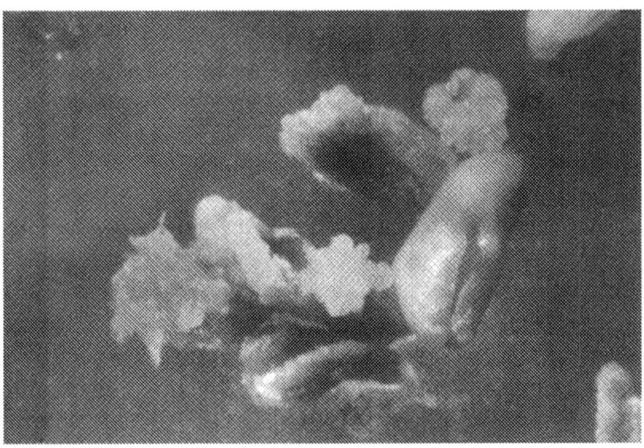

Figure 2. Undifferentiated, white and haploid callus and embryos emerging from inside a *Citrus clementina* (Nules) anther.

Green and compact callus emerging from inside the anthers has often been observed in *Poncirus* (Fig. 4) and *C. limon*.

Sporophytic divisions in pollen grains of various species start at different intervals after first pollen mitosis and after the trauma of wounding and culture, depending on the degree of repression of the sporophytic gene programme (Heberle-Bors, 1985). Hidaka & Omura (1989b) described cytologically the development of embryos from microspores in *C. aurantium* and

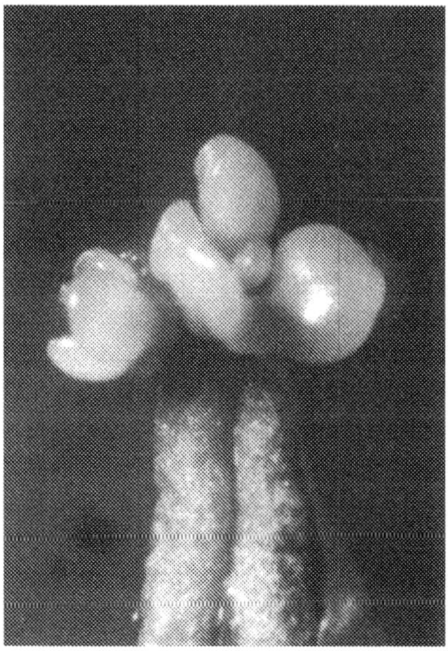

Figure 3. Embryos from anthers of *Poncirus trifoliata* after six weeks of culture.

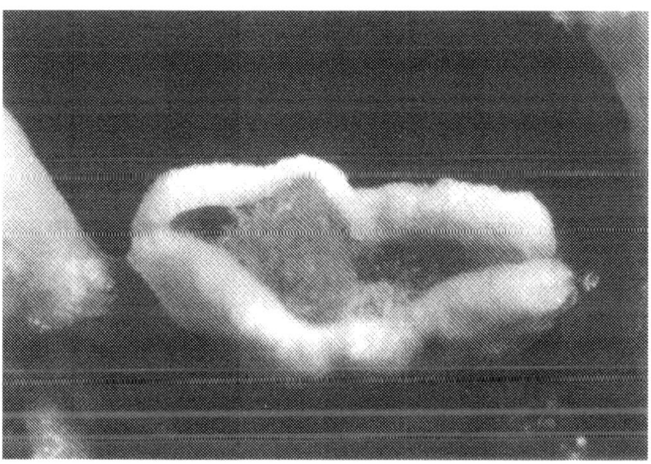

Figure 4. Green and compact callus emerging from inside a *Poncirus trifoliata* anther.

P. trifoliata. They observed three main routes of microspore development (Fig. 5): in route A, the microspores lose their contents; in route B, microspores develop as *in vivo*, i.e., a normal nuclear division results in one vegetative and one generative nucleus, but this route is rarely observed; in route

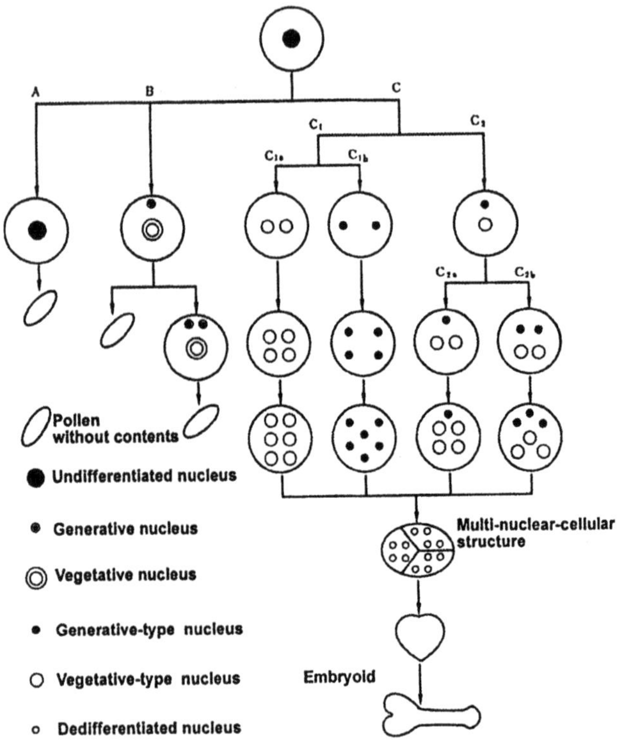

Figure 5. Developmental pathways of microspores in anther culture of *Citrus* (from Hidaka & Omura, 1989).

C, there are two pathways, neither of which resembles normal microspore development. In route C1, two types of morphologically similar nuclei are observed: "vegetative-type" (C1a) or "generative-type" (C1b). In route C2, both "vegetative-type" and "generative-type" nuclei are present in the same cell; repeated divisions of only the "vegetative-type" nucleus or both nuclei independently results in embryogenesis. Most embryos in *Citrus* anther culture have been attributed to the C2 route.

When the nucleus divides without cell division, a multinucleate pollen grain is initially formed that later gives rise to a multicellular structure that develops into a proembryo (Fig. 6) and finally an embryo. Moreover, nuclear fusion between the vegetative and generative nuclei has also been observed and this can result in increased ploidy level among anther-derived plants.

Haploid micropropagation by *in vitro* tissue culture is particularly important in woody species, where the frequency of haploid induction by anther culture is low. For multiplying haploid plants in large numbers, shoot meristems or axillary bud cultures may be used. Numerous haploid plants can

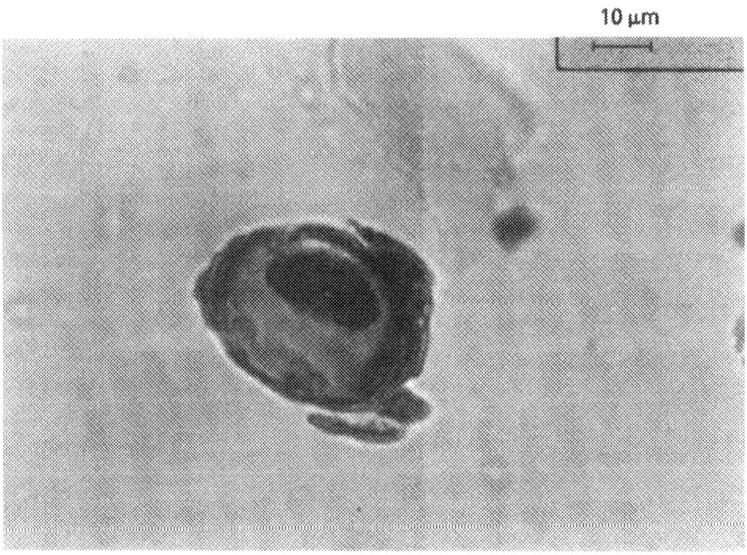

Figure 6. A proembryo in *Citrus aurantium* (A.A. C.N.R. 10) anther culture.

easily be obtained when highly embryogenic callus is produced by anther culture in *C. clementina*.

7. Microspore culture

Microspore culture involves isolation of microspore from somatic tissue of the anthers. Although it is more difficult and laborious than anther culture, it is well-suited to study the mechanisms of pollen embryogenesis because it eliminates the unknown effects of the anther tissues, thereby, allowing greater control over the culture process. However, there has been no success with isolated microspore culture of plants that are also recalcitrant to anther culture (Heberle-Bors, 1989). So far, there have been no published reports on isolated microspore culture of *Citrus*. However, we have conducted microspore cultures of several *Citrus* species in our laboratory.

Anthers containing microspores at the uninucleate stage were excised from cold-pretreated (4°C for ten days) and surface-sterilized flower buds that were precultured on either liquid or solid medium in Petri dishes at 27 ± 1°C in the dark. Medium for isolated microspore culture is more complex than that for anther culture. Our unpublished results were obtained by using NTH (Nitsch & Nitsch, 1969) or N_6 (Chu, 1978) liquid filter-sterilized medium supplemented with casein (500 mg l^{-1}), glutamine (800 mg l^{-1}), malt extract (500 mg l^{-1}), biotin (0.5 mg l^{-1}), myo-inositol (5 g l^{-1}), glycine (2 mg l^{-1}),

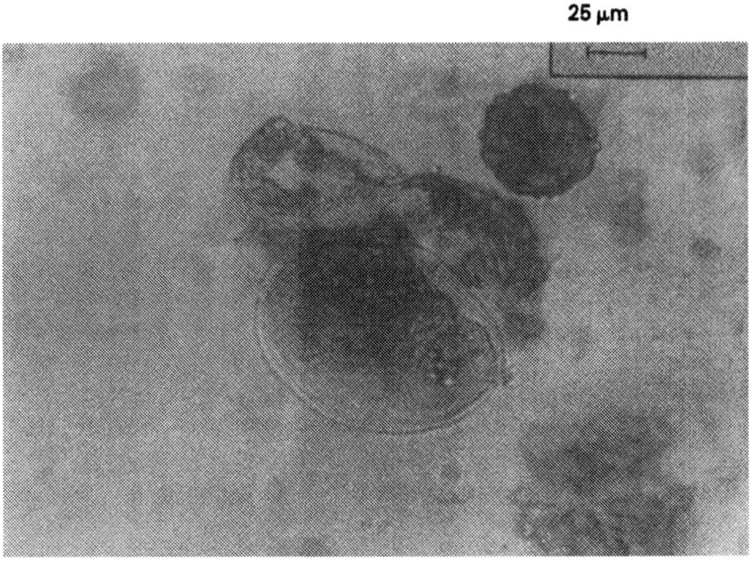

Figure 7. Microspore culture in *Citrus clementina*: different stages.

pyridoxine (5 mg l^{-1}), thiamine (5 mg l^{-1}), serine (100 mg l^{-1}), coconut water (5–10%), ascorbic acid (500 mg l^{-1}), with galactose (18 g l^{-1}) and sorbitol (18 g l^{-1}) as carbon source and a complex combination of growth regulators: 2,4–D (0.5 mg l^{-1}) + kin (0.5 mg l^{-1}) + zeatin (0.5 mg l^{-1}) + gibberellic acid (0.05 mg l^{-1}). After 5–15 days of culture, anthers were gently squeezed with a glass rod in approx. 2 ml liquid medium. The anther mixture was filtered through a sterile nylon sieve (40 μm) and microspores were centrifuged (1000 rpm for 10 min) and washed twice with fresh medium. Finally, isolated microspores were resuspended in the fresh liquid medium at a density of 10^3 to 10^4 grains per ml, then cultured in Petri dishes (6 cm diam) containing a thin layer (3 ml) of liquid medium, sealed with parafilm and incubated at 27 ± 1°C under cool white fluorescent lamps with a photosynthetic photon flux density of 35 μmol m^{-2} s^{-1} under a 16 h photoperiod. The isolated microspores of almost all investigated *Citrus* species produced multinucleate structures that developed into small proembryos within 1–4 months (Fig. 7). However, these proembryos failed to develop further, although several media and different culture conditions (double layer, liquid medium, soft agar, solidified agar, etc.) were employed.

8. Ploidy of androgenic plants

Haploid (Fig. 8), diploid, aneuploid and mixoploid calluses and plantlets have been produced from *Citrus* and its relatives by anther culture (Table

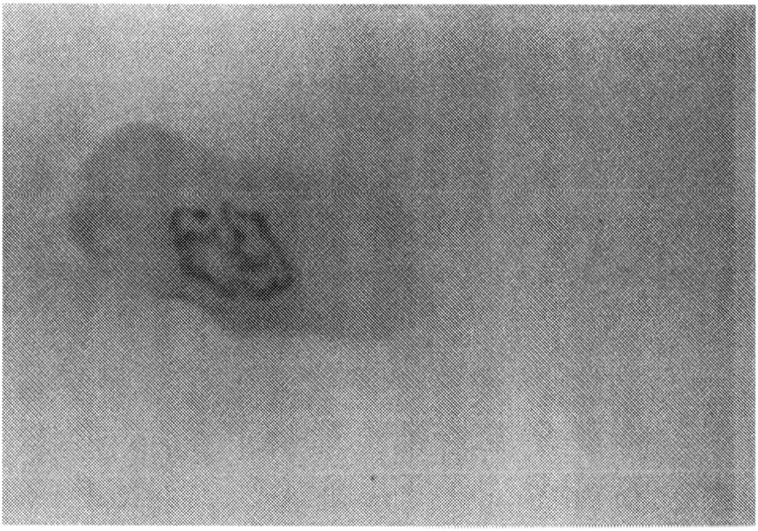

Figure 8. A haploid set of chromosomes ($2n = 1x = 9$) from a shoot tip of Nules clementine.

1). Non-haploids may arise from: a) somatic tissues of anther walls, b) fusion of nuclei, c) endomitosis, and d) irregular microspores formed by meiotic irregularities (D'Amato, 1977; Sangwan-Norreel, 1983; Sunderland & Dunwell, 1977; Narayanaswamy & George, 1982). Ploidy variation seems to depend on the developmental stage of the pollen at the time of culture. In general, the older the stage, the higher the ploidy of the resulting embryos (Maheshwari *et al.*, 1980). In cases of embryogenesis from unreduced microspores, resulting regenerated plants are expected to be heterozygous (Wenzel *et al.*, 1977).

9. Androclonal variation and its utilization

Generally, a large amount of variation is generated by androgenesis due to gene mutation and chromosome variation (Hu & Huang, 1987). Sometimes undesirable gametoclonal variation, e.g., lower vigor, has been observed among anther-derived doubled haploids (Veilleux, 1994). An unexpected elevation of DNA content per cell has been also observed (De Paepe *et al.*, 1982; Dhillon *et al.*, 1983).

Although haploid embryos of clementine germinated vigorously, haploid plantlets grew slowly in soil, presumably due to the expression of harmful recessive genes expressed in homozygosity (inbreeding depression). These haploid plantlets have always died as a result of fungal infection. However, better results were obtained by grafting them onto Citrange Troyer seedlings.

Figure 9. A haploid Nules shoot grafted on Citrange Troyer seedling compared with two haploid plantlets of the same age.

At the moment, these grafts of haploid clementine and several ungrafted haploid clementine (very slowly growing, Fig. 9) are the only *Citrus* haploids successfully transferred to the greenhouse. Such haploid plantlets have sometimes shown differences in leaf morphology and in plant habit.

10. Potential applications of haploids in *Citrus*

Angiosperm microspores may be easily manipulated to improve the efficiency of conventional plant breeding. By culturing isolated microspores of elite plants, haploid and doubled haploid plants can be recovered via embryogenesis. Because microspores are the products of genetic recombination, it may be possible to obtain new allelic combinations. By exerting selective stress on microspore cultures, it may also be possible to isolate physiological and biochemical mutants (Evans *et al.*, 1990). Such selection assumes that genes expressed by the gametophyte are also expressed by the sporophyte. However, success of microspore culture needs a well-defined procedure of regeneration through embryogenesis.

Microspores may also be considered as targets for gene transfer by cocultivation with *Agrobacterium tumefaciens*, by biolistic methods followed by pollen culture, or by microinjection of DNA into pollen embryos (Heberle-

Bors *et al.*, 1990; Sangwan & Sangwan-Norreel, 1990). Haploid protoplasts obtained from microspores can be fused with somatic cell protoplasts to produce triploid hybrids (gametosomatic fusion; Deng *et al.*, 1992), or with other haploid protoplasts to obtain diploid hybrids between distant species.

Another application of haploids is the diploidization of haploid plants derived from disease resistant heterozygous species, e.g., L CNR 26 (Monachello lemon) contains resistant genes against Italian lemon disease caused by *Deuterophoma tracheiphila*. In this case, as a result of gene dosage, a higher level of resistance may be expected in doubled haploids. Moreover, *C. clementina* cv. Nules is characterized by monoembryony and self-incompatibility; therefore, production of highly embryogenic haploid callus and plantlets could be of particular importance. In fact, it is difficult to obtain nucellar callus from monoembryonic *Citrus* species, which is essential for genetic improvement with biotechnology. Haploids from self-incompatible *C. clementina* plants would be desirable for fixing quantitative characters in completely homozygous self-incompatible diploid lines (de Nettancourt & Devreux, 1977).

11. Conclusions

Haploids have significant potential in *Citrus* breeding for developing new cultivars. Although anther and microspore culture has become more or less standard tools for haploid production in many crops, this is not yet the case for *Citrus*. Several other methods of haploid production, such as spontaneous haploid seedlings, use of irradiated pollen, hybridization between diploid and triploid plants, embryo rescue and ovary culture, have similarly failed to produce large numbers of haploids in *Citrus* spp.

Concerted effort is needed to develop and modify various aspects of biotechnology in order for it to be applied successfully in improving *Citrus*. A clear understanding of different steps required for *in vitro* regeneration and microspore embryogenesis of *Citrus* is essential to improve the frequency of embryos from pollen grains and the number of plantlets per anther. Therefore, in our opinion, this type of research has great prospects and needs further support and attention, connected with a study in depth on the acclimatization process of the regenerated haploid plants in order to insure their survival from *in vitro* conditions to greenhouse and field conditions.

12. Acknowledgements

The author wishes to thank E. Rugini for critically reading the manuscript.

13. References

Bajaj, Y.P.S., 1990. *In vitro* production of haploids and their use in cell genetics and plant breeding. In: Y.P.S. Bajaj (Ed.), Biotechnology in Agriculture and Forestry, Vol. 12: Haploids in Crop Improvement, Part I, pp. 1–44. Springer-Verlag, Berlin.

Blakeslee, A.F., J. Belling, M.E. Farnham & A.D. Bergner, 1922. A haploid mutant in the Jimson weed, *Datura stramonium*. Science 55: 646–647.

Blakeslee, A.F. & J. Belling, 1924. Chromosomal mutations in the Jimson weed, *Datura stramonium*. J. Hered. 15: 195–206.

Bouvier, L., Y.X., Zhang & Y. Lespinasse, 1993. Two methods of haploidization in pear, *Pyrus communis* L.: greenhouse seedling selection and *in situ* parthenogenesis induced by irradiated pollen. Theor. Appl. Genet. 87: 229–232.

Cameron, J.W. & H.B. Frost, 1968. Genetic, breeding and nucellar embryony. In: W. Reuther, L.D. Batchelor & H.J. Webber (Eds.), The Citrus Industry, Vol. 2., pp. 325–370. Div. Agr. Sci., Univ. Calif., Berkeley.

Chang, K., 1992. The evolution of Citrus demand and supply. Proc. Intl. Soc. Citriculture 3: 1153–1155.

Chaturvedi, H.C. & G.C. Mitra, 1974. Clonal propagation of *Citrus* from somatic callus cultures. Hort. Sci. 9: 118–120.

Chaturvedi, H.C. & A.K. Sharma, 1985. Androgenesis in *Citrus aurantifolia* (Christm.) Swingle. Planta 165: 142–144.

Chen, Z., H. Wang & H. Liao, 1980. The induction of *Citrus* pollen plants in artificial media. Acta Genet. Sin. 7: 189–192.

Chen, Z., 1985. A study on induction of plants from *Citrus* pollen. Fruit Varieties J. 39: 44–50.

Chen, Z., 1986. Induction of androgenesis in woody plants. In: H. Han & Y. Hongyuan (Eds), Haploids of Higher Plants *in Vitro*, pp. 42–66. China Academic Publishers, Beijing/Springer-Verlag, Berlin.

Chu, C., 1978. The N6 medium and its applications to anther culture of cereal crops. In: Proc. Symp. Plant Tissue Culture, pp. 43–50. Science Press, Peking.

D'Amato, F., 1977. Cytogenetics of differentiation in tissue and cell cultures. In: J. Reinert & Y.P.S. Bajaj (Eds.), Applied and Fundamental Aspects of Plant Cell, Tissue, and Organ Culture, pp. 343–356. Springer-Verlag, Berlin.

de Nettancourt, D. & M. Devreux, 1977. Incompatibility and *in vitro* cultures. In: J. Reinert & Y.P.S. Bajaj (Eds.), Applied and Fundamental Aspects of Plant Cell, Tissue and Organ Culture, pp. 426–441. Springer-Verlag, Berlin.

Deng, X.X., Z.A. Deng, S.Y. Xiao & W.C. Zhang, 1992. Pollen derived plantlets from anther culture of Ichang papeda hybrids No. 14 and Trifoliate orange. Proc. Intl. Soc. Citriculture 1: 190–192.

De Paepe, R., D. Prat & T. Huguet, 1982. Hereditable nuclear DNA changes in doubled haploid plants obtained by pollen culture of *Nicotiana sylvestris*. Plant Sci. Lett. 28: 11–28.

Dhillon, S.S., E.A. Wernsman & J.P. Miksche, 1983. Evaluation of nuclear DNA content and heterochromatin changes in anther-derived dihaploids of tobacco (*Nicotiana tabacum*) cv. Coker 139. Can. J. Genet. Cytol. 25: 169–173.

Drira, N. & A. Benbadis, 1975. Analyse, par culture d'anthères *in vitro*, des potentialités androgenetiques de deux espèces de *Citrus* (*Citrus medica* L. et *Citrus limon* L. Burm.). C.R. Hebdomadaires Sci. Acad. Sci. 281(18): 1321–1324.

Dunwell, J.M., 1986. Pollen, ovule and embryo culture, as tools in plant breeding. In: L.A. Withers & P.G. Alderson (Eds.), Plant Tissue Culture and its Agricultural Applications, pp. 375–404. Butterworths, London.

Esen, A. & R.K. Soost, 1972. Unexpected triploids in *Citrus*: their origin, identification, and possible use. J. Hered. 62: 329–333.

Evans, D.E., M.B. Singh & R.B. Knox, 1990. Pollen development: applications in biotechnology. In: S. Blackmore & R.B. Knox (Eds.), Microspores: Evolution and Ontogeny. pp. 309–338. Academic Press, New York.

FAO, 1994. Citrus fruit fresh and processed. Annual Statistics.

Frau, A.M., 1973. Osservazioni sulla quantità e sulla germinabilità del polline di alcune cultivar di Agrumi. Studi Sassaresi Sez. III. Ann. Fac. di Agraria. XXI: 3–10.

Frost, H.B., 1925. The chromosomes of *Citrus*. J. Washington Acad. Sci. 15: 1–3.

Fusurato, K., 1957. Studies on polyembryony in *Citrus*. Rep. Kihara Inst. Biol. Res. 8: 40–48.

Gamborg, O.L., R.A. Miller & K. Ojima, 1968. Nutrient requirements of suspension cultures of soybean root cells. Exptl. Cell. Res. 50: 151–158.

Geraci, G. & A. Starrantino, 1990. Attempts to regenerate haploid plants from "*in vitro*" cultures of *Citrus* anthers. Acta Hort. 280: 315–320.

Germanà, M.A., Yu Ying Wang, G. Chironi, G. Viglianisi & F.G. Crescimanno, 1990. Indagine sull'androgenesi in *Citrus aurantium* (L.) e *C. sinensis* (L.) Osbeck. In: XXXIV Congresso Annuale Società Italiana di Genetica Agraria, Poster.

Germanà, M.A., F.G. Crescimanno, F. De Pasquale & Yu Ying Wang, 1991. Androgenesis in five cultivars of *Citrus limon* L. Burm. f. Acta Hort. 300: 315–324.

Germanà, M.A., Yu Ying Wang, M.G. Barbagallo, G. Iannolino & F.G. Crescimanno, 1994. Recovery of haploid and diploid plantlets from anther culture of *Citrus clementina* Hort. ex Tan. and *Citrus reticulata* Blanco. J. Hort. Sci. 69: 473–480.

Gmitter F.G. Jr., X.B. Ling & X.X. Deng, 1990. Induction of triploid *Citrus* plants from endosperm calli *in vitro*. Theor. Appl. Genet. 80: 785–790.

Grosser, J.W., F.G. Gmitter Jr. & J.L. Chandler, Jr., 1988. Intergeneric somatic hybrid plants from sexually incompatible woody species: *Citrus sinensis* and *Severina disticha*. Theor. Appl. Genet. 75: 397–401.

Grosser J.W. & F.G. Gmitter, Jr., 1990. Protoplast fusion and citrus improvement. Plant Breed. Rev. 8: 339–374.

Guha, S. & S.C. Maheshwari, 1964. *In vitro* production of embryos from anthers of *Datura*. Nature (London) 204: 497–498.

Guha, S. & S.C. Maheshwari, 1966. Cell division and differentiation of embryos in the pollen grains of *Datura in vitro*. Nature (London) 212: 97–98.

Heberle-Bors, E., 1980. Isolated pollen cultures and pollen dimorphism. Naturwissenschaften 67: 311–312.

Heberle-Bors, E., 1982. *In vitro* pollen embryogenesis in *Nicotiana tabacum* L. and its relation to pollen sterility, sex balance and floral induction of the pollen donor plants. Planta 156: 396–401.

Heberle-Bors, E., 1983. Induction of embryogenic pollen grains and subsequent embryogenesis in *Nicotiana tabacum* L. by treatments of the pollen donor plants with feminizing agents. Physiol. Plant. 59: 67–72.

Heberle-Bors, E., 1985. *In vitro* haploid formation from pollen: a critical review. Theor. Appl. Genet. 71: 361–374.

Heberle-Bors, E., 1989. Isolated pollen culture in tobacco: plant reproductive development in a nutshell. Sex. Plant Reprod. 2: 1–10.

Heberle-Bors, E., R.M. Benito Moreno, A. Alwen, E. Stöger & O. Vincente, 1990. Transformation of pollen. In: H.J.J. Nijkamp, L.H.W. van der Plas & J. van Aartijk (Eds.), Progress in Plant Cellular and Molecular Biology, pp. 244–251. Kluwer Academic Publishers, Dordrecht.

Hidaka, T., Y. Yamada & T. Shichijo, 1979. *In vitro* differentiation of haploid plants by anther culture in *Poncirus trifoliata* (L.) Raf. Jpn. J. Breed. 29: 248–254.

Hidaka, T., Y. Yamada, & T. Shichijo, 1981. Plantlet formation from anthers of *Citrus aurantium* L. Proc. Intl. Soc. Citriculture 1: 153–155.

Hidaka, T., 1984a. Effects of sucrose concentration, pH of media, and culture temperature on anther culture of *Citrus*. Jpn. J. Breed. 34: 416–422.

Hidaka, T., 1984b. Induction of plantlets from anthers of "Trovita" orange. J. Jpn. Soc. Hort. Sci. 53: 1–5.

Hidaka, T. & M. Omura, 1989a. Control of embryogenesis in *Citrus* cell culture. Bull. Fruit Tree Res. Stn. 16: 1–17.
Hidaka, T. & M. Omura, 1989b. Origin and development of embryoids from microspores in anther culture of *Citrus*. Jpn. J. Breed. 39: 169–178.
Hidaka, T. & M. Omura, 1993. Transformation of *Citrus* protoplasts by electroporation. J. Jpn. Soc. Hort. Sci. 62: 371–376.
Hu, H. & Y. Hongyuan, 1986. Haploids of Higher Plants *in Vitro*. China Acad. Publ., Beijing/ Springer-Verlag, Berlin.
Hu, H. & B. Huang, 1987. Application of pollen-derived plants to crop improvement. In: K.L. Giles & J. Prakash (Eds.), Pollen: Cytology and Development, Intl. Rev. Cytology, Vol. 107, pp. 293–314. Academic Press, New York.
Kaneyoshi (Hiramatsu), J., S. Kobayashi, Y. Nakamura, N. Shigemoto & Y. Doi, 1994. A simple and efficient gene tranfer system of trifoliate orange (*Poncirus trifoliata* Raf.). Plant Cell Rep. 13: 541–545.
Karasawa, K., 1971. On the occurrence of haploid seedlings in *Citrus natsudaidai* Hayata. Bull. Sakushingakuin Junior Col. for Women 1: 1–2.
Kasha, K.J. (Ed.), 1974. Haploids in Higher Plants: Advances and Potential. Univ. Guelph Press, Guelph, Ontario.
Keller, W.A., T. Rajhathy, & J. Lacapra, 1975. *In vitro* production of plants from pollen in *Brassica campestris*. Can. J. Genet. Cytol. 17: 655–666.
Khuroshvili, K.G., Sh.K. Goliadze & Ts.S. Kashakashvili, 1982. Use of anther culture to obtain haploid plants of citrus crops. Subtropicheskie-Kul'tury. 5: 72–76.
Kostoff, D., 1929. An androgenic *Nicotiana* haploid. Z. Zellforsch. Mikrosk. Anat. 9: 640–642.
Ling, J., M. Iwamasa, & N. Nito, 1988. Plantlet regeneration by anther culture of Calamondin (*C. madurensis* Lour.). Proc. Sixth Intl. Citrus Congr. 1: 251–256.
Maheshwari, S.C., A.K. Tyagi, K. Malhotra & S.K. Sopory, 1980. Induction of Haploidy from pollen grains in angiosperms – the current status. Theor. Appl. Genet. 58: 193–206.
Moore, G.A., C.C. Jacono, J.L. Neidigh, S.D. Lawrence & K. Cline, 1992. *Agrobacterium*-mediated tranformation of *Citrus* stem segments and regeneration of transgenic plants. Plant Cell Rep. 11: 238–242.
Murashige, T. & F. Skoog, 1962. A revised medium for rapid growth and bioassays with tobacco tissue cultures. Physiol. Plant. 15: 473–497.
Murashige, T. & D.P.H. Tucker, 1969. Growth factor requirement of *Citrus* tissue culture. Proc. 1st Intl. Citrus Symp. 3: 1155–1161.
Narayanaswamy, S. & L. George, 1982. Anther culture. In: B.M. Johri (Ed.), Experimental Embryology of Vascular Plants, pp. 79–103. Springer-Verlag, Berlin.
Nitsch, C., 1977. Culture of isolated microspores. In: J. Reinert & Y.P.S. Bajaj (Eds.), Applied and Fundamental Aspects of Plant Cell, Tissue and Organ Culture, pp. 268–278. Springer-Verlag, Berlin.
Nitsch, J.P. & C. Nitsch, 1969. Haploid plants from pollen grains. Science 163: 85–87.
Nuti Ronchi V., L. Giorgetti, M. Tonelli & G. Martini, 1992. Ploidy reduction and genome segregation in cultured carrot cell lines. II. Somatic meiosis. Plant Cell Tiss. Organ Cult. 30: 115–119.
Oiyama, I.I. & S. Kobayashi, 1993. Haploid obtained from diploid X triploid crosses of *Citrus*. J. Jpn. Soc. Hort. Sci. 62: 89–93.
Raghavan, V., 1990. From microspore to embryoid: faces of the angiosperm pollen grain. In: H.J.J. Nijkamp, L.H.W. van der Plas & J. van Hartrigik (Eds.), Progress in Plant Cellular and Molecular Biology, pp. 213–221. Kluwer Academic Publishers, Dordrecht.
Rangan, T.S., T. Murashige & W.P. Bitters, 1969. *In vitro* studies of zygotic and nucellar embryogenesis in *Citrus*. In: H.D. Chapman (Ed.), Proc. Intl. Citrus Symp., Univ. California, Riverside, pp. 225–229.
Reinert, J. & Y.P.S. Bajaj, 1977. Anther culture: Haploid production and its significance. In: J. Reinert & Y.P.S. Bajaj (Eds.), Applied and Fundamental Aspects of Plant Cell, Tissue and Organ Culture, pp. 251–267. Springer-Verlag, Berlin.

Sangwan, R.S. & B.S. Sangwan-Norreel, 1990. Anther and pollen culture. In: S.S. Bhojwani (Ed.), Plant Tissue Culture: Applications and Limitations, pp. 220–241. Elsevier, Amsterdam.
Sangwan-Norreel, B.S., 1983. Male gametophyte nuclear DNA content evolution during androgenic induction in *Datura innoxia*. Z. Pflanzenphysiol. 111: 47–54.
Sangwan-Norreel, B.S., R.S. Sangwan & J. Pare, 1986. Haploïdie et embryogenèse provoquée *in vitro*. Bull. Soc. Bot. Fr. 133. Actual. Bot. 4:7–39.
Schenk, R.U. & A.C. Hildebrandt, 1972. Medium and techniques for induction and growth of monocotyledonous and dicotyledonous plant cell cultures. Can. J. Bot. 50: 199–204.
Shull, J.K. & M.Y. Menzel, 1977. A study of the reliability of synchrony in the development of pollen mother cells of *Lilium longiflorum* at the first meiotic prophase. Am. J. Bot. 64: 670–679.
Soost, R.K. & J.W. Cameron, 1980. 'Oroblanco', a triploid pummelo-grapefruit hybrid. Hort Science 15: 667–669.
Soost, R.K. & J.W. Cameron, 1985. 'Melogold', a triploid pummelo-grapefruit hybrid. Hort Science 20: 1134–1135.
Starrantino, A., 1986. Produzione di aploidi androgenetici di agrumi. In: Il recente contributo della ricerca allo sviluppo dell'agrumicoltura italiana, Delfino, Cagliari, pp.31–37.
Stuart, D.A., S.G. Strickland & K.A. Walker, 1987. Bioreactor production of alfalfa somatic embryos. HortScience 22: 800–809.
Sunderland, N., 1971. Anther culture: a progress report. Sci. Prog. 59: 527–549.
Sunderland, N., 1974. Anther culture as means of haploid induction. In: K.J. Kasha (Ed.), Haploids in Higher Plants: Advances and Potential, pp. 91–122. Univ. Guelph Press, Guelph, Ontario.
Sunderland, N., 1978. Strategies on the improvement of yields in anther culture. In: Proceedings of the Symposium on Plant Tissue Culture, pp. 65–86. Science Press, Peking.
Sunderland, N., 1983. The concept of morphogenic competence with reference to anther and pollen culture. In: S.K. Sen & K.C. Giles (Eds.), Plant Cell Culture in Crop Improvement, pp. 125–129. Plenum Press, New York.
Sunderland, N. & J.M. Dunwell, 1977. Anther and pollen culture. In: H.E. Street (Ed.), Plant Tissue and Cell Culture, pp. 223–265. Blackwell, Oxford.
Vardi, A., 1988a. Application of recent taxonomical approaches and new techniques to *Citrus* breeding. Proc. Sixth Intl. Citrus Congress 1: 303–308.
Vardi, A., 1988b. Progressi nel miglioramento genetico delle piante coltivate. Tecnica agricola 40. 341–355.
Vardi, A., S. Bleichman & D. David, 1990. Genetic transformation of *Citrus* protoplasts and regeneration of transgenic plants. Plant Sci. 69: 199 206.
Vardi, A., 1992. Conventional and novel approaches to *Citrus* breeding. Proc. Intl. Citrus Congress 1: 39–43.
Vasil, I.K., 1967. Physiology and cytology of anther development. Biol. Rev. 42: 327–373.
Vasil, I.K., 1980. Androgenic haploids. Intl. Rev. Cytol. Suppl. 11A: 195–223.
Veilleux, R.E., 1994. Development of new cultivars via anther culture. HortScience 29: 1238–1241.
Wenzel, G., F. Hoffmann & E. Thomas, 1977. Anther culture as a breeding tool in rape. I. Ploidy level and phenotype of androgenetic plants. Z. Pflanzenzüchtg. 78: 149–155.
Wenzel, G., U. Frei, A. Jahoor, A. Graner & B. Foroughi-Wehr, 1995. Haploids: An integral part of applied and basic research. In: M. Terzi, R. Cella & A. Falavigna (Eds.), Current Issues in Plant Molecular and Cellualar Biology, pp. 127–135. Kluwer Academic Publishers, Dordrecht.
Zhang, Y.X., Y. Lespinasse & E. Chevreau, 1990. Induction of haploidy in fruit trees. Acta Hort. 280: 293–304.

12. Haploid plantlet production in tomato

WILLIAM L. SUMMERS

Contents

1. Introduction

One of the routine problems facing plant breeders is how to rapidly develop inbred lines for either direct release or for use as parents of hybrids. One way to accomplish inbreeding efficiently is to produce haploid or dihaploid plants from pollen or anthers of F_1 hybrids. Although much research has been conducted on tomato (*Lycopersicon esculentum* Mill.) pollen or anther culture, at least three problems have emerged as roadblocks to the successful application of these techniques. First, because both the number of anthers producing calli and the number of plants regenerated per callus have been limited, few plants have been recovered. Secondly, very few tomato anther genotypes have been evaluated under a variety of experimental conditions to rate their callus forming and regeneration potential. Third, various media/-methods have been used to produce plantlets of unknown ploidy. In many of these systems closer evaluation suggests that plantlets developed as a product of organogenesis from anther wall tissue rather than via androgenesis (Zamir *et al.*, 1981). This chapter reviews published reports of tomato anther and microspore culture, contrasts procedures that produce anther-derived calli and plantlets (Fig. 1) with those that do not, and explores factors that need optimization or further study prior to routine haploid plantlet production via anther and/or microspore culture.

2. Tomato haploids

The first recorded description of a tomato haploid grown from seed was provided by Lindstrom (1929). He noted that the plant was dwarfed with many axillary branches. Cuttings readily rooted allowing vegetative propa-

S.M. Jain, S.K. Sopory & R.E. Veilleux (eds.), In Vitro *Haploid Production in Higher Plants, Vol. 5,* 219–231.

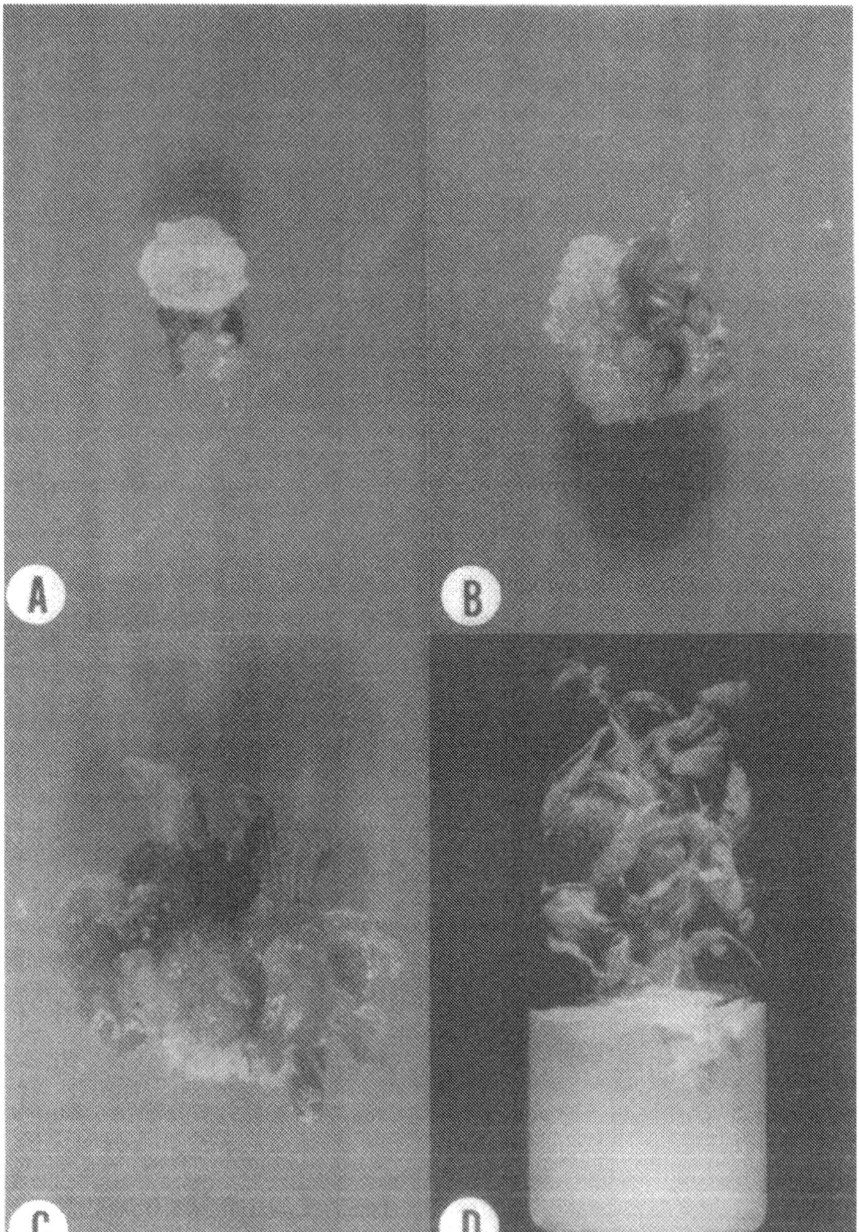

Figure 1. Regeneration of tomato anther callus. (A) Brown anther wall tissue and green callus. (B) Small anther callus with a single plantlet. (C) Larger anther callus producing many plantlets. (D) Rooted plantlet prior to transfer to a soil medium.

gation of the original plant. Haploid flowers were described as smaller than diploid, with shrunken anthers. When used as a male parent on diploid tomato no seed were produced but, when used as a female parent, 11% of the crosses with diploid pollen produced seed. Of interest are Lindstrom's notes on meristematic root tip cells. While most of the cells contained 12 chromosomes, some were aneuploid or diploid. Cook (1936) described the use of spontaneously occurring haploids of Marglobe, Gulf State Market, Globe, and Earliana by the Ferry-Morse Seed Breeding Institute to produce doubled haploid tomato cultivars for general use. Further descriptions of commercial doubled haploids were provided by Morrison (1932).

3. Nurse culture

In order to minimize culturing remnants of the anther wall or filament which might lead to diploid/polyploid plantlets, Sharp *et al.* (1972) developed a nurse culture system for pollen culture. In this system microspores from 1 cm long Rutgers anthers were placed on a filter paper disk which rested on whole tomato anthers. Twelve percent of the pollen produced plantlets while none of the pollen plated directly on an agar-based basal media grew.

4. Anther culture

Direct androgenesis, i.e., the formation of microspore-derived embryos in anther or microspore culture without an intervening callus phase has been difficult in tomato. The few reports of anther-derived embryos have not resulted in the regeneration of haploid plants. Formation of callus by tomato anther tissue has also been difficult to obtain and very few plantlets have been regenerated (Sink & Reynolds, 1986). We now report on research aimed at maximizing callus induction from tomato anthers. Because of the nature of the tomato inflorescence (monochasial cyme), microspores within tomato anthers exhibit various stages of development at any fixed length (Picken *et al.*, 1985). Anthers in smaller buds that occur towards the apex of an inflorescence typically contain more advanced microspores than similar sized anthers at the basal portion of an inflorescence. This hampers the use of anther size as a macroscopic gauge of microspore developmental stage. Hence, the most appropriate developmental stage for anther culture has been difficult to determine.

In a study using genetic stocks that carried one of 15 different male sterile tomato loci used to disrupt normal anther/microspore development over a range of stages, Zamir *et al.* (1980) found that "only anthers containing *ms*10^{35}, which disrupts development in the meiotic vs. the premeiotic or tetrad stages, consistently produced callus". Arrested development seemed to occur in anthers that were 1.5 to 2.5 mm long. In these studies, 26% of the 2.5 mm long *ms*10^{35} anthers produced callus while only 0.2% of anthers of similar length from corresponding isogenic fertile lines produced callus. These studies led Summers *et al.* (1992) to investigate whether fertile anthers were less able to form calli or alternatively, whether fertile anthers were plated at a suboptimal developmental stage. We compared the number of anthers forming callus and mean callus diameter from anthers plated at each of five developmental stages (leptotene, zygotene to diakinesis, metaphase I to telophase II, uninucleate, or binucleate) from three tomato lines (Licato, Ailsa Craig, or L-680A). Callus could be induced from all five stages of anther development in tomato although some cultivars did not produce calli after metaphase I. The optimal stage for callus induction and growth was prophase I. These results agreed with those of Gresshoff & Doy (1972). However, our results showed that there was greater callus induction from microspores plated at early prophase I (leptotene) than at late prophase I (zygotene). Generally, fewer calluses were produced as anther length increased and microspore development proceeded. Few calluses were induced from uninucleate or binucleate microspores. These observations did not agree with those of Gulshan *et al.* (1981) and Levenko *et al.* (1977) who found that only 3 mm long tomato anthers containing uninucleate microspores produced callus. In contrast to Zamir *et al.* (1980), 33% of the fertile anthers plated produced callus in our study. This demonstrated that anther development did not need to be genetically disrupted to obtain increased callus induction rates if fertile anthers were plated at optimal developmental stages and appropriate cultivars were used.

5. Media solidification agent

Although semisolid media have been used extensively in plant tissue culture (George & Sherrington, 1984; Dunwell, 1986) agar may contain substances that inhibit embryo production from microspores. In some instances liquid media have been shown to be significantly better than agar-solidified media for anther culture, e.g., *Nicotiana tabacum* (Kohlenbach & Wernicke, 1978).

In tissue culture applications agar concentration may generally range from <3 to $15\,g\,l^{-1}$ (Debergh, 1983; George & Sherrington, 1984; Miller & Murashige, 1976; Singha, 1982). Gelrite, a hydrocolloid gellan gum has also been used at concentrations of $1.7\,g\,l^{-1}$ and up. For tomato anther culture, the most commonly used solidifying agent has been Difco Bacto-Agar at concentrations of 7 to $8\,g\,l^{-1}$ (Cappadocia & Sree Ramulu, 1980; Dao & Shamina, 1978; Gresshoff & Doy, 1972; Zamir *et al.*, 1980); however, little data exist to determine the optimal concentration of each solidifying agent. Jaramillo & Summers (1990) determined that the optimal concentration of Noble agar was $6\,g\,l^{-1}$; Bacto-agar $8\,g\,l^{-1}$; Gelrite $3\,g\,l^{-1}$; and Phytagar $5\,g\,l^{-1}$ for tomato anther callus studies. When callus production was examined at each of these rates and compared to a liquid treatment of DBM_1 medium, 38% fewer tomato anthers on Bacto agar produced callus than the liquid treatment and 42% fewer than Noble agar. Tomato anther culture systems may not require solidified culture media; however, gelling agents and concentrations should be evaluated.

6. Basal media composition

Much of the research on tomato anther culture has focused on increasing the efficiency of callus formation and plantlet production by optimizing media composition (Table 1). Murashige & Skoog's (1962) basal (MS_b) or revised (MS_r) media have been widely used for anther and microspore culture (Dunwell, 1985). However, variations of White's (1934) media (Sharp *et al.*, 1972), i.e., Modified White's (MW), Gamborg & Eveleigh's (1968) medium, i.e., Doy's basal medium 1 (DBM_1), and Blaydes (1966) medium, i.e., Doy's basal medium 2 (DBM_2), have been used to induce callus and plantlet formation from tomato anthers (Gresshoff & Doy, 1972, 1974). The absence of replicated studies which compare media effects on anther culture across a number of tomato genotypes has led to confusion about which medium produces the most anther-derived calli. Jaramillo (1988) conducted a study where five media, MS_b, MS_r, MW, DBM_1, and DBM_2, were compared for their ability to form callus from leptotene stage anthers of Licato, Ailsa Craig and L-680A. Both media and cultivar effects were significant. More anthers produced callus on DBM_1 medium (Table 2) while the calluses grew largest on either DBM_1 or MS_r media (Table 3). Forty-five percent more L-680A anthers produced callus and the calluses grew to a larger size by 30% than those of other cultivars in the study. In other studies using tetrad stage anthers, DBM_1 and DBM_2 media promoted significantly more callus growth (75%) than the other three media.

7. Carbohydrate type and concentration

Carbohydrates serve as both a carbon source and as an osmotic stabilizer in tissue culture media (Dunwell, 1985; Limberg *et al.*, 1979). Generally, sucrose is used as a carbohydrate source in culture media although glucose has also been used. Both sugars have been used as a carbon source in anther culture studies (Gresshoff & Doy, 1974). In *Petunia*, another solanaceous species where embryogenic response has been low (Maizonnier, 1984), it has been demonstrated that media containing maltose (0.167 M) produced more embryos than those with lactose, sucrose, glucose + fructose, or fructose + - galactose (Raquin, 1983). Jaramillo (1988) found that DBM_1 media supplemented with 30 g l^{-1} maltose, 20 g l^{-1} sucrose, or a combination of 10 g l^{-1} of sucrose and glucose produced significantly more tomato anther calluses

Table 1. Culture media used to produce haploid tomato callus

Media Component	MS REVISED 1962 mg/l MS_r	MS BASAL 1962 mg/l MS_b	ZAMIR 1980 mg/l	ZAMIR 1981 mg/l	WHITE 1963 mg/l MW	SHARP DOUGALL 1971 mg/l	SHARP RASKIN 1972 mg/l	GRESSHOFF & DOY 1972 mg/l DBM1	GRESSHOFF & DOY 1972 mg/l DBM2
NH_4NO_3	1650	400	400	400					1000
KNO_3	1900	80	80	80	80	80	80	1000	1000
$CaCl_2 \cdot 2H_2O$	440							150	
$MgSO_4 \cdot 7H_2O$	370	72	72	72	720	750	750	250	35
KH_2PO_4	170	12.5	12.5	12.5					300
$(NH_4)_2SO_4$								200	
$NaH_2PO_4 \cdot H_2O$					16.5	16.5	16.5	90	
Na_2HPO_4								30	
$Ca(NO_3)_2 \cdot 4H_2O$		144	144	144	300	300	300		347
KCl		65	65	65	65	65	65	300	65
Na_2SO_4					200	200	200		
KI	0.83	0.75	0.75	0.75	0.75	0.75	0.75	0.75	0.80
H_3BO_3	6.2	1.6	1.6	1.6	1.5	1.5	1.5	3.0	3.0
$MnSO_4 \cdot 4H_2O$ or	22.3	6.5	6.5	6.5	7.0	7.0	7.0		
$MnSO_4 \cdot H_2O$	16.9							10	10
$ZnSO_4 \cdot 4H_2O$ or	8.6								
$ZnSO_4 \cdot 7H_2O$		2.7	2.7	2.7	3.0	3.0	3.0	3.0	3.0
$Na_2MoO_4 \cdot 2H_2O$	0.25							0.25	0.25
MoO_3					0.0001	0.0001	0.0001		
$CuSO_4 \cdot 5H_2O$	0.025				0.001	0.001	0.001	0.25	0.25
$CoCl_2 \cdot 6H_2O$	0.025							0.25	0.25
$Fe_2(SO_4)_3$					2.5				
$FeSO_4 \cdot 7H_2O$	27.8	27.8	27.8	27.8		27.8	27.8	27.8	27.8
Na_2 EDTA	37.3	37.3	37.3	37.3		37.3	37.3	37.3	37.3
Thiamine-HCl	0.1	0.1	0.1	0.1	0.1	4.0	4.0	10.0	10.0
Pyridoxine-HCl	0.5	0.5	0.5	0.5	0.1	1.0	1.0	1.0	1.0
Nicotinic Acid	0.5	0.5	0.5	0.5	0.5	5.0	5.0	1.0	1.0
Biotin (Optional)								2.0	2.0
Panthothenate-Ca (Optional)					1.0	0.1	0.1	1.0	1.0
Sucrose	30000	20000	20000	10000	20000	40000	40000	20000	20000
Myo-Inositol	100	100			100	100	100	100	100
Agar	10000	10000	8000	8000	8000	8000	8000	8000	8000
Glycine	2	2			3			4	4
Edamin	1000	1000							
Kinetin	0.04-10	0.2				0.128[a]		5	1
2,4-D					6	0.32	6		
NAA						1[a]		2	2
trans - Zeatin			0.25	0.25					
IAA	1-30	2	0.50	0.50					
Detergent						YES	YES		
ETOH Percent			70	70				100	100
ETOH Time, Seconds			20	20				10	10
Chlorox Percent			0.25	0.25		7	1	-	-
Chlorox Time, Minutes		5	5		3	5	-	-	
Medium pH		5.8	5.8	5.8				5.8	5.8
Temp Minimum, °C		27	27	27		25	25	27	27
Temp Maximum, °C		27	27	27		25	25	27	27
Photoperiod-Induction		16	16		12	12	0	0	
Photoperiod-Differentiation			-	-				16	16
Light Level, Foot-Candles			278	278		200	200	370	370

[a] Use either 2,4-D or NAA, not both.

Table 2. Mean number of callus producing anthers per dish

			Media[a]			
Cultivar	DBM_1	DBM_2	MS_r	MS_b	MW	Mean[b]
L-680A	3.7[c]	2.6	2.7	1.4	1.1	2.3
Ailsa Craig	3.1	1.5	1.8	1.1	0.4	1.6
Licato	2.3	1.6	1.3	1.0	0.6	1.4
Mean[d]	3.1	1.9	1.9	1.1	0.7	

[a] Dunnett 0.05 = N.S.
[b] Dunnett 0.05 = 0.5, cv% 56.7.
[c] Five anthers per dish, 14 dishes per treatment (70 anthers per treatment mean).
[d] Dunnett 0.05 = 0.4, cv% 53.1.

Table 3. Mean callus diameter derived from cultured tomato anthers

			Media[a]			
Cultivar	MS_r	DBM_1	DBM_2	MS_b	MW	Mean[b]
L-680A	2.1[c]	2.0	1.2	1.0	0.9	1.4
Ailsa Craig	1.7	1.7	0.9	0.9	0.4	1.1
Licato	1.6	1.6	1.0	0.8	0.5	1.1
Mean[d]	1.8	1.8	1.1	0.9	0.6	

[a] Dunnett 0.05 = N.S.
[b] Dunnett 0.05 = 0.4, cv% 52.7.
[c] Mean diameter of calli per dish in mm. Five anthers per dish, 14 dishes per treatment, 70 anthers per treatment mean.
[d] Dunnett 0.05 = 0.5, cv% 54.9.

than similar media supplemented with 10 or 20 $g\,l^{-1}$ maltose. Studies by Yelle *et al.* (1988) have shown that *L. esculentum* fruit accumulate fructose and glucose while *L. chmielewskii* fruit accumulate sucrose. The effects of substituting hexoses for sucrose in DBM_1 anther culture media remain to be determined.

8. Media-growth regulators type and concentration

Since root culture of various vegetable crops, including tomato, was a major focus of tissue culture in the 1920s and 30s as evidenced by the work of Kotte (1922) and Robbins (1922) and the continuous root culture system of White (1934) it should not be surprising that shoot regeneration in tomato tissue culture was first reported on callus derived from roots (Norton & Boll, 1954) and later from stem internode (DeLanghe, 1973). Attempts to optimize shoot regeneration began with studies of the effects of NAA, IAA and kinetin. A medium containing 22.8 μM IAA and 18.6 μM kinetin was found to be optimal (Padmanabhan *et al.*, 1974). Later Behki & Lesley (1976) proposed that shoot regeneration occurred more frequently when 0.5 or 2.5 μM NAA and 1.0 to 20 μM BA were used. Additional studies found that zeatin could be substituted for BA. Although optimal rates vary with tissue and genotype, a suitable starting point was proposed as 0.5 μM NAA and 50 μM zeatin. Throughout the latter half of the 1970s much work focused on optimizing growth regulator type and concentration. Kartha *et al.* (1977) found that shoots emerged earlier on medium with BA compared to zeatin and that 0.1 μM kinetin was effective at inducing callus. De Langhe & DeBruijne (1976) found that when combinations of zeatin + IAA, BA + IAA, zeatin + NAA, or BA + NAA were compared for their shoot regeneration potential, the combination of 1.0 μM zeatin + 0.1 or 1.0 μM IAA was best. Cassells (1979) found that calluses derived from explants of greenhouse tomatoes grown in the winter could be induced to produce more shoots, (31 vs. 1.7 per plant) if cut and subcultured on a medium containing 0.2 μM TIBA. Shoot regeneration from calluses derived from similar explants taken from tomatoes grown in the summer was not significantly increased by the treatment.

9. NO_3^-/NH_4^+ ratio

Behki & Lesley (1980) found that when nitrogen sources in MS medium were compared, NH_4^+ inhibited shoot regeneration compared to NO_3^-. Additional studies showed that NO_3^- could be substituted for NH_4^+ in MS media with no reduction in the number of shoots regenerated. Our studies focused on the effects of NO_3^-/NH_4^+ ratio on callus induction and growth on DBM_1 medium. When total nitrogen was provided by either NO_3^- or NH_4^+ on DBM_1 medium, callus induction doubled and growth increased 50% on NO_3^- compared to NH_4^+ plates. Increasing the nitrate ratio to 26:1 produced significant improvement in induction and growth compared to basal DBM_1 (Jaramillo, 1988).

10. Culture environment

Light is known to influence the rates of cell division (Fraser *et al.*, 1967; George & Sherrington, 1984) and ethylene evolution (Huxter *et al.*, 1981) which, in turn, may influence caulogenesis and rhizogenesis (Cornejo-Martin *et al.*, 1979). Therefore, the duration and timing of light application play an important role in explant morphogenesis (Noth & Abel, 1971). An alternating period of dark-light exposure using white light was superior to a continuous photoperiod or monochromatic red light for embryo induction of dark-treated *Datura* anthers (Sopory & Maheshwari, 1976). Rice (*Oryza sativa* L.) anthers must be incubated in the dark for three weeks before light exposure (Niizeki & Oono, 1968; Noth & Abel, 1971). However, in some species dark exposure may be associated with negative effects. In *Nicotiana tabacum*, anthers exposed to dark treatments produced less callus and fewer embryos than anthers exposed to light (Noth & Abel, 1971; Sunderland, 1971). Typically tomato anthers have been exposed to cool white light (2,000 to 20,000 lux) for 12 or 16 h photoperiods (Kut *et al.*, 1984), or placed in the dark for 4 to 8 weeks before light exposure (Gresshoff & Doy, 1972; Gulshan *et al.*, 1981). Jaramillo & Summers (1991) studied the effects of increasing the dark period from 2 to 10 weeks followed by culture under a 16 h photoperiod (88 $\mu mol \cdot s^{-1} \cdot m^{-2}$) provided by cool-white fluorescent lamps. The number of anthers producing callus increased significantly during the 4th, 6th and 7th weeks of dark treatment. No significant increase in callus induction was observed after eight weeks of dark treatment. In addition, calluses grew faster in the dark; calluses grown for ten weeks in the dark were 3.4 times larger than those exposed to five weeks of darkness followed by five weeks of light. In general, each week of dark exposure increased callus growth by 0.27 mm. Some tomato lines produced significantly more calluses that also grew at a faster rate. The line "L-680A" produced 1.8 times more calluses which grew 1.6 times larger than other lines used in this study. In contrast, the callus diameter of both "Ailsa Craig" and "Licato" was reduced if more than eight weeks of dark treatment was applied. This study suggests that calluses should be transferred to regeneration media after eight weeks of dark treatment.

11. Ploidy

Because it has been difficult to produce tomato anther callus and regenerate plantlets we wondered if these difficulties were due to gene dosage dependent products expressed in haploid tissue. Sharp *et al.* (1971) reported that haploid cell lines derived from tetraploid *L. esculentum* grew five times faster than haploid lines from diploid tomato. In addition, Ohki *et al.* (1978) found that hybrid tomato hypocotyls exhibited greater shoot production than comparable inbred lines. In a comparison of callus formation and growth among anthers of dihaploid, autotetraploid, and diploid "San Marzano" we found that the number and diameter of calluses did not differ significantly. Further comparisons between diploid vs. tetraploid "Break O'Day", "Ohio WR Globe", "Pearson", and "Stokesdale" yielded similar results. These data suggest that neither heterozygosity nor gene dosage play a significant role in stimulating the induction and/or growth of tomato anther callus (Jaramillo, 1988).

12. Temperature treatments

In many plant species traumatic shocks promote induction of embryogenesis and increase the number of pollen grains that develop into embryos (Sink & Padmanabhan, 1977). Temperature stress pretreatments typically produce the most significant effects (Maheshwari *et al.*, 1980). Debergh & Nitsch (1973) demonstrated that cold treatments increased the number of tomato embryos produced in culture. Levenko *et al.* (1977) found that 70–90% of tomato anthers treated at 4°C for 24 h produced callus. We wanted to determine whether cold or heat pretreatments of different duration significantly affected the number and growth of tomato anther callus compared to a control without pretreatment. Temperatures of 3, 11, 19, 27, 31, or 35°C were provided for 0, 2, 4, or 6 days. Callus number was significantly reduced for anthers treated at 11°C. Callus growth was significantly greater at 3 or 27°C when treated for 0, 2 or 4 days. When anthers were treated longer than four days callus induction was also reduced. Significant reductions in callus number and growth were observed when anthers were treated at 35°C or when heat treatments extended beyond two days (Jaramillo, 1988).

13. Future directions

Anthers contain cells of varying ploidy. Callus derived from tomato anthers, may originate from haploid or somatic cells that result in polysomatic cultures where the growth of 2*n* cells maybe favored. We are attempting to use flow cytometry to separate anther derived tomato cells based on ploidy. Such sorting systems offer the potential of selecting exclusively haploid cells from which new cultures can be established. We hope that the use of cultures enriched for haploid cells will enhance the likelihood of regenerating haploid tomatoes.

One difficulty in characterizing anther derived plants is to determine which are haploids, doubled haploids, or heterozygous diploids. We have found that growth chambers and growth regulators may disrupt leaf morphology and pigmentation marker systems. Early selections based on morphology or color frequently revert to a normal phenotype when grown under greenhouse conditions. We believe that the use of cultures established from exclusively haploid cells derived from hybrid anthers heterozygous for isozyme banding patterns, cell reporter systems (GUS, etc.), or differential northern blot behavior could prove useful in determining the origin of anther-derived plantlets.

Finally, we are convinced that modification of the gas headspace within a culture system holds great promise in stimulating cell growth, removing ethylene, and perhaps moving to a gas based growth environment.

14. References

Behki, R.M. & S.M. Lesley, 1976. *In vitro* plant regeneration from leaf explants of *Lycopersicon esculentum* (tomato). Can. J. Bot. 54: 2409–2414.

Behki, R.M. & S.M. Lesley, 1980. Shoot regeneration from leaf callus of *Lycopersicon esculentum*. Z. Pflanzenphysiol. 98: 83–87.

Blaydes, D.F., 1966. Interaction of kinetin and various inhibitors in the growth of soybean tissue. Physiol. Plant. 19: 748–753.

Cappadocia, M. & K. Sree Ramulu, 1980. Plant regeneration from *in vitro* cultures of anthers and stem internodes in an interspecific hybrid, *Lycopersicon esculentum* × *L. peruvianum* Mill. and cytogenetic analysis of the regenerated plants. Plant Sci. Lett. 20: 157–166.

Cassells, A.C., 1979. The effect of 2,3,5-triiodobenzoic acid on caulogenesis in callus cultures of tomato and *Pelargonium*. Physiol. Plant. 46: 159–164.

Cook, R.C., 1936. A haploid Marglobe tomato. J. Hered. 27: 433–435.

Cornejo-Martin, M.J., A.M. Mingo-Castel & E. Primo-Millo, 1979. Organ redifferentiation in rice callus: effects of C_2H_4, CO_2 and cytokinins. Z. Pflanzenphysiol. 94: 117–123.

Dao, N.T. & Z.B. Shamina, 1978. Cultivation of isolated tomato anthers. Soviet Plant Physiol. 25: 120–126.

Debergh, P.C. 1983. Effects of agar brand and concentration on the tissue culture medium. Physiol. Plant. 59: 270–276.

Debergh, P.C. & C. Nitsch, 1973. Premières resultats sur la culture *in vitro* de grains de pollen isolés chez la tomate. C.R. Acad. Sci. Paris 276: 1281–1284.

DeLanghe, E., 1973. Vegetative multiplication *in vitro*. Intl. Assoc. Plant Tiss. Cult. Newsl. 9: 7.

DeLanghe, E. & E. DeBruijne, 1976. Continuous propagation of tomato plants by means of callus culture. Sci. Hortic. 4: 221–227.

Dunwell, J.M., 1985. Haploid cell cultures. In: R.A. Dixon (Ed.), Plant Cell Cultures: A Practical Approach, pp. 21–36. IRL Press Ltd., Boca Raton.

Dunwell, J.M., 1986. Pollen, ovule, and embryo culture as tools in plant breeding. In: L.A. Withers & P.G. Alderson (Eds.), Plant Tissue Culture and its Agricultural Applications, pp. 374–404. Butterworth, London.

Fraser, R.S., U.E. Loeinig & M.M. Yeoman, 1967. Effect of light on cell division in plant tissue cultures. Nature (London) 215: 873.

Gamborg, O.L. & D. Eveleigh, 1968. Culture methods and detection of glucanases in suspension cultures of wheat and barley. Can. J. Biochem. 46: 417–421.

George, E.F. & P.D. Sherrington, 1984. Plant Propagation by Tissue Culture. Exegetics Ltd., Hants.

Gresshoff, P.M. & C.H. Doy, 1972. Development and differentiation of haploid *Lycopersicon esculentum* (tomato). Planta 107: 161–170.

Gresshoff, P.M. & C.H. Doy, 1974. Derivation of a haploid cell line from *Vitis vinifera* and the importance of the stage of meiotic development of anthers for haploid culture of this and other genera. Z. Pflanzenphysiol. 73: 132–141.

Gulshan, T., T.M. Varghese & D.R. Sharma, 1981. Studies on anther cultures of tomato *Lycopersicon esculentum* Mill. Biol. Plant. 23: 414–420.

Huxter, T.J., T.A. Thorpe & D.M. Reid, 1981. Shoot initiation in light and dark-grown tobacco callus. The role of ethylene. Physiol. Plant 53: 319–326.

Jaramillo, J., 1988. Anther culture in the tomato (*Lycopersicon esculentum* Mill.) callus production and plantlet production studies. Ph.D. Dissertation, Iowa State University, Ames.

Jaramillo, J. & W.L. Summers, 1990. Tomato anther callus production: Solidifying agent and concentration influence induction of callus. J. Amer. Soc. Hort. Sci. 115: 1047–1050.

Jaramillo, J. & W.L. Summers, 1991. Dark-light treatments influence induction of tomato anther callus. HortScience 26: 915–916.

Kartha, K.K., S. Champoux, O.L. Gamborg & K. Pahl, 1977. *In vitro* propagation of tomato by shoot apical meristem culture. J. Amer. Soc. Hort. Sci. 102: 346–349.

Kohlenbach, H.W. & W. Wernicke, 1978. Investigations on the inhibitory effect of agar and the function of active carbon in anther culture. Z. Pflanzenphysiol. 86: 463–472.
Kotte, W., 1922. Wurzelmeristem in Gewebekultur. Ber. Deutsch. Bot. Ges. 40: 269–272.
Kut, S.A., J.E. Bravo & D.A. Evans, 1984. Tomato. In: P.V. Ammirato, D.A. Evans, W.R. Sharp & Y. Yamada (Eds.), Handbook of Plant Cell Culture: Crop Species. Vol. 3, pp. 247–289. Macmillan, New York.
Levenko, B.A., V.A. Kunakh & G.N. Yurkova, 1977. Studies on callus tissue from anthers. I. Tomato. Phytomorphology 27: 377–383.
Limberg, M., D. Cress & K. Lark, 1979. Variants of soybean cells which can grow in suspension with maltose as a carbon energy source. Plant Physiol. 63: 718–721.
Lindstrom, E.W., 1929. A haploid mutant in the tomato. J. Hered. 20: 23–30.
Maheshwari, S.C., A. Rashid & A.K. Tyagi, 1980. Physiology of pollen haploid formation. The current status. In: P. Sala, B. Parisi, R. Cella & O. Ciferri (Eds.), Plant Cell Cultures: Results and Perspectives, pp. 393–398. Biomedical Press, Elsevier, Amsterdam.
Maizonnier, D., 1984. Haploidy. In: K. Sink (Ed.), Petunia, pp. 151–154. Springer-Verlag, New York.
Miller, C. & T. Murashige, 1976. Tissue culture propagation of tropical foliage plants. In Vitro 12: 797–813.
Morrison, G., 1932. The occurrence and use of haploid plants in the tomato with special reference to the variety Marglobe. Proc. 6th Intl. Congr. Genet. 2: 137–139.
Murashige T. & F. Skoog, 1962. A revised medium for rapid growth and bioassays with tobacco tissue cultures. Physiol. Plant. 15: 473–497.
Niizeki, H. & K. Oono, 1968. Induction of haploid rice plant from anther culture. Proc. Jpn. Acad. 44: 554–557.
Norton, J.P. & W.G. Boll, 1954. Callus and shoot formation from tomato roots *in vitro*. Science 119: 220–221.
Noth, M.H. & W.O. Abel, 1971. Zur Entwicklung haploider Pflanzen aus unreifen Mikrosporen verschiedener *Nicotiana* Arten. Z. Pflanzenzüchtg. 65: 277–284.
Ohki, S., C. Bigot & J. Mousseau, 1978. Analysis of shoot forming capacity *in vitro* in two lines of tomato (*L. esculentum* Mill.) and their hybrids. Plant Cell Physiol. 19: 27–42.
Padmanabhan, V., E.F. Paddock & W.R. Sharp, 1974. Plantlet formation from *Lycopersicon esculentum* leaf callus. Can. J. Bot. 52: 1429–1432.
Picken, A.J.F., R.G. Hurd, & D. Vince-Prue, 1985. *Lycopersicon esculentum*. In: A.H. Halevy (Ed.), Handbook of Flowering, Vol. III, pp. 330–346. CRC Press Inc., Boca Raton.
Raquin, C., 1983. Utilization of different sugars as carbon source for *in vitro* anther culture of petunia. Z. Pflanzenphysiol. 111: 453–457.
Robbins, W.J., 1922. Wurzelmeristem in Gewebekultur. Ber. Deutsch Bot. Ges. 40: 269–272.
Sharp, W.R., D.K. Dougall & E.F. Paddock, 1971. Haploid plantlets and callus from immature pollen grains of *Nicotiana* and *Lycopersicon*. Bull. Torrey Bot. Club 98: 219–222.
Sharp, W.R., R.S. Raskin & E. Sommer, 1972. The use of nurse culture in the development of haploid clones in tomato. Planta 104: 357–361.
Singha, S., 1982. Influence of agar concentration on *in vitro* shoot proliferation of *Malus* sp. "Almey" and *Pyrus communis* Seckel. J. Amer. Soc. Hort. Sci. 107: 657–660.
Sink, K.C. & J.F. Reynolds, 1986. Tomato (*Lycopersicon* L.), In: Y.P.S. Bajai (Ed.), Biotechnology in Agriculture and Forestry Vol. 2: Crops I, pp. 319–344. Springer-Verlag, Berlin.
Sink, K.C. & V. Padmanabhan, 1977. Anther and pollen culture to produce haploids: progress and application for the plant breeder. HortScience 12: 143–147.
Sopory, S.K. & S.C. Maheshwari, 1976. Development of pollen embryoids in anther culture of *Datura innoxia*. I. General observations and effects of physical factors. J. Exp. Bot. 27: 49–57.
Summers, W., J. Jaramillo & T. Bailey, 1992. Microspore developmental stage and anther length influence the induction of tomato anther callus. HortScience 27: 838–840.
Sunderland, N., 1971. Anther culture: a progress report. Sci. Prog. 59: 527–549.

Yelle, S., J.D. Hewitt, N.L. Robinson, S. Damon & A. Bennett, 1988. Sink metabolism in tomato fruit. Plant Physiol. 87: 737–740.
White, P.R., 1934. Potentially unlimited growth of excised tomato root tips in a liquid medium. Plant Physiol. 9: 585–600.
Zamir, D., R. Jones & N. Kedar, 1980. Anther culture of male-sterile tomato (*Lycopersicon esculentum* Mill.) mutants. Plant Sci. Lett. 17: 353–361.
Zamir, D., S.D. Tanksley & R.A. Jones, 1981. Genetic analysis of the origin of plants regenerated from anther tissue of *Lycopersicon esculentum* Mill. Plant Sci. Lett. 21: 223–227.

13. Somatohaploid production by anther culture of interspecific somatic hybrids and their prospects in potato breeding

VELI-MATTI ROKKA, JARI P.T. VALKONEN and EIJA PEHU

Contents

1. Introduction

Potato is world's fourth most important food crop after wheat, maize and rice (Hawkes, 1990). However, it is generally felt that the nutrition and breeding potential of potato have not yet been fully exploited. The nutritional constituents of potato tubers are well-balanced for human requirements. In developing countries, where the immediate demand for food is often crucial, potato outcompetes rice and wheat in production of edible energy per hectare and is nearly as efficient as groundnut, lentil, cabbage, rice, and wheat in production of edible protein (Horton, 1986). In most parts of the world, except the natural habitats of potato species in Latin America, a single potato species (*Solanum tuberosum* L.) is cultivated and just a minor part of existing potato germplasm (ca. 230 species) has been used in potato breeding and production (Ross, 1986). Many useful traits that are yet to be transferred to cultivated potato lines have been identified in wild potato germplasm; still, many traits and useful genotypes await identification (Hawkes, 1990). The unexploited traits identified in the non-cultivated tuber-bearing species include resistance to various stresses, such as plant pathogens, insect pests and frost, and also tolerance to heat under tropical climates (Hawkes, 1990).

Potato breeding by conventional methods has been limited, mainly due to two factors: i) the narrow genetic base of the cultivated tetraploid potato (*S. tuberosum*; $2n = 4x = 48$) (Hawkes, 1990); and ii) the difficulties encountered in the incorporation of specific traits in crosses among tetraploid cultivars. These difficulties include a high level of heterozygosity, complex segregation at the tetraploid level, frequently encountered pollen sterility, germplasm storage and transport, and the accumulation of viruses through vegetative

S.M. Jain, S.K. Sopory & R.E. Veilleux (eds.), In Vitro *Haploid Production in Higher Plants, Vol. 5,* 233–243.

propagation cycles (Howard, 1978). Thus, breeding at the tetraploid ($4x - 4x$) level is very slow and labor intensive, as large populations of seedlings need to be evaluated (Ross, 1986). Breeding at $2x$ level can offer significant advantages (Hougas & Peloquin, 1958; Ortiz & Peloquin, 1994). The majority of naturally occurring potato species are diploid ($2n = 2x = 24$) or tetraploid ($2n = 4x = 48$). However, other polyploid potato species, such as triploids, pentaploids and hexaploids, also occur. In all species, the genome is assumed to consist of multiples of the monoploid ($2n = 1x = 12$) set of chromosomes (Hawkes, 1990). Crosses between many wild and cultivated potato species are possible, including interploidy crosses between diploid and tetraploid species, as a result of production of unreduced gametes (Hougas & Peloquin, 1958; Mendiburu & Peloquin, 1976; Ortiz & Peloquin, 1994; Tai, 1994). However, several species cannot be readily crossed due to an inherent inability to achieve the required balance of a genetic factor called the endosperm balance number (EBN); deviation from a 2:1 ratio of maternal:paternal EBN causes failure in endosperm development and abortion of embryos (Johnston & Hanneman, 1982). Crosses which do not readily result in hybrid seeds can be obtained using a combination of rescue pollination and subsequent isolation and culture of embryos *in vitro* (embryo rescue; Watanabe *et al.*, 1995). However, new breeding methods and strategies are required for the enhanced utilization of wild potato germplasm.

Successful regeneration of potato plants from *in vitro* cultures of commercial cultivars reported by Lam (1975) led to various approaches to incorporate tissue culture techniques into traditional potato breeding programs. These approaches can broaden the gene pool of cultivated potato by incorporation of wild germplasm and enable selection for genotypes with agronomically desirable traits. Such techniques include haploid production by anther culture *in vitro* (Dunwell & Sunderland, 1973; Wenzel *et al.*, 1979) and somatic hybridization (Jones, 1988). Micropropagation and haploid production are techniques which are well-established in potato (Bajaj & Sopory, 1986) and can, therefore, be efficiently employed for the development of new breeding schemes. We have produced haploids by anther culture of interspecific somatic hybrids of potato (Rokka *et al.*, 1995) which will be described below, including the prospects of using these haploids in potato breeding. The term "haploid" has been used to describe a sporophyte with the gametic chromosome number (Ortiz & Peloquin, 1994). As haploids derived from somatic hybrids are unique from the viewpoint of applied methodology and genetic constitution, we propose the term *somatohaploid* to be used for haploids derived from somatic hybrids as distinct from the hybrids derived from sexual hybrids.

2. Analytical-synthetic breeding schemes

The anther culture method was first reported by Guha & Maheswari (1966) in *Datura innoxia*. Haploid production by anther culture offered a quick way to obtain homozygous lines. Therefore, the method has subsequently been extended to several crop species of diverse taxonomic origins (Bajaj, 1990). Dunwell & Sunderland (1973) were the first to regenerate haploid plants through anther culture of potato. A potato improvement program based on cellular and traditional breeding methods was first proposed by Wenzel *et al.* (1982). Their "analytical-synthetic" breeding scheme included anther culture and protoplast fusion in combination with hybrid production through sexual crosses and was based on stepwise reduction of the ploidy of tetraploid potato to the monoploid level ($2n = 1x = 12$) (the analytical part), followed by controlled resynthesis of a new heterozygous tetraploid clone (the synthetic part). This scheme has been applied to incorporate virus and nematode resistance genes as well as *in vitro* culturability (Wenzel & Uhrig, 1981). Uhrig & Salamini (1987) suggested a similar breeding scheme which included reduction of ploidy of tetraploid potato by anther culture, followed by crossing with superior (di)haploid lines or diploid wild species, selection of spontaneously doubled 4x lines, and their further crossing with tetraploid cultivars.

Naturally occurring unreduced gametes offer a possibility to take advantage of variation in ploidy level when utilizing wild germplasm (Vidal-Martinez & Garcia-Velazquez, 1986; Zimnoch-Guzowska & Wasilewicz, 1987; Watanabe & Peloquin, 1991). Veilleux *et al.* (1985) developed a scheme, the first step of which included production of monoploid plants ($2n = 1x = 12$) by anther culture of two unrelated *S. phureja* clones ($2n = 2x = 24$) which produce unreduced gametes. Protoplasts of the monoploid clones could be fused to produce highly heterozygous, diploid plants expected to produce unreduced gametes. The diploids could then be used as pollen parents in subsequent $4x \times 2x$ crosses.

There are several reports on anther culture of interspecific sexual hybrids produced through sexual crosses between *Capsicum annuum* and *C. chinense* (Morrison *et al.*, 1986); *Paspalum dilatatum* and *P. regnellii* (Bovo *et al.*, 1985); *S. phureja* and *S. berthaultii* or *S. phureja* and *S. microdontum* (Singsit & Veilleux, 1989); *Helianthus annuus* and *H. tuberosus* (Pugliesi *et al.*, 1993); and *Oryza sativa* and *O. rufipogon* (Rout *et al.*, 1989). There are, however, few prior reports on anther culture of somatic interspecific hybrids (Iwai *et al.*, 1981; Morrison *et al.*, 1985). So far, the interest of all schemes has been to reduce the ploidy level in order to improve selection of agronomically desirable parents or protoplast fusion partners. None of the schemes has, however, pursued an anther culture step of an interspecific somatic hybrid.

3. A case study on *Solanum brevidens*

At the Department of Plant Production of the University of Helsinki and the Plant Breeding Section, Agricultural Research Center, Finland, a major interest has been to incorporate the virus resistances (Valkonen *et al.*, 1992) of a wild, non tuber-bearing diploid potato species, *S. brevidens* ($2n = 2x =$

24) into cultivated *S. tuberosum* lines. We have carried out symmetric somatic hybridization (Rokka *et al.*, 1994) followed by a backcrossing program, and also asymmetric somatic hybridization (Xu *et al.*, 1993; Valkonen *et al.*, 1994b). In both approaches, reduced fertility of the hybrids has hampered further breeding.

An alternative approach that we have recently attempted is to culture anthers of interspecific somatic hybrids between *S. brevidens* and *S. tuberosum* to produce virus resistant lines with reduced ploidy. Such lines could be subsequently used for protoplast fusion with superior (di)haploid lines ($2n = 2x = 24$) of *S. tuberosum*. Furthermore, it permits study of the fate of the chromosomes of the two species after chromosome reduction by anther culture and the possible tissue culture-induced genetic instability which may lead either to preferential elimination of certain chromosomes or enhancement of recombination between the two species.

The interest in pursuing anther culture of the somatic hybrids of *S. brevidens* and *S. tuberosum*, even though their pollen fertility was low, was based on our success in anther culture of cultivated potato (Rokka *et al.*, 1996). Furthermore, Dai *et al.* (1993) observed that monohaploid potato plants could be produced from a male sterile (di)haploid line of *S. tuberosum* which indicated that pollen production and anther culture response were not necessarily correlated.

4. Anther culture of somatic hybrids between *S. brevidens* and *S. tuberosum*

Anthers were cultured from 11 tetraploid and five hexaploid symmetric somatic hybrids between *S. brevidens* ($2n = 2x = 24$) and *S. tuberosum* ($2n = 2x = 24$) grown in the greenhouse as described by Rokka *et al.* (1995). Five of the 16 somatic hybrids produced a total of 34 embryos. Four of the five hybrids that responded produced shoots and a total of 17 shoots regenerated (i.e., about one shoot per 100 cultured anthers). However, only five shoots survived the *in vitro* culture (Rokka *et al.*, 1995). All of the responsive anther-donor hybrids were hexaploid (Table 1), except hybrid 0201 which was tetraploid (data not shown). Morphology of the regenerants was clearly intermediate between the (di)haploid *S. tuberosum* "Pito dh. 45/4" and *S. brevidens* (Fig. 1).

5. Characterization of the somatohaploids

Species-specific RAPD (randomly amplified polymorphic DNA) patterns were consistent with morphological analysis and confirmed that the anther-derived lines contained DNA from both species (Fig. 2). Chromosome counts also revealed that the anther-derived plants had a chromosome number of 34–36 which was approximately half that of the hexaploid ($2n = 6x = 72$)

Table 1. Characterization of the fusion parents, *S. brevidens* and (di)haploid *S. tuberosum* "Pito dh.45/4", three hybrids produced by somatic hybridization between them, and five somatohaploids produced by anther culture of the somatic hybrids

Plant	Chromosome number/ ploidy level	Mean 2C value (pg)	Virus resistance
Fusion parental lines:			
S. brevidens CPC 2451	24 / 2x	1.64	PLRV, PVA, PVY
S. tuberosum 'Pito dh.45/4'	24 / 2x	1.78	PLRV
Somatic hybrids:			
0502	68±4 / 6x	4.00, 4.52[1]	PLRV, PVA, PVY
0507	* / 6x	*	*
0603	64±4 / 6x	3.73, 4.47	*
Somatohaploids:			
0502.1.1.1	34 / 3x	2.24	PLRV
0507.1.2.1	34 / 3x	2.26	PLRV
0603.1.5.1	34 / 3x	2.26	*
0603.1.5.2	35, 36 / 3x	2.21	*
0603.1.5.4	35, 36 / 3x	2.18	*

* Not determined.
[1] Two distinct peaks.
[2] PLRV, potato leaf roll luteovirus; PVA, potato A potyvirus; PVY, potato Y potyvirus.

anther-donor hybrids (Table 1). Nuclear DNA content (2C values) was similar in all of the somatohaploids as estimated by flow cytometry (2C = 2.15–2.30 pg; Table 1; Rokka *et al.*, 1995).

Successful plant regeneration from cultured anthers of the interspecific somatic hybrids between *S. brevidens* and *S. tuberosum* (Rokka *et al.*, 1995) showed that the somatic hybrids possessed androgenic capacity. Since *S. brevidens* is recalcitrant in anther culture (Rokka *et al.*, unpublished results), it is likely that the anther culture capability of the somatic hybrids came from *S. tuberosum*. Our study also confirmed that pollen production and anther culture responses were not necessarily associated because the frequency of viable (stainable) pollen in the somatic hybrids was low (Rokka *et al.*, 1994). This finding agrees with the previous studies of Uhrig & Salamini (1987) and Dai *et al.* (1993). The report by Rokka *et al.* (1995) is one of the few reports on anther culture of interspecific somatic hybrids (Iwai *et al.*, 1981; Morrison *et al.*, 1985), and is the first study on potato.

As indicated earlier, only the hexaploid somatic hybrids of our study demonstrated androgenic capacity, whereas the tetraploid somatic hybrids

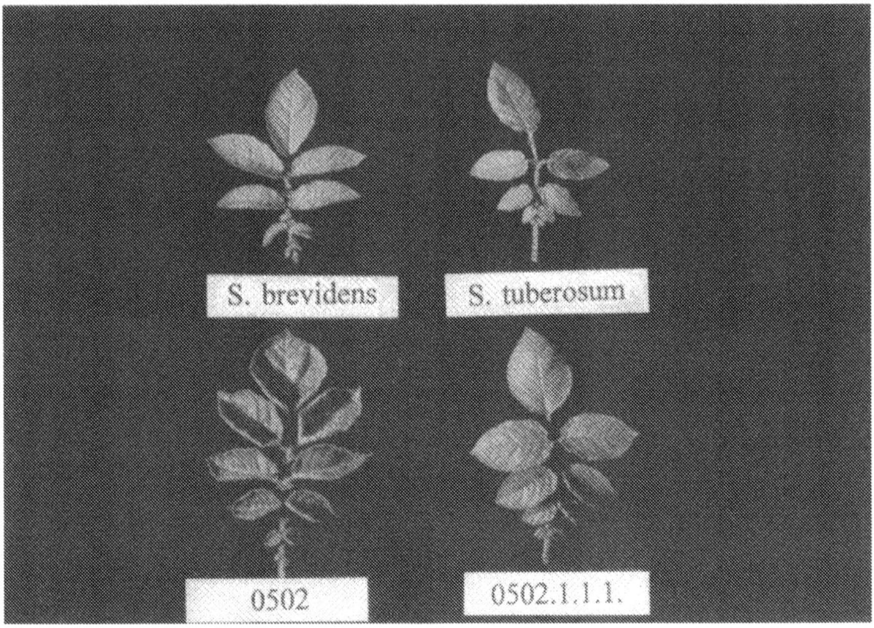

Figure 1. Leaf morphologies of the somatic fusion parents, *S. brevidens* and (di)haploid *S. tuberosum* "Pito dh.45/4", the hexaploid somatic hybrid 0502, and the somatohaploid 0502.1.1.1 derived from the somatic hybrid 0502.

formed embryos but failed to regenerate shoots. In our backcrossing program, using *S. brevidens* + *S. tuberosum* somatic hybrids, we found that the hexaploid hybrids were more effective as male parents than the tetraploids (Jacobsen *et al.*, 1993; Rokka *et al.*, 1994). It is likely that the hexaploid somatic hybrids responded better to anther culture because they contained more anther culture competence genes from *S. tuberosum*. Most probably, because the hexaploids were more successful pollinators, their meiosis and chromosome segregation were more likely to be normal and therefore result in more viable microspores (Rokka *et al.*, 1994; R. Veilleux, personnel communication).

We have observed previously that highly chimeric plants that can be detected by flow cytometry of isolated nuclei frequently occur among somatic hybrids (Valkonen *et al.*, 1994a). It may be relevant that the hexaploid somatic hybrids that responded in anther culture were chimeric according to chromosome counts and flow cytometry (Rokka *et al.*, 1995). However, somatohaploids derived from these chimeric somatic hybrids had approximately 50% of the nuclear DNA content of the anther-donor hybrids and showed no intracellular variability in chromosome numbers. Genome stability of the somatohaploids is an advantage for further protoplast fusions.

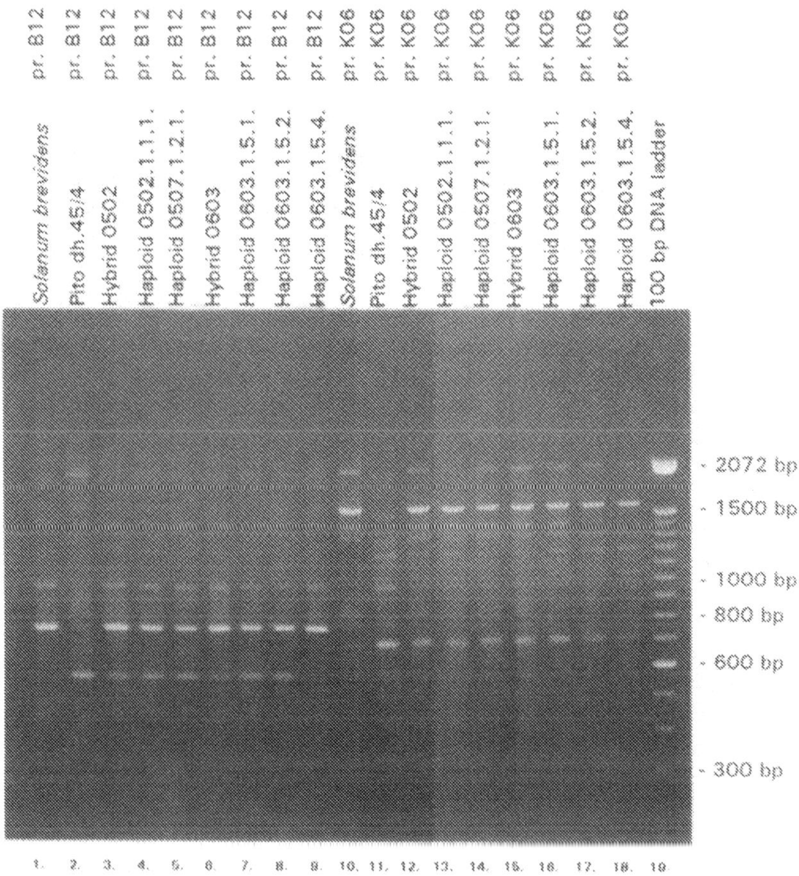

Figure 2. RAPD profiles produced using the primers B-12 (lanes 1–9) and K-06 (lanes 10–18) (Operon Technologies). *S. brevidens* and (di)haploid *S. tuberosum* "Pito dh.45/4" were used as fusion parents in somatic hybridization to produce the somatic hybrids 0502, 0507 and 0603 that were used for anther culture to obtain the somatohaploids (called haploids in the figure; the somatic hybrid 0507 was not included in the RAPD analysis). Note that RAPDs of 750 bp and 950 bp, and 1600 bp and 2070 bp amplified with primers B-12 and K-06, respectively, appeared in *S. brevidens* and all somatohaploids but not in "Pito dh.45/4". In turn, the RAPDs of 550 bp and 700 bp amplified by the primers B-12 and K-06, respectively, appeared in "Pito dh.45/4" and all somatohaploids, except 0603.1.5.4, but not in *S. brevidens*.

The parental genomic composition of the somatohaploids analysed by RAPD patterns showed that the somatohaploids retained at least parts of the genomes of both *S. brevidens* and *S. tuberosum*. Morphologically the somatohaploid plants were intermediate between the two parental species. These observations were consistent with the results reported by others who have shown transmission of *S. brevidens* chromosomes through meiosis of somatic hybrids (Williams *et al.*, 1993) and that all of the *S. brevidens* chromosomes were capable of recombining with the potato chromosomes (McGrath *et al.*, 1994).

Transfer of virus resistance from *S. brevidens* was of special interest in our breeding program. Therefore, the somatohaploid plants were challenged by graft-inoculation with potato leaf roll virus (PLRV), potato virus A (PVA) and potato virus Y (PVY), as previously described (Valkonen *et al.*, 1994b). Two anther-derived somatohaploids expressed extreme resistance to PLRV (similar to *S. brevidens*) but only some resistance to PVA and PVY. In contrast, *S. brevidens* is extremely resistant to PVA and PVY (Valkonen *et al.*, 1992). These results showed that PLRV resistance can be retained during haploidization via anther culture and thus transferred into the resulting anther-derived plants (Rokka *et al.*, 1995).

6. Potential of the somatohaploids

Production of somatohaploids from the interspecific somatic hybrids between *S. brevidens* and *S. tuberosum* offers a new potato breeding scheme including ploidy reduction by anther culture, selection of superior anther-derived lines, and somatic hybridization for reconstruction of improved tetraploid genotypes. A major advantage would be to overcome sexual crossing and follow cycles of anther culture and protoplast fusion to resynthesize an agronomically acceptable cultivar carrying desirable traits from the wild species. Androgenesis of an interspecific somatic hybrid can be an additional tool for enhancing recombination between the two different genomes and thus contribute to retaining the desired traits in subsequent cellular and other breeding steps. Our on-going *in situ* hybridization experiments using species-specific probes (Pehu *et al.*, 1990), genomic *in situ* hybridization (GISH) (Xu *et al.*, 1994) and direct labelling (Xu & Earle, 1994) will yield new data to evaluate the potential of the anther culture process in intergenomic recombination.

The present work has shown that anther culture of somatic hybrids is one way to simplify the genetics of somatic hybrids and the incorporation of wild germplasm into potato breeding programmes. However, the direct use of somatohaploids ($n = 3x = 34–36$) in sexual crosses may be limited as a result of reduced male fertility. Therefore, it would be interesting to do anther culture of the fertile tetraploid somatic hybrids reported by Austin *et al.* (1985) and Ehlenfeldt & Helgeson (1987). Anther culture could also be expanded to a wider range to interspecific and even intergeneric, symmetric and asymmetric somatic hybrids creating new genetic combinations and producing doubled haploids. These doubled haploids could be utilized in potato breeding programmes. All these and the above described approaches are expected to enhance utilization of the yet widely unexploited natural resources of wild potato species.

7. Acknowledgements

We gratefully acknowledge Profs. M.G.K. Jones, J.G. Helgeson, and the editors of this book for their valuable suggestions and comments on the manuscript.

8. References

Austin, S., M.A. Baer and J.P. Helgeson, 1985. Transfer of resistance to potato leaf roll virus from *Solanum brevidens* into *Solanum tuberosum* by somatic fusion. Plant Sci. 39: 75–82.

Bajaj, Y.P.S., 1990. *In vitro* production of haploids and their use in cell genetics and breeding. In: Y.P.S. Bajaj (Ed.), Biotechnology in Agriculture and Forestry 12, Haploids in Crop Improvement I, pp. 1–44. Springer-Verlag, Berlin.

Bajaj, Y.P.S. & S.K. Sopory, 1986. Biotechnology of potato improvement. In: Y.P.S. Bajaj (Ed.), Biotechnology in Agriculture and Forestry, Vol. 2, pp. 429–454. Springer-Verlag, Berlin.

Bovo, O.A., C.L. Quarin & L.A. Mroginski, 1985. Regeneration of plants from anther-derived calli of *Paspalum quadrifarium*. Agronomie 5: 865–870.

Dai, C., P. Yu, Y. Ran & X. Qu, 1993. Induction of monohaploid plants from a male sterile dihaploid strain of *Solanum tuberosum* L. by anther culture. Acta Genet. Sin. 20: 141–146.

Dunwell, J.M. & N. Sunderland, 1973. Anther culture of *Solanum tuberosum* L. Euphytica 22: 317–323.

Ehlenfeldt, M.K. & J.P. Helgeson, 1987. Fertility of somatic hybrids from protoplast fusions of *Solanum brevidens* and *S. tuberosum*. Theor. Appl. Genet. 73: 395–402.

Guha, S. & S.C. Maheswari, 1966. Cell division and differentiation of embryos in the pollen grains of *Datura in vitro*. Nature 212: 97–98.

Hawkes, J.G., 1990. The Potato: Evolution, Biodiversity and Genetic Resources. Smithsonian Institution, Washington, D.C.

Horton, D.E., 1986. Potatoes: truly a world crop. Span 30: 116–118.

Hougas, R.W. & S.J. Peloquin, 1958. The potential of potato haploids in breeding and genetic research. Am. Potato J. 35: 701–707.

Howard, H.W. 1978. The production of new varieties. In: P.M. Harris (Ed.), The Potato Crop: The Scientific Basis for Improvement, pp. 607–646. Chapman and Hall, London.

Iwai, S., K. Nakata, T. Nagao, N. Kawashima & S. Matsuyama, 1981. Detection of *Nicotiana rustica* chloroplast genome coding for the large subunit of Fraction I protein in a somatic hybrid in which only the *N. tabacum* chloroplast genome appeared to have been expressed. Planta 152: 478–480.

Jacobsen, E., R. Malvar, D.J. Huigen, J.E.M. Bergervoet & M.S. Ramanna, 1993. Isolation and characterisation of somatic hybrids of diploid *Solanum tuberosum* and *Solanum brevidens* and the use of amylose-free starch mutation for detection of introgression. Euphytica 69: 191–201.

Johnston, S.A. & R.E. Hanneman, 1982. Manipulations of endosperm balance number overcome crossing barriers between diploid *Solanum* species. Science 217: 446–448.

Jones, M.G.K., 1988. Fusing plant protoplasts. Trends Biotech. 6: 153–158.

Lam, S.L. 1975. Shoot formation from potato tuber discs *in vitro*. Am. Potato J. 52: 103–106.

McGrath, J.M., S.M. Wielgus, T.F. Uchytil, H. Kim-Lee, G.T. Haberlach, C.E. Williams &

J.P. Helgeson, 1994. Recombination of *Solanum brevidens* chromosomes in the second backcross generation from a somatic hybrid with *S. tuberosum*. Theor. Appl. Genet. 88: 917–924.
Mendiburu, A.P. & S.J. Peloquin, 1976. Sexual polyploidization and depolyploidization: some terminology and definitions. Theor. Appl. Genet. 48: 137–143.
Morrison, R.A., R.E. Koning & D.A. Evans, 1986. Anther culture of an interspecific hybrid of *Capsicum*. J. Plant Physiol. 126: 1–10.
Morrison, R.A., D.A. Evans & R.E. Koning, 1985. Anther culture of *Nicotiana* somatic hybrid. Am. J. Bot. 72: 877.
Ortiz, R. & S.J. Peloquin, 1994. Use of 24–chromosome potatoes (diploids and dihaploids) for genetical analysis. In: J.E. Bradshaw & G.R. Mackay (Eds.), Potato Genetics, pp. 133–154. CAB International, Wallingford.
Pehu, E., M. Thomas, T. Poutala, A. Karp & M.G.K. Jones, 1990. Species-specific sequences in the genus *Solanum*: identification, characterization and application to study somatic hybrids of *S. brevidens* and *S. tuberosum*. Theor. Appl. Genet. 80: 693–698.
Pugliesi, C., P. Megale, F. Cecconi & S. Baroncelli, 1993. Organogenesis and embryogenesis in *Helianthus tuberosus* and in the interspecific hybrid *Helianthus annuus* × *Helianthus tuberosus*. Plant Cell Tiss. Organ Cult. 33: 187–193.
Rokka, V-M., L. Pietilä & E. Pehu, 1996. Enhanced production of dihaploid lines via anther culture of tetraploid potato (*Solanum tuberosum* L. spp. *tuberosum*) clones. Am. Potato J. 73: 1–12.
Rokka, V-M., J.P.T. Valkonen & E. Pehu, 1995. Production and characterization of haploids derived from somatic hybrids between *Solanum brevidens* and *S. tuberosum* through anther culture. Plant Sci. 112: 85–95.
Rokka, V-M., Y-S. Xu, J. Kankila, A. Kuusela, S. Pulli & E. Pehu, 1994. Identification of somatic hybrids of dihaploid *Solanum tuberosum* lines and *S. brevidens* by species specific RAPD patterns and assessment of disease resistance of the hybrids. Euphytica 80: 207–217.
Ross, H, 1986. Potato breeding – problems and perspectives. J. Plant Breed. Suppl. No. 13. Paul Parey, Berlin.
Rout, J.R., N.P. Sarma & G.J.N Rao, 1989. Effect of potato-2 medium on anther culture of interspecific rice hybrids. Ann. Bot. 63: 621–624.
Singsit, C. & R.E. Veilleux, 1989. Intraspecific and interspecific transmission of androgenetic competence in diploid potato species. Euphytica 43: 105–112.
Tai, G.C.C., 1994. Use of 2*n* gametes. In: J.E. Bradshaw & G.R. Mackay (Eds.), Potato Genetics, pp. 109–132. CAB International, Wallingford.
Uhrig, H. & F. Salamini, 1987. Dihaploid plant production from 4x-genotypes of potato by the use of efficient anther plants producing tetraploid strains (4x EAPP-clones) – proposal of a breeding methodology. Plant Breed. 98: 228–235.
Valkonen, J.P.T., G. Brigneti, L.F. Salazar, E. Pehu & R.W. Gibson, 1992. Interactions of *Solanum* spp. of the Etuberosa group and nine potato infecting viruses and a viroid. Ann. Appl. Biol. 120: 301–313.
Valkonen, J.P.T., K.N. Watanabe & E. Pehu, 1994a. Analysis of correlation between nuclear DNA content, chromosome number, and flowering capacity of asymmetric somatic hybrids of diploid *Solanum brevidens* and (di)haploid *S. tuberosum*. Jpn. J. Genet. 69: 525–536.
Valkonen, J.P.T., Y-S. Xu, V-M. Rokka, S. Pulli & E. Pehu, 1994b. Transfer of resistance to potato leafroll virus, potato virus Y and potato virus X from *Solanum brevidens* to *S. tuberosum* through symmetric and designed asymmetric somatic hybridisation. Ann. Appl. Biol. 124: 351–362.
Veilleux, R.E., J. Booze-Daniels & E. Pehu, 1985. Anther culture of a 2*n* pollen producing clone of *Solanum phureja*. Can. J. Genet. Cytol. 27: 559–564.
Vidal-Martinez, V.A. & A. Garcia-Velazquez, 1986. Cytological characterization and gamete production in dihaploids and hybrids of potato (*Solanum* sp.). Agrociencia 66: 203–218.
Watanabe, K. & S.J. Peloquin, 1991. Occurrence and frequency of 2*n* pollen in 2*x*, 4*x* and 6*x* wild tuber-bearing *Solanum* species from Mexico, and Central and South Americas. Theor. Appl. Genet. 82: 621–626.

Watanabe, K.N., M. Orillo, S. Vega, J.P.T. Valkonen, E. Pehu, A. Hurtado, & S.D. Tanksley, 1995. Overcoming crossing barriers between non tuber-bearing and tuber-bearing *Solanum* species: towards potato germplasm enhancement with a broad spectrum of Solanaceous genetic resources. Genome 38: 27–35.

Wenzel, G., C. Meyer, T. Przewozny, H. Uhrig, O. Schieder & G. Melchers, 1982. Incorporation of microspore and protoplast techniques into potato breeding programs. In: E.D. Earle & Y. Demarly (Eds.), Variability in Plants Regenerated from Tissue Culture, pp. 290–302. Praeger Pub., New York.

Wenzel, G., O. Schieder, T. Przewozny, S.K. Sopory & G. Melchers, 1979. Comparison of single cell culture derived *Solanum tuberosum* plants and a model for their application in breeding programs. Theor. Appl. Genet. 55: 49–55.

Wenzel, G. & H. Uhrig, 1981. Breeding for nematode and virus resistance in potato via anther culture. Theor. Appl. Genet. 59: 333–340.

Williams, C.E., S.M. Wielgus, G.T. Haberlach, C. Guenther, H. Kim-Lee & J.P. Helgeson, 1993. RFLP analysis of chromosomal segregation in progeny from an interspecific hexaploid somatic hybrid between *Solanum brevidens* and *S. tuberosum*. Genetics 135: 1167–1173.

Xu, J., R.L. Conner & A. Laroche, 1994. C-banding and fluorescence *in situ* hybridization studies of the wheat-alien hybrid "Argotana". Genome 37: 477–481.

Xu, J. & E.D. Earle. 1994. Direct and sensitive fluorescence *in situ* hybridization of 45s rDNA on tomato chromosomes. Genome 37: 1062–1065.

Xu, Y-S., M. Murto, R. Duncley, M.G.K. Jones & E. Pehu, 1993. Production of asymmetric hybrids between *Solanum tuberosum* and irradiated *S. brevidens*. Theor. Appl. Genet. 85: 729–734.

Zimnoch-Guzowska, E. & I. Wasilewicz, 1987. Unreduced gametes in the breeding of potatoes at the diploid level. In: G.J. Jellis & D.E. Richardson (Eds.), The Production of New Potato Varieties, pp. 222–224. Cambridge University Press, Cambridge.

In Vitro Haploid Production in Higher Plants
Volume 5 – Oil, Ornamental and Miscellaneous Plants

Y.P.S. Bajaj, c/o Director, Lady Irwin College, Sikandra Road, New Delhi 110001, India.

R. Bergmann, Institute of Crop Science and Plant Breeding I, Justus-Liebig-University, Ludwigstr. 23, D-35390 Giessen, Germany.

N. Bhalla-Sarin, School of Life Sciences, Jawaharlal Nehru University, New Mehrauli Road, New Delhi-110067, India. Fax: +91-11-686 5886.

R.W. van den Bulk, Department of Developmental Biology and Department of Ornamental Cops, Center for Plant Breeding and Reproduction Research (CPRO-DLO), P.O. Box 16, 6700 AA, Wageningen, The Netherlands.

A. Cersosimo, Istituto Sperimentale per II Tobacco, 35071 Bovolone (VR) 8, Febbraio 1994, Via Canton 14, Italy. Fax: ++39 45 710 1097.

W. Friedt, Institut fur Pflanzenbau und Pflanzenzüchtung I, Ludwigstr. 23, DW-6300 Giessen, Germany. Fax: +49 641 702 5997.

M.A. Germana, Centro Studio Miglioramento Genetico Agrumi, Vialle delle Scienze, C.N.R., 90128, Palermo, Italia. Fax: ++39 91 652 1098.

M.S. Gill, Punjab Agricultural University, Ludhiana, 141 004, India.

S. Mohan Jain, Department of Plant Production, University of Helsinki, SF-00014, Box 27, Helsinki, Finland. Fax: 358-9-708 5582.

P.B. Kavi Kishor, Department of Genetics, Osmania University, Hyderabad-500 007, India. Fax: ++91 40 8688 46.

H. Köhler, Institute of Crop Science and Plant Breeding I, Justus-Liebig-University, Ludwigstr. 23, D-35390, Giessen, Germany.

G. Lakshmi Sita, Department of Microbiology & Cell Biology, Indian Institute of Science, Banglore – 560 012, India. Fax: ++91 80 336 7789.

R.G. Leonardo, Istituto Sperimentale per L'Orticoltura sezione Di Montanaso Lombardo, via Paullese, 28-20075, Montanaso Lombardo (Milano), Italy.

T. Nurhidayah, Institute of Crop Science and Plant Breeding I, Justus-Liebig-University, Ludwigstr. 23, D-35390, Giessen, Germany.

M.S. Pais, Departamento de Biologia Vegetal, Faculdade de Ciencias, Universidade de Lisboa, Bloc C2, Piso 1, Campo Grande, P-1700 Lisboa, Portugal.

M.C. Pedroso, Universidade de Lisboa, Faculdade de Ciencias, Department of Biologia Vegetal, Bloco, Piso 1, Campo Grande 1700 Lisboa, Portugal. Fax: 351 1 759 77 16.

E. Pehu, Department of Plant Production, University of Helsinki, FIN-00014, Box 27, Helsinki, Finland. Fax: +358 9 708 5582.

T.P. Reddy, Department of Genetics, Osmania University, Hyderabad-500 007, India.

V.-M. Rokka, Agricultural Research Center, Institute of Crop and Soil Science, Plant Breeding Section, FIN-31600, Jokioinen, Finland.

A. Sarvesh, Department of Animal biology, School of Veterinary Medicine, 3800 Spruce Street, University of Pennsylvania, Philadelphia, PA 19104, USA.

C. Sorrentino, Istituto Sperimentale per IL Tabacco Sezione di Genetica, Via P. Vitiello 66, 84018, Scafati (SA), Italy.

B. Summers, Department of Horticulture, Iowa State University, Room 251, Ames, IA 50011-1100, USA. Fax: 515-294-0730.

H.S. Tsay, Department of Agronomy, Taiwan Agricultural Research Institute, 289 Chang-Chung Road, Wufeng, Taichung 41301, Taiwan. Fax: ++886 4 333 8162.

J. van Tuyl, CPRO-DLO, Droevendaalsesteeg, P.O. Box 16, NL 6700 AA, Wageningen, The Netherlands. Fax: +31 8370 18094.

J.P.T. Valkonen, Department of Plant Production, University of Helsinki, FIN-00014, Box 27, Helsinki, Finland. Fax: +358 9 708 5463.

I.K. Vasil, Laboratory of Plant Cell and Molecular Biology, Department of Horticultural Sciences, 1143 Fifield Hall, University of Florida, Gainesville, FL 32611-0690, USA. Fax: 904 392 9366.

G. Venkatesham, Department of Genetics, Osmania University, Hyderabad – 500 007, India.

Species and subject index

Current Plant Science and Biotechnology in Agriculture

1. H.J. Evans, P.J. Bottomley and W.E. Newton (eds.): *Nitrogen Fixation Research Progress.* Proceedings of the 6th International Symposium on Nitrogen Fixation (Corvallis, Oregon, 1985). 1985 ISBN 90-247-3255-7
2. R.H. Zimmerman, R.J. Griesbach, F.A. Hammerschlag and R.H. Lawson (eds.): *Tissue Culture as a Plant Production System for Horticultural Crops.* Proceedings of a Conference (Beltsville, Maryland, 1985). 1986 ISBN 90-247-3378-2
3. D.P.S. Verma and N. Brisson (eds.): *Molecular Genetics of Plant-microbe Interactions.* Proceedings of the 3rd International Symposium on this subject (Montréal, Québec, 1986). 1987 ISBN 90-247-3426-6
4. E.L. Civerolo, A. Collmer, R.E. Davis and A.G. Gillaspie (eds.): *Plant Pathogenic Bacteria.* Proceedings of the 6th International Conference on this subject (College Park, Maryland, 1985). 1987 ISBN 90-247-3476-2
5. R.J. Summerfield (ed.): *World Crops: Cool Season Food Legumes.* A Global Perspective of the Problems and Prospects for Crop Improvement in Pea, Lentil, Faba Bean and Chickpea. Proceedings of the International Food Legume Research Conference (Spokane, Washington, 1986). 1988 ISBN 90-247-3641-2
6. P. Gepts (ed.): *Genetic Resources of* Phaseolus *Beans.* Their Maintenance, Domestication, Evolution, and Utilization. 1988 ISBN 90-247-3685-4
7. K.J. Puite, J.J.M. Dons, H.J. Huizing, A.J. Kool, M. Koorneef and F.A. Krens (eds.): *Progress in Plant Protoplast Research.* Proceedings of the 7th International Protoplast Symposium (Wageningen, The Netherlands, 1987). 1988 ISBN 90-247-3688-9
8. R.S. Sangwan and B.S. Sangwan-Norreel (eds.): *The Impact of Biotechnology in Agriculture.* Proceedings of the International Conference The Meeting Point between Fundamental and Applied in vitro Culture Research (Amiens, France, 1989). 1990. ISBN 0-7923-0741-0
9. H.J.J. Nijkamp, L.H.W. van der Plas and J. van Aartrijk (eds.): *Progress in Plant Cellular and Molecular Biology.* Proceedings of the 8th International Congress on Plant Tissue and Cell Culture (Amsterdam, The Netherlands, 1990). 1990 ISBN 0-7923-0873-5
10. H. Hennecke and D.P.S. Verma (eds.): *Advances in Molecular Genetics of Plant–Microbe Interactions.* Volume 1. 1991 ISBN 0-7923-1082-9
11. J. Harding, F. Singh and J.N.M. Mol (eds.): *Genetics and Breeding of Ornamental Species.* 1991 ISBN 0-7923-1094-2
12. J. Prakash and R.L.M. Pierik (eds.): *Horticulture – New Technologies and Applications.* Proceedings of the International Seminar on New Frontiers in Horticulture (Bangalore, India, 1990). 1991 ISBN 0-7923-1279-1
13. C.M. Karssen, L.C. van Loon and D. Vreugdenhil (eds.): *Progress in Plant Growth Regulation.* Proceedings of the 14th International Conference on Plant Growth Substances (Amsterdam, The Netherlands, 1991). 1992 ISBN 0-7923-1617-7
14. E.W. Nester and D.P.S. Verma (eds.): *Advances in Molecular Genetics of Plant–Microbe Interactions.* Volume 2. 1993 ISBN 0-7923-2045-X
15. C.B. You, Z.L. Chen and Y. Ding (eds.): *Biotechnology in Agriculture.* Proceedings of the First Asia-Pacific Conference on Agricultural Biotechnology (Beijing, China, 1992). 1993 ISBN 0-7923-2168-5

Current Plant Science and Biotechnology in Agriculture

16. J.C. Pech, A. Latché and C. Balagué (eds.): *Cellular and Molecular Aspects of the Plant Hormone Ethylene*. 1993 ISBN 0-7923-2169-3
17. R. Palacios, J. Mora and W.E. Newton (eds.): *New Horizons in Nitrogen Fixation*. Proceedings of the 9th International Congress on Nitrogen Fixation (Cancún, Mexico, 1992). 1993 ISBN 0-7923-2207-X
18. Th. Jacobs and J.E. Parlevliet (eds.): *Durability of Disease Resistance*. 1993 ISBN 0-7923-2314-9
19. F.J. Muehlbauer and W.J. Kaiser (eds.): *Expanding the Production and Use of Cool Season Food Legumes*. A Global Perspective of Peristent Constraints and of Opportunities and Strategies for Further Increasing the Productivity and Use of Pea, Lentil, Faba Bean, Chickpea, and Grasspea in Different Farming Systems. Proceedings of the Second International Food Legume Research Conference (Cairo, Egypt, 1992). 1994 ISBN 0-7923-2535-4
20. T.A. Thorpe (ed.): In Vitro *Embryogenesis in Plants*. 1995 ISBN 0-7923-3149-4
21. M.J. Daniels, J.A. Downie and A.E. Osbourn (eds.): *Advances in Molecular Genetics of Plant-Microbe Interactions*. Volume 3. 1994 ISBN 0-7923-3207-5
22. M. Terzi, R. Cella and A. Falavigna (eds.): *Current Issues in Plant Molecular and Cellular Biology*. Proceedings of the VIIIth International Congress on Plant Tissue and Cell Culture (Florence, Italy, 1994). 1995 ISBN 0-7923-3322-5
23. S.M. Jain, S.K. Sopory and R.E. Veilleux (eds.): In Vitro *Haploid Production in Higher Plants*. Volume 1: Fundamental Aspects and Methods. 1996 ISBN 0-7923-3577-5
24. S.M. Jain, S.K. Sopory and R.E. Veilleux (eds.): In Vitro *Haploid Production in Higher Plants*. Volume 2: Applications. 1996 ISBN 0-7923-3578-3
25. S.M. Jain, S.K. Sopory and R.E. Veilleux (eds.): In Vitro *Haploid Production in Higher Plants*. Volume 3: Important Selected Plants. 1996 ISBN 0-7923-3579-1
26. S.M. Jain, S.K. Sopory and R.E. Veilleux (eds.): In Vitro *Haploid Production in Higher Plants*. Volume 4: Cereals. 1996 ISBN 0-7923-3978-9
27. I.A. Tikhonovich, N.A. Provorov, V.I. Romanov and W.E. Newton (eds.): *Nitrogen Fixation: Fundamentals and Applications*. 1995 ISBN 0-7923-3707-7
28. N.L. Taylor and K.H. Quesenberry: *Red Clover Science*. 1996 ISBN 0-7923-3887-1
29. S.M. Jain, S.K. Sopory and R.E. Veilleux (eds.): In Vitro *Haploid Production in Higher Plants*. Volume 5: Oil, Ornamental and Miscellaneous Plants. 1996 ISBN 0-7923-3979-7
30. R.H. Ellis, M. Black, A.J. Murdoch and T.D. Hong (eds.): *Basic and Applied Aspects of Seed Biology*. 1997 ISBN 0-7923-4363-8

KLUWER ACADEMIC PUBLISHERS – DORDRECHT / BOSTON / LONDON

GPSR Compliance
The European Union's (EU) General Product Safety Regulation (GPSR) is a set of rules that requires consumer products to be safe and our obligations to ensure this.

If you have any concerns about our products, you can contact us on

ProductSafety@springernature.com

In case Publisher is established outside the EU, the EU authorized representative is:

Springer Nature Customer Service Center GmbH
Europaplatz 3
69115 Heidelberg, Germany

www.ingramcontent.com/pod-product-compliance
Ingram Content Group UK Ltd.
Pitfield, Milton Keynes, MK11 3LW, UK
UKHW020136160926
475607UK00003B/68
* 9 7 8 0 7 9 2 3 3 9 7 9 3 *